HOT SPOTTER'S REPORT

HOT SPOTTER'S REPORT

MILITARY FABLES OF TOXIC WASTE

SHILOH R. KRUPAR

A Quadrant Book

University of Minnesota Press
Minneapolis • London

Quadrant, a joint initiative of the University of Minnesota Press and the Institute for Advanced Study at the University of Minnesota, provides support for interdisciplinary scholarship within a new, more collaborative model of research and publication.

http://quadrant.umn.edu

Sponsored by the Quadrant Environment, Culture, and Sustainability group (advisory board: Bruce Braun and Christine Marran), the Quadrant Health and Society group (advisory board: Susan Craddock, Jennifer Gunn, Alexander J. Rothman, and Karen-Sue Taussig), and the University of Minnesota's Center for Bioethics and Institute on the Environment.

Quadrant is generously funded by the Andrew W. Mellon Foundation.

An earlier version of chapter 1 was previously published as "Where Eagles Dare: Remediating the Rocky Mountain Arsenal," *Environment and Planning D: Society and Space* 25, no. 2 (2007): 194–212; reprinted with permission of Pion Ltd., London. An earlier version of chapter 2 was previously published as "Alien Still Life: Distilling the Toxic Logics of the Rocky Flats National Wildlife Refuge," *Environment and Planning D: Society and Space* 29, no. 2 (2011): 268–90; reprinted with permission of Pion Ltd., London. Portions of chapters 2 and 4 were previously published as "Transnatural Ethics: Revisiting the Nuclear Cleanup of Rocky Flats, Colorado, through the Queer Ecology of Nuclia Waste," *Cultural Geographies* 19, no. 3 (2012): 303–27; reprinted with permission of Sage Publications, Ltd.

Published by the University of Minnesota Press
111 Third Avenue South, Suite 290
Minneapolis, MN 55401-2520
http://www.upress.umn.edu

Library of Congress Cataloging-in-Publication Data

Krupar, Shiloh R.
Hot spotter's report : military fables of toxic waste / Shiloh R. Krupar.
"A quadrant book."
Includes bibliographical references and index.
ISBN 978-0-8166-7638-5 (hb) — ISBN 978-0-8166-7639-2 (pb)
1. Hazardous waste site remediation—Political aspects—United States.
2. Radioactive waste sites—Cleanup—United States. 3. Biopolitics—United States.
4. Environmental responsibility—United States. I. Title.
TD1050.P64K78 2013
363.72'8760973—dc23 2013007995

Printed in the United States of America on acid-free paper

The University of Minnesota is an equal-opportunity educator and employer.

20 19 18 17 16 15 14 13 10 9 8 7 6 5 4 3 2 1

For my parents, Karen and Joe, on a dare,
after many phone calls, and with much laughter.

Leopards break into the temple and drink to the dregs
what is in the sacrificial pitchers; this is repeated over
and over again; finally it can be calculated in advance,
and it becomes a part of the ceremony.

FRANZ KAFKA, "LEOPARDS IN THE TEMPLE,"
PARABLES AND PARADOXES

And for Allan Pred,
who showed me a positive nuclear metaphor.

Everything which fell under the scrutiny of his words
was transformed, as though it had become radioactive.
His capacity for continually bringing out new aspects,
not by exploding conventions through criticism,
but rather by organizing himself so as to be able to relate
to his subject-matter in a way that seemed beyond all
convention . . . Had all analogies drawn from physics not
become profoundly suspect in an age which has been
characterized by the radical divergence of social and
scientific consciousness, his intellectual energy might well
be described as a kind of mental atomic fission.

THEODOR ADORNO ON WALTER BENJAMIN, *PRISMS*

CONTENTS

PREFACE

TO: The Reader
FROM: The Principal Investigator (Shiloh Krupar et al.)
CC: Office of Academia Accountability (OOAA)
SUBJECT: *Hot Spotter's Report: Military Fables of Toxic Waste*

Hot Spotter's Report: Military Fables of Toxic Waste is concerned with the remains of former World War II and Cold War military production, owing to the restructuring of the U.S. military and the closure and remediation of nuclear facilities. "Remains" permeate the project: the term serves as a broad rubric for the range of residues generated by the domestic production of war, including nuclear and hazardous wastes, perforated land, flourishing wildlife, and sick former nuclear workers. The investigation offers case studies of the conversion of former military arsenals and nuclear facilities to nature refuges; it details the administration of land, the concessions to wildlife, and the compensation program for former workers now taking place within the reorganization of industry and ongoing domestic toxicity of twentieth-century U.S. war making. The project examines the fraught relations between humans and waste, nature and waste, the body and environment, with respect to "post–Cold War" cleanup, remediation, and concessions of the nation. The book tracks some of the subtle shifts in memory, rhetoric, and politics needed to naturalize military–industrial territories as "pure" wild spaces, and the reframing of military practices as compatible with environmental protection. It also explores the management of waste and strategies of governing the population and territory—the "life" of the nation—through powerful governing rationalities, value-generating uncertainties, and absurd efficiencies in response to toxicity, exposures, and occupational illnesses.

Hot Spotter's Report seeks to draw connections between labor and landscape, memory and bureaucracy, contamination and genetic preservation, military wastes and aesthetic practices. To that aim, the investigation poses the following questions:

- How is a military base or nuclear facility converted into a nature refuge?
- What relations between human and animal, bureaucracy and uncertainty, body and territory, ecology and containment, are mobilized by such conversions?
- How does the spectacle of nature, organized as separate from the human, external and pure, allow for and obscure toxicity, and contribute to the denial of responsibility for negative legacies of war?
- What kinds of ethical responses do such radioactive natures and military-to-wildlife conversions demand? What tactics are—and might be—employed by impacted populations to live in the remains of domestic war production?
- How would an environmental ethics that takes waste as inspiration operate? What forms of subjectivity, ethical practices, and aesthetics might emerge? What experimental coalitions and potential politics could be put into play by attending to the domestic material remains and ambiguities of so-called green war?
- More generally, what opportunities to practice criticism and documentation might be cultivated under conditions of toxicity, uncertainty, holes of data, partial knowledge, and secrecy?

This line of inquiry has no clear authorial origins. *Hot Spotter's Report* emerges from a series of solo performances and articles, ongoing collaborative projects and collective work, intellectual mentoring and a rich network of people who have generously supported the author's efforts to understand her life within U.S. war making and the military–industrial complex. Many of the ideas of the project first took shape in conversation with Jenna Loyd and within the Activist Geographers Group (AGG), which collectively explored the relations between militarism and geographic education, and organized around the possibilities of life "beyond the welfare/warfare nexus." Experimentations with performative techniques and creative methodologies of geographical research were encouraged by my late adviser Allan Pred and the community of geographers, scholars, and artists that filled his absence and pushed the project forward in various ways: C. Greig Crysler, Trevor Paglen, Laura Levin, Jason Moore and Diana Gildea, Gill Hart, Cindi Katz, Derek Gregory, Neil Smith, David Pinder, Sarah Kanouse and the National Toxic Land/Labor Conservation Service, and my geography department cohort and intellectual community at the University of California–Berkeley and beyond. A special thanks to Geraldine Pratt, who, as former editor of *Society and Space,* saw merit in the project from the beginning and created the space for me to advance the work's strategies rather than regressively justify its existence; her support has continued into the present, providing invaluable feedback on the project's current book form. Special thanks, too, to Michael Crang, who engaged with and challenged the early work on its own grounds, and supported my efforts to channel my proclivity for live performance scripts toward a more scholarly "mess." I am also grateful to Stuart Elden and the current editorial team at *Society and Space* for recognition of the scholarly value of this project.

The writing of *Hot Spotter's Report* has been intimately connected with my faculty position at Georgetown University; the making of the book took place largely within the diverse multidisciplinary curriculum, public scholarship atmosphere, and steady stream of remarkable students in the Culture and Politics (CULP) program within the School of Foreign Service at Georgetown. I am grateful to Katrin Sieg, in particular, for seeing the importance and relevance of this project, to the School of Foreign Service for financial support and time off, and to the many CULPies who have enriched my research through seminars, reading groups, thesis projects, and documentary activisms. A supplementary but no less crucial site for the production of *Hot Spotter's Report* has been the University of Minnesota in Minneapolis, where the innovative Quadrant program stimulated the book project. As a Quadrant fellow, I benefited from repeated intellectual engagements with other university fellows, University of Minnesota faculty, and the University of Minnesota Press. Many thanks to the Institute for Advanced Study for fostering a collegial work environment; to the generosity of

Bruce Braun, who provided me an indispensable opportunity to reconnect with the field of geography under different terms and to advance the project's framing and stakes; and to my editor, Jason Weidemann at the University of Minnesota Press, who has given thoughtful direction and crucial feedback over the long haul and helped me to envision the project in a printed form congruent with its analysis.

Additional thanks to Whitman College's Global Studies Symposium "Global Media, Global Spectacles," which was not only intellectually invigorating but brought me back to Richland and Hanford, Washington, after twenty-three years—a well-timed homecoming; the DOE Visual Archive; Bob McAllister; Nuclia Waste; Arthur S. Aubry; Ellen Davis and Lucas Adams at Southwestern University; Valerie Kuletz; and the various nuclear workers with whom I have spoken. I hope to contribute something to the efforts of individuals and communities in the wider public concerned with the remains of war; indeed, the book would not have been possible without those dedicated citizens, watchdog groups, reporters, artists, writers, environmental groups, academics, scientists, historians, activists and organizers, worker unions and groups who, in different ways, refuse the rhetorical cleanup operations of the "post–" in "post–Cold War" and "postnuclear."

My deepest gratitude to Nadine Ehlers, whose intellectual companionship and sharp insights I find everywhere in this project's current form; Joshua McDonald, for his queer humor, intellect, musical consorting, and editorial skills, and with whom life has become what Walter Benjamin called a "one-way street"; and my nuclear family—the historical rigor and documentary ethics of Jason Krupar, the inexhaustible knowledge, generosity, and delight in the absurd of Joseph Krupar, and the dynamic mind and captivating drive of Karen Krupar.

When the book is at its best, it is because of this generous support network. Any inaccuracies, elisions, or misunderstandings are the fault of the author alone.

ABBREVIATIONS

AEC	Atomic Energy Commission
AFL-CIO	American Federation of Labor and Congress of Industrial Organizations
AGG	Activist Geographers Group
AIDS	acquired immunodeficiency syndrome
ANWAG	Alliance of Nuclear Worker Advocacy Groups
AS	Assigned Share
ASAPs	Accelerated Site Action Projects
AWE	Atomic Weapons Employers
BCTD	Building and Construction Trades Department (of the AFL-CIO)
BRAC	Base Realignment and Closure program
***BRD**	Bureaucracy Reduction Director
CDC	Centers for Disease Control and Prevention
CDPHE	Colorado Department of Public Health and Environment
CERCLA	Comprehensive Environmental Response, Compensation, and Liability Act
CHWA	Colorado Hazardous Waste Act
CO	Colorado State
COPEEN	Colorado People's Environmental and Economic Network
CULP	Culture and Politics program
D & D	decommissioning and decontamination
DDT	Dichlorodiphenyltrichloroethane
DEEOIC	Department of Energy Employees Occupational Illness Compensation (part of DOL)
DOD	U.S. Department of Defense

* fictitious office or organization

DOE	U.S. Department of Energy
DOI	U.S. Department of Interior
DOJ	U.S. Department of Justice
DOL	U.S. Department of Labor
***E.A.G.L.E.**	Environmental Artist Garbage Landscape Engineers
EECAP	Energy Employees Claimant Assistance Project
EEOICPA	Energy Employees Occupational Illness Compensation Program Act
***EGO**	Endgame of Government Oversight
EM	Office of Environment Management (part of DOE)
EPA	U.S. Environmental Protection Agency
FBI	Federal Bureau of Investigation
FFTF	Fast Flux Test Facility (located in Hanford, Washington)
FWS	U.S. Fish and Wildlife Service (part of DOI)
GAO	Government Accountability Office (formerly General Accounting Office)
GIS	geographic information systems
HAZ-MAP	National Library of Medicine's Occupational Exposure to Hazardous Agents Database
HHS	U.S. Department of Health and Human Services
IREP	Interactive RadioEpidemiological Program
LM	Office of Legacy Management (part of DOE)
LSS	Japanese Life Span Study
NCI	National Cancer Institute
NIOSH	National Institute for Occupational Safety and Health (part of CDC)
OECD	Organisation for Economic Co-operation and Development
OMB	Office of Management and Budget
***OOAA**	Office of Academia Accountability
ORAU	Oak Ridge Associated Universities
OWCP	Office of Workers' Compensation Programs (part of DOL)
POC	probability of causation
POW	prisoner of war

* fictitious office or organization

RCRA	Resource Conservation and Recovery Act
RECA	Radiation Exposure Compensation Act
RERF	Radiation Effects Research Foundation
RMA	Rocky Mountain Arsenal
RMANWR	Rocky Mountain Arsenal National Wildlife Refuge
RSAL	Radionuclide Soil Action Level
SEC	special exposure cohort
SEM	Site Exposure Matrices
SOARS	System Operation and Analysis at Remote Sites
VA	U.S. Department of Veterans Affairs
WIPP	Waste Isolation Pilot Plant

INTRODUCTION

The farewell address of U.S. President Dwight D. Eisenhower on January 17, 1961, delivered a sober warning to a nation absorbed by its newfound prosperity: Keep watch of a rising and permanent armaments industry of vast proportions and unforeseeable influence.[1] As early as 1953, Eisenhower's "Chance for Peace" inaugural speech made reference to such a future, a "world in arms" that was spending not money alone but "the sweat of its laborers, the genius of its scientists, the hopes of its children." This total mobilization of society for war, he cautioned, was "not a way of life at all"; humanity would be sacrificed by the extension of warfare and military influence into all spheres of society, by binding national economic strength, productive capacities, and general welfare and security *to war making*.[2] Eisenhower's words still resonate with contemporary concerns over the problem of defense: How far can you go without destroying from within what you are trying to defend from without?[3]

The domestic organization of life for national defense has left the marks of war within the nation; while differentially experienced, war has unequivocally happened "at home," involving countless harms to the health, safety, and environment of the nation. Massive contamination remains from the production, maintenance, and disposal of weapons systems, decades of improper and unsafe handling, storing, and discarding of hazardous materials, and the privileging of production goals over ecological concerns.[4] Military activities have generated numerous toxic troubles, among them fuel spills and solvent leaks, compromised groundwater and soil, unexploded ordnance (artillery shells, bombs, mines), heavy metals pollution and radioactive wastes, and genetic impacts on humans and nonhumans alike. These domestic legacies of war are often hard to see. Toxicity can be difficult to physically detect; a history of secrecy and misinformation makes it difficult to weigh evidence and substantiate claims; military sites are scattered across the country, largely in rural areas that do not command public attention; the burden of inhabiting the risks and attritional hazards of such sites falls largely on the poor and marginalized communities.

The current restructuring of the military toward "green war" and the sanitizing language of "post–Cold War" and "postnuclear" suggest that war is not here, within the nation—that the residually negative impacts of war have not contaminated the nation. Just as land withdrawals for military activity have been inextricably connected with American ideas about nature's bounty, freedom, and the open frontier in the American West, "nature" continues to play a central ideological role in sustaining American exceptionalism today. Nature serves to contain the residues of war; such containment avers that the nation is clean and properly bounded, pure and environmentally stabile—an exceptional Eden untainted by war. The U.S. Department of Defense (DOD) has invested significant resources in environmental matters, from ecological audits and compliance with environmental laws to enhancing efficiencies with respect to the daily business of military bases. The U.S. Department of Energy (DOE), too, has created new management offices supposedly accountable for stewardship of nuclear hazards in perpetuity. Both the DOD and the DOE have organized cleanups of redundant facilities and transferred excess land to the private sector or to the public, from shopping malls and commercial spaces to recreational opportunities and nature refuges. Furthermore, environmental protection is now a key component of national security.

Yet war lives on, as Eisenhower prophetically warned, and we are still living in war. The leftovers of past war making are everywhere; moreover, the current organizing of war as "green"—as "green war," wherein war is waged under the aegis of the environment—actively remakes and maintains certain lands and populations of the nation as residues, relegated to the past or deemed superfluous. New efficiencies, rationalities, and forms of life attendant to the rise of green war have resulted in various forms of displacement, abandonment, and disposal—a politics "of dislocation and deferral that cut through the nation-state by delimiting interior frontiers as well as exterior ones."[5]

Hot Spotter's Report endeavors to show green war as a basis of the social, economic, and material organization of the nation, and to delineate the perforated land, depleted citizens, and subjugated knowledge that are the legacy of that organization of life. *This approach contends that the nation lives in and as the residues of war.* Its U.S. focus is one effort to excavate the material, residual impacts of war within the nation, as a contribution to delineating longer and larger—historical and transnational—geographies of war.

Hot Spotter's Report takes a broad social-cultural approach to the military and war. It is not intended to be a multisited cultural history of the military–industrial complex and its restructuring per se. However, revisiting the conceptual terrain of "military–industrial complex" and, by extension, "militarization" and "militarism," positions the book and lays out important background information on the contemporary organization of green war as an underlying basis of the nation. A malleable term, "military–industrial complex" generally refers to the economic and spatial structuring of the U.S. economy for mass industrial warfare, with subsidized employments and research, industrial policies, institutions, and land uses that normalize war and preparations for war. The concept, often used interchangeably with "military Keynesianism," points beyond weapons production toward industrial capitalism and the ordering of the nation from World War II through the Cold War period; it instantiates the organization of military production as the industrial basis of the United States and, therefore, the extension of warfare into all aspects of the social life of the nation.[6] The related term "militarization" refers to the social processes by which society composes itself for the production of weapons and national defense, such as the material organization of labor and resources, military land withdrawals, military–industrial corporate power, and social geographies of classification and secrecy.[7] Subsuming internal conflicts under the demands for national unity and the nation's survival, militarization has molded social institutions seemingly separate from battle, weaponry, or the formal military, from the creation and advancement of certain fields of knowledge (at the expense of other potentials) to the organization of suburbs and domestic relations of masculinity, femininity, and sexuality.[8]

Denoting the extension of military influence into civilian, social, political, and economic spheres, "militarism" even more broadly captures the processes that normalize war. Accounts of militarism have sought to explain how the military makes geographies—arranges social relations and spaces—either as a deliberate extension of the production and reproduction of military capabilities and the "warfare state," or as a by-product of those processes and of prioritizing military institutions.[9] From the political economy of weapons production and environmental impacts of military training or industrial nuclear processes, to the unremarkable commonplace extension of military objectives, technologies, and capabilities into everyday civilian life, militarism works to legitimate military control over environments and landscapes. It also naturalizes social conflict over those who benefit from and those who bear the costs of war—to the point that war is largely unrecognized as a building block of the nation. War is relegated to specific places, or blamed on particular institutions, industries, or people, rather than viewed as *a geography of power integral to the nation*.

These terms capture the complex structural aspects—the industries, institutions, and so forth—and social extensions of twentieth-century U.S. war making. *Hot Spotter's Report* employs a diffuse understanding of the state and the military—not as monolithic actors with intention, but as a disparate condensation of projects that are dynamic, never fixed, both organized and dispersed. By doing so, the book endeavors to delineate the contours of green war and excavate some of the logics and practices. *Green war is a powerful reorganization of social life that occludes the domestic impacts of war through tactical spectacles of nature, and that produces a slew of military remains caused by restructuring, remilitarization, and new demands for military efficiency and accountability.* "The half decade since the end of the Cold War has seen not a purging of organized violence from the international state system, but a restructuring of the ways in which it is organized and supplied. Not demilitarization but remilitarization."[10]

The sudden redundancy of the Cold War apparatus of military capacity and the industries that sustained it can be attributed, in part, to several national changes and global transitions.[11] The internationalization of the U.S. military–industrial core and globalized defense industry, combined with reduced U.S. social spending and the reconfiguration of the state around coercive and social-managerial functions—commonly ascribed to the political project of neoliberalism, all have contributed to military restructuring and post–Cold War efforts to create a more efficacious military. Defense has contracted, concentrated, shed jobs, shut down many national arsenals, and sought to establish consistency between discourses of security, good economic management, and professionalized forms of accountability.[12] Austerity and restructuring have impacted the nuclear weapons complex most dramatically, leading to the massive decommissioning of facilities, redundancy of labor, large-scale remediation and waste management projects, and the collapse of much of the moral, ideological, and symbolic legitimacy of the Cold War. In addition to the downsizing and privatizing of work, with scientific and engineering labor far surpassing manufacturing labor in importance, many landholdings have been liquidated. Military expenditures are used to justify such actions. "Peace dividends" are also given out in the form of returning land—often heavily contaminated—to the public for recreation or redevelopment.[13] The release of this land allows the military to consolidate terrain elsewhere, with more air space and territory for long-range weaponry.

The management of this liquidation within the larger reconstitution of U.S. defense has contributed to the "environmental turn" of the military and the reorientation of military practices and principles to those of green war.[14] For example, toxicity and environmental hazards—often by-products of military activities and nuclear weapons production—can help rationalize new military response efforts; to preempt potential disasters and to address environmental damage, former surveillance networks are revamped as ecological monitoring systems.[15] Green war also preserves research, weapons infrastructures, and contracts by redirecting military capacities toward sustainable development, alternative energy, and other long-standing environmental goals, excepting the end of the military or war. For example, the Pentagon has not only received an energy-efficient makeover, but the DOD, as the largest energy user in the nation, has heavily invested in the development of alternative energy sources, with the goal of advancing defense under conditions of global warming and fossil fuel scarcity.

The DOE has explored new weapons systems and "green upgrades" of the existing stockpile by integrating environmental considerations into the life cycles of weapons and by replacing highly toxic components with environmentally friendly substances supposedly "clean enough to eat."[16]

If the recent complicity of military activities with ecological restoration is a strategy of pursuing war through environmentalist means and mobilizing power through environmental monitoring and feedback, then the DOD and DOE both have also integrated environmental stewardship into their operations as a way to mitigate, contain, and defer the economic and social consequences of domestic war: the environmental hazards, cleanup costs, regulatory demands, the pressure to extend compensation for the negative health impacts of military activities, and so forth. This is most visible in the cleanup and management of former military arsenals and nuclear facilities as wildlife refuges. The DOD and DOE have been remediating large swaths of land and transferring former bases and landholdings to the Department of Interior (DOI) to be administered as part of the national wildlife refuge system; that which remains is repackaged as ecological improvement. Such land conversions demonstrate that military activities have not just destroyed nature but have actively produced it, and that nature—ideologically and materially—serves to contain the risks resulting from decades of domestic mobilization for war.

Green war not only annexes environmental concerns to the mobilizing and maintenance of war but also constructs truths about what is natural and therefore rational, incontrovertible, and/or removed from political context and responsibility. Negative legacies of war are relegated to the past through sanitizing rhetoric and by emphasizing economic rationalities, stewardship, and clean technologies in place of former industrial processes and negligent environmental safety. The DOD and DOE currently acknowledge many of the negative environmental impacts of national defense, in some cases providing concessions to U.S. populations harmed by fallout or hazardous materials, and making appearances of responsibility for the toxic residues of war. Yet residual hazards are often eclipsed through such visibility; military-to-wilderness conversions of land, environmental audits, and memorials to former Cold War workers can instrumentally diminish controversy and defer accountability for the accumulated domestic collateral damages of war. Green war effectively absorbs challenges to the legitimacy of the ongoing organization of war in the United States through the remediation and turnover of sites to the public, compensation efforts, and linkages to everyday environmental concerns. Moreover, military realignment actively produces remnants. From cost-benefit logics inherent to the assessment of risk or exposure, to "sustainability" as a rationale for streamlining oversight of long-term hazards, green war makes military remains in order to make better, cleaner, more efficient and responsible war. Inherent to this are processes of displacement, abandonment, disposal, and remaindering—the ongoing, attritional aftermath of war as a condition for the life of the nation itself.

Hot Spotter's Report investigates the biopolitics of green war—*the ways that war is produced and imagined within biopolitical methods of governing*. Such an approach scrutinizes green war as an underlying *biopolitical* organization of society. The book draws on the insights of poststructural theorist and activist Michel Foucault to define biopolitics thusly: "Biopolitics is the series of governmental strategies centered on this new concept called 'life.'"[17] Biopolitics refers to the historical rise of biological life as an object of state calculation and strategy—in other words, the securing and standardization of a population.[18] For Foucault, biopolitics is modern and productive; historically, this entailed the establishment of regulatory controls over biological processes, such as births, mortality, public health, and so forth. Biopolitics operates through the regularization of the population; it aims to curtail unpredictable events, achieve stability, and eradicate internal threats, which often entails judgments about the kinds of life that present a danger to the population.[19] It involves the power "to *make* something live or to let it die, the power to regularize life, the authority to *force* living not just to happen but to endure and appear in particular ways . . . where living increasingly becomes a scene of administration, discipline, and recalibration of what constitutes health."[20] This is crucial to understanding the way biopolitics operates today: fostering capacities to control, manage, engineer, reshape, and modulate "life."

Biopolitics in effect saturates ordinary material existence with power; it arranges, concentrates, and mobilizes living energies according to particular rationales, logics, and efficiencies.[21] It can distribute the nonhuman, environment, species, and so on, in the domain of value and utility by increasing productivity, or it may define environments or contain materials as dangerous in the drama of safety.[22] Green war is biopolitical in that life is organized through the administration of the environment—namely, through the supposed separations of human and environment, human and waste. *Hot Spotter's Report* emphasizes the regulations, procedures, and calculations that proliferate and diffuse state power, that target the population and the environment as principal objects of knowledge. The book focuses largely on institutional practices—that government is a doing or making—and *green war's arrangement of relationships between humans and nature, and waste and humans.* The biopolitical organization of waste, for example, is a form of social management that delimits spaces and relationships with materials, to control biological, material threats, secure general well-being, and generate economic opportunities.[23]

While the function of biopolitics is to sustain rather than take life, one of the contradictions at the heart of green war, brought into relief from the perspective of the nation domestically, is that this regime of biopolitics must incorporate the threat of violence and death within the regulation of life. This means normalizing the exceptional conditions of war and rationalizing forms of violence and death as inevitable, necessary for the protection and security of the population, and/or required by American power and modernity.[24] Not only do residues of war persist, as environmental threats and inhabited or embodied risks of exposure to toxic chemicals and radiation, but green war insistently remainders: it detaches remainders of the living, such as "othered" populations, economically redundant workers, exposed and cancer-ridden communities, as well as contaminated land, nuclear and hazardous wastes, abandoned facilities, and repudiated industrial material culture.[25] *Hot Spotter's Report* concentrates on such *military remains.* Drawing on the category of the residual—the abandoned and discarded, leftovers, and by-products—the book investigates the broad range of what might be considered the residues of war and military restructuring. It seeks to show how military remains are made, managed, and, furthermore, how they linger on and persist as potential in spite of rationales in operation that would deny their existence.[26]

Hot Spotter's Report considers *spectacle* and *uncertainty* to be central to the biopolitical operations of green war and the conditions of exposure, displacement, and abandonment that underpin it. The book understands spectacle to be a technology of power and pedagogy; spectacle functions as an architecture of separation—an arrangement of division—between humans and nature, humans and waste that resuscitates ideas about purity positioned against contamination. Chapter 1 examines some of the ways the spectacle of nature as pure wilderness at a former DOD arsenal enables new forms of value, such as real-estate development; it demonstrates how environmental remediation can involve "predatory" forms of preservation of animals, through genetic preservation, wilderness education, and racialized resuscitations of the frontier. Chapter 2 explores how the accelerated cleanup of a former nuclear site and creation of a postnuclear nature refuge mobilize purifying auditing systems, designed to keep up the nature spectacle and make appearances of responsibility; the heavy bureaucratization of land and waste can govern in such a way that the legacies of war and actual ecologies become superfluous to management, as indicated by the DOE's new office of "legacy management" to monitor nuclear hazards, supposedly in perpetuity. Combined with the popular consumption of nature spectacle—the tours, viewing blinds, and children's programs promoted at the military-to-wildlife refuges explored in this book—spectacular hypervisibility is shown to function as a means of disposal; spectacle obscures any lingering toxicity, the relations between human and nonhuman, and the way the past still shapes the present not merely theoretically but materially. Spectacle arranges for unaccountability and perpetuates exposure rather than controversy.

This marks a significant shift in the way that spectacle is methodologically delineated. Instead of a limited visual understanding of spectacle as repressive of reality—as a lie or false image, as "greenwashing"—*Hot Spotter's Report* understands spectacle to be institutionally embedded, relational (even if that relation is one of separation), constitutive of the subject and population, and, therefore, relevant to investigations of aesthetics, politics, affect, and biophysical effects. The postmilitary (and postnuclear) nature refuge can be seen as spectacle, as a tactical ontological break between humans and nature—a "species line" that monetizes environmental damage only to maintain the industrial paradigm of nature as free disposal and buffer zone.[27] Waste becomes embodied in material and representational practices associated with the production and valorization of nature—and together these allow displacement and/or abandonment. Chapters 1 and 2 portray how the postmilitary nature refuge fixes waste to wilderness, instrumentalizes certain animals as biomonitoring devices in the service of bureaucratized waste management, and implements surveillance technologies as the means to secure human health. These nature refuges are symbolic redefinitions of the state in terms of environmental protection, and, more fundamentally, they embed a tactical ontology—meaning a truth-telling, world-making strategy—that powerfully presents nature as recuperated by technology and separate from the human in order to conceal the fact that exposures to toxicity and contamination continue.

The blurring of sacrifice zone and protection zone embodied by the postmilitary nature refuge, then, is a means of governing by contingency. Risks are essentially security elements—measurable uncertainties—placed around mutable, unfathomable, and potentially dangerous elements in the population and environment so as to optimize life; they are not merely epiphenomena of industrialization but epistemic and material objects of rule. Risk assessments circumscribe postmilitary land use as well as other forms of remediation and rehabilitation. Aporias, impurifications, and uncertainty also operate as tools of political management. Chapter 3 focuses on the ways that holes in data can be used to rationalize "doing nothing" for the remains of war and to turn accountability into a bewildering maze. Technical emphasis on the monitoring of waste disregards legacies of secrecy and denial, and dislocates political responsibility for ecological and embodied effects. Uncertainty emerges as that which is amassed and calculated—as "certain uncertainty"—so as to maintain possible emergencies but, more often, to take no action and subjugate any claims to know. For example, in the case of populations harmed by military activity, illness and suffering from exposures to hazardous materials are treated as a rational-technical process: legal and bureaucratic structures interact with scientific practices that assign medical risk and liability to individual claimants, ultimately assigning value to people's lives against the backdrop of a culture of denial and granting primacy to data over personal testimony and experience. Acknowledgment of environmental harm caused by military activity, recognition of responsibility to former nuclear workers, and concessions made to affected populations can result in the honing of uncertainty in order to maintain the status quo. Through applications of epidemiology that create mystifying abstraction, compensation can obscure the embodied effects of decades of domestic war preparation and weapons production. As chapter 3 broadly outlines, an industry of lucrative obscurantism cleanses the land of evidence and converts domestic—and transnational—geographies of war into spectacular data. The tactical spectacle of data production dehumanizes nature and divides labor from landscape, splitting workers from former work sites and from each other.

Together, the first three chapters show how green war tactically uses *the ontology of nature as separate,* external, pure, and/or available for containing waste and promoting consumption—nature *as spectacle.* The arrangement of nature spectacle can dispose of land and responsibility, legitimate remilitarization through environmental security and uncertainty, and even displace residual hazards *in place* through environmental monitoring and auditing systems. This binary division of nature and the human also works to abandon former workers who are living the industrial past in their bodies, in the form of exposures and ingested hazardous materials. Such workers are in fact doubly remaindered—by the return of former sites to nature and by the procedures of compensation legislation. Compensation effectively obscures historical geographies of war through spectacular calculations of uncertainty with respect to dosages and exposure assessments.

Hot Spotter's Report excavates such key aspects of the biopolitics of green war in order to cultivate a series of ethical responses. The book seeks to develop *an ethics and method appropriate to the conditions of green war.* Ethics, here, does not refer to abstract moral principles, separated from the moral problem and to be applied by an individual moral agent. Nor does it necessarily advance a liberal conception of ethical agency, as an individualized human ethics wherein the body merely serves as a vehicle for the territory of the ethical mind and Cartesian human subject. Instead, this is "an active and practical form of ethics founded in an evaluative sensibility arising from the concrete experience of specific situations."[28] The book considers the ethical to be tactical and relational; accordingly, ethical responses do not exist in a position of distance but rather are born out of the oppression that they seek to contest, unhinge, or dismantle. Foucault set in motion a "genealogical" approach to ethics, attuned to the microdynamics of power rather than responding to problems and oppression with new ontologies of the world or claims to the real.[29] The point of this form of ethics is to figure out how to live and how to die; such an ethics pursues particular ways of living that are not based on universal moral assertions but on the quest for "ways to live otherwise." The practices of relational ethics envisioned in this book seek to affirm the slow critical work of generating new conditions of existence from within existing institutions and the biopolitical organization of life, and, in doing so, support forms of persistence that already take place inside a culture of green war that does not provide sustenance. Such work humbly—maybe radically—endures, to defy "rationalistic conceits which establish the officially sanctioned good evidence of what's obviously possible and what's not." The overall purpose of *Hot Spotter's Report,* then, is to excavate "diagnostic *insights* and the imaginative *means to render society adequate to human [and other forms of] life.*"[30]

Centrally, *Hot Spotter's Report* is concerned with how criticism can be produced under conditions of toxicity, uncertainty, partial knowledge, and secrecy. A primary motivation is that criticism designed to "expose the truth" cannot adequately address the conditions of everyday toxicity, banal forms of death, and underlying organization of war. Such toxicity and contamination are here and now, and will persist long into the future; there can be no recourse to pure nature or myths of purity, such as unblemished land or the clean and proper body. The cost of trying to return to nature is grave; there are no people in the Edenic nightmare that is the postnuclear nature refuge. Nuclear scholar Valerie Kuletz puts it simply: "[there is only] Silence."[31]

Environmental critique tends to perpetuate relations with nature that make nature the passive victim of contamination or a realm of lost purity, that pit human culture against nonhuman nature, and that treat waste as that which contaminates the purity of both sides of the human/nonhuman opposition.[32] A myth of containment and creation undergirds such approaches: first, that nature can be contained and preserved through delimiting protection zones, and second, that humans can remain separate, unaffected, and behind the scenes as creators. Whether defined as utopia or dystopia, the bounding of nature as the inert remains of war or other forms of human development does not alter or challenge the contemporary organization of life; it in fact entrenches biopolitical governing relations and limits critique to the problem of visibility, that is, uncovering toxicity, revealing contamination, exposing the leakages and injustices. *Hot Spotter's Report* does not deny the importance of this work; however, it is essential to recognize that visibility does not guarantee that anything different will be done. In fact, visibility is often the problem. The biopolitical organization of nature spectacle or the mastery of genetic preservation of certain animals in chapter 1 allows for subtle forms of suffering, exposure, and death to continue. As detailed in chapter 2, the DOE's new Office of Legacy Management, through relentless organizational self-auditing in the name of ecological efficiencies, essentially produces an oversight of the negative legacies of nuclear weapons production through its own oversight. The spectacle of former nuclear worker illnesses in chapter 3 is yet another example of disposal and denial under the highly visible sign of benevolence of a federal compensation program. In the realm of popular culture, criticism and consumption of the present as apocalyptic spectacle rely on and reproduce the fear and paranoia endemic to a culture of exposure and toxicity.

One response is to affirm "indeterminacy," with respect to contamination and military natures—that such indeterminacy opens up new possibilities and new life-forms as opposed to retroactive nostalgia for nature as purity. However, such an approach potentially loses sight of the way risk and contingency are used to govern; the ontology of an indeterminate world is often strategically evoked so as to legitimate doing nothing, which effectively maintains the way things are rather than interrupting it and/or demanding change. *Hot Spotter's Report* does *not* assume that there is anything inherently hopeful about indeterminate futures or the uncertain natures resulting from domestic war making; this can too easily obscure the terror of the "ticking time bomb" of cancerous mutation, as described by a former nuclear worker exposed to radiation at work.[33] It also potentially misses the importance of attending to the residual, to that which materially remains and persists in the present, unsupported by present modes of life, but that might show concrete ways to live otherwise.[34] *Hot Spotter's Report*, therefore, resolves that the critical operations of "making visible" and openness to indeterminacy, uncertainty, ambiguity, and the like *must remain ethically attuned to the larger problematic of biopolitical governing:* the ways life and death are organized and experienced as anything but a pure opposition, that is, depleting livelihoods, spectacular suffering, and residual vulnerabilities. To that end, the book pursues what chapter 4 refers to as "transnatural ethics."

Hot Spotter's Report explores a relational "transnatural ethics" that deontologizes spectacle and plays with uncertainty. Transnatural ethics musters deontologizing force by unraveling truth claims, exploring the residual, rearranging material social relations, and thereby opening the present to other possibilities and *compositions*. Chapter 4 explicates the ways that nature and humans, nature and waste, are always already imbricated; they cannot be categorically separated or pure in existence. By refusing these separations and pure categories instrumental to the perpetuation of green war, such a transnatural ethics attends to the residual—not simply to affirm it but to explore the ways that remains persist within the conditions of the present. These residues encompass anything from long-lasting materials, such as nuclear waste, to humans who are compelled to live within conditions of death, often dealing with uncertainty in creative ways, such as irreverent grief or humor. The whole book attends to the outmoded, worn-out, and cast-off, in order to support ethical-political play with the potential of the negative.

This effort is *not* invested in arguments about the "end of nature" or the "social construction of nature" as the end point of critique, but in *nature's ongoing composition*. Chapter 4 regards practices of transnatural ethics as resolutely materialist; because they are tactical and relational, they examine the historical and geographical conditions of the problem at hand. Transnatural ethics practices truth as a relation and responsibility, and welcomes pedagogical play and aesthetic practices as central to political responses and coalitions. Furthermore, when confronted with the widespread conditions of exposure that characterize green war and contemporary society, transnatural ethics tries to grapple with the material connections between human bodies and other bodies—the material "stuff" of humans and that of the world. The emphasis on composition, therefore, is crucial; ethics, aesthetics, and politics are all conceived as integral aspects of the same relational ethical approach. It challenges humans to take responsibility yet relinquish the certainty of knowing what justice looks like beforehand, in order to explore more experimental coalitions.[35]

Although this might be unsettling for many environmentalists, and unseat ethical approaches that rely on arguments about nature or truth, the effort rejects totalizing views of power to foster different sensibilities and styles.[36] This can be of use to people studying landscapes and environments. Social processes, political institutions, and cultural values all affect not only how landscapes are perceived but also how they are physically reproduced; a less dichotomous and more relational ethic toward landscapes and subjects—contaminated or otherwise—can elucidate the ways landscapes are managed and conceptualized *and* cultivate different engagements with them and different environmental practices. The approach also raises questions about the sovereignty and agency of the "human" in occupational illness issues, environmental justice struggles, and human rights violations centered on the environment; if what is human is relational to the environment, then the arrangement of populations and places—bodies of land, people, animals, and so on—rises in importance over notions of autonomous personhood and traditional understandings of "rights" as being connected to solo, supposedly free, agents. Acknowledging the difficulty and uneasiness of this approach for communities that inhabit toxicity or that live as the remains of military activities, *Hot Spotter's Report* sees opportunity for calling out taken-for-granted assumptions and understandings of landscapes, nature, or the human in order to potentially free up other possibilities. It offers ways to think about the relations between nature and technology, social life and waste, the human and nonhuman, life and death, ecology and politics—*and* the multiple truth-telling practices, counterspectacles, and coalitions that might respond to green war.

Hot Spotter's Report puts into play various technologies of form and representation. While still structured by the conventions of academic book-length publishing, the chapters of the book employ relational aesthetics, such as theatrical collage, and combine satirical performance with empirical and theoretical writing. Fictional elements are used to reveal truth as a practice, to risk the certainty of empirical research and thereby hopefully to direct attention to *knowledge as doing, making, and arranging*. Each chapter has a documentary aspect, including a forensics front company, faux documents—a congressional PowerPoint testimonial and report—that "tell the truth," official documents that are barely credible, and seriously ridiculous figures that cluster real and fictive elements together in ways that proliferate controversy and perform counterspectacles to militarism, waste, and war. The visual-textual format of each chapter endeavors to convey key aspects, rhetorical contrivances, and technological interfaces encountered and/or engaged during the researching of *Hot Spotter's Report*.[37] Most important, each chapter features a particular figure that is either a composite drawn together from fieldwork, interviews, or documentary work *or* is drawn out from field investigations. Inspired by the material figuration strategies of Donna Haraway and (eco)feminist science studies and fiction, the figures in this report, which range from a collective of nuclear workers to a "radioactive" drag queen, are all creatures of surprisingly ordinary reality that mark places of transformative potential.[38] Through their creative implementation and exposition, *Hot Spotter's Report* endeavors to practice cultural criticism and environmental education as a diagnostic art of connection with an absurdist documentary sensibility: as "hot spotting."

"Hot spot" refers to both contamination and, more generally, an area where something resides unassimilated, unmetabolized, and thus remains productively troubling and politically open. This figure of an ambiguous, impure material and semiotic site, discussed in more detail in the Conclusion, informs the practice of hot spotting, which aspires to "bury" truths, regulations, and rationalities back into social and material relations, thereby setting forth relations hitherto unseen.

Such activity disrupts current arrangements of knowledge, truth, and affect; politicizes uncertainty; excavates the absurdities of current forms of governing and regimes of truth; and plays with the residual to develop tactical ethics that reject "ways of doing things" by rearranging materials and selves differently. Although hot spotting could initially evoke the visual operation of "spotting," of locating the truth by making it visible, the method imagined here and performed throughout the book is to proliferate ambiguity through absurdist methods and irreverence toward categories, containers, and endgames of governing. Essentially, hot spotting questions the rationality of the absurdism immanent to governing. It looks for empirical contradictions inherent to green war, and, using absurdist documentary methods, attempts to show how many of the rationalities at work are in fact absurd.

The Conclusion argues that the absurdist approach is a transnatural one that refuses the finality of truth to proliferate ambiguity and controversy in the present. This is not to say that ambiguity is somehow good or liberating, but rather that ambiguity is what must be explored to contest routinized recitations of evidence and established truths. In place of a sole focus on reparations for injury and the politics of recognition or victimhood, which recites violence and retroactively links the self to the state, hot spotting attempts to multiply documentation of the same information in different arenas for the purposes of multiplying publics. It is not antitruth, antiscience, or antievidence; it favors documentary excess in order to contest the ways that truths and practices become sedimented. The book incorporates multiple registers of the same material—multiple pedagogies and rhetorics—in order to encourage the ambiguity necessary for new and different coalitions. The entire volume models hot spotting.

More experimental diagnostic arts of connection attempt to rearrange the usual cartographies of conventional, bounded landscapes, namely, territories, bodies, and regions. This makes gestures toward new coalitional possibilities, between labor and environmentalists, nuclear scientists and artists, or disability rights activists, environmental justice movements, and environmental health and illness-as-politics approaches. It could also generate new and uncanny intimacies between people, nonhumans, even inorganic materials. By foregrounding the materiality of various bodies and the material linkages between humans and more-than-human natures, the degradation of environments can also be seen as forms of destruction to particular bodies, communities, and cultures. The vulnerability of ecosystems might inspire a more embodied ecological politics that puts the human within ecology as part of the missing matter of landscapes.[39] The ambiguities of what counts as environmental damage—and what responses are needed—are openings for thinking about things tactically, differently, and more intensely, as not only a question of meaning and the organization of knowledge but also affect, relational intensities, collective energies, and experimental coalitions.

Site Map and Chapter Tour

Hot Spotter's Report introduces a key former DOD facility in the urban fringes of Denver, Colorado: the Rocky Mountain Arsenal (RMA), now officially called the Rocky Mountain Arsenal National Wildlife Refuge (RMANWR) and administered by the U.S. Fish and Wildlife Service. This leading case of military environmental reuse and reinterpretation has served as an important model for the remediation and conversion to refuge of a second featured site in the report: the Rocky Flats Plant, also located near Denver. The now Rocky Flats National Wildlife Refuge is currently managed under the same administration as that of the Arsenal refuge.[40] Together, these two cases have influenced the decommissioning, decontamination, and remediation of military and nuclear lands across the United States. They are frequently cited as success stories in congressional testimonies, DOD, DOE, and DOI reports, environmental regulatory documents, and news venues.[41]

The sites exist within a broader historical-geographical context that requires brief elaboration: the state of Colorado's long-standing support of military enclosures along the Front Range. Denver was rapidly developed in the mid–twentieth century when federal dollars flooded the state to establish a series of military facilities, among them the $40 million Lowry Air Force Base and airfield, established in 1937 approximately fifteen miles southeast of Denver; the more than $40 million RMA chemical weapons factory, where the Army began producing chemical weapons and incendiary bombs in 1943; and the estimated $45 million Rocky Flats nuclear facility sixteen miles northwest of Denver—first announced in 1951—for the purposes of recovering and recycling scrap plutonium, manufacturing plutonium and highly enriched uranium components, and producing nuclear bomb pits in coordination with the United States' nationally distributed infrastructure of nuclear weapons facilities. These sites were key components of Denver's and, more generally, Colorado's militarization, which included the full life cycle of military activities, for instance, uranium mining, plutonium processing, underground defense posts and labs, the U.S. Air Force Academy, nuclear testing and bombing ranges, and waste storage. The interaction between Denver and communities along the Front Range with the remains of weapons production exemplifies the confluence of ecological attitudes and practices surrounding contaminated military land uses; the ascendancy of post–Cold War environmental remediation efforts is readily apparent.[42] Following national environmental regulations and federal mandates to promote redevelopment and public reuse of military lands, and abetted by private industry, the former arsenal and former plutonium factory are now wildlife refuges, while Lowry has undergone "brownfield" rehabilitation to become a mixed-income housing project.[43]

The replacement of the RMA and Rocky Flats missions of weapons production with environmental stewardship has effectively prompted reinterpretation of the history of domestic military occupation and industrial production along with that of nature preservation; nature plays a significant role in green war, aligning the wildlife refuge with military enclosure and disposing of liability for ongoing contamination and military–industrial ecology. Bureaus and offices for the purposes of managing the legacies of such sites—remaining wastes, hazards, *and* histories—separate out and subjugate the concerns, workplace/environmental knowledge, and occupational illnesses of former workers. Compensation legislation for sick former nuclear workers scattered across the country further installs an intransigent division of human from environment, labor from landscape, by channeling individual claims through a complicated procedure of dose reconstruction and exposure assessment. With many claims denied, these workers are but one example of the many remains of military restructuring, separated and abandoned through the process of environmental remediation and wildlife refuge conversion. However, residues of the industrial past—for example, the buildings of former military sites or the knowledge of former workers, as well as the stubborn materiality and abiding presence of military and transuranic wastes—can also function as resources for popular culture, criticism, and experimental coalitions.

This review of the empirical and analytic terrain of *Hot Spotter's Report* discloses the overall narrative arc of the book. While the chapters may be read separately as stand-alone cases, they also present, together, a biopolitical analysis of increasing complexity. Each chapter builds on important questions and picks up analytic threads from the previous. It begins by focusing on the animals at the RMANWR in chapter 1, using the format of an ethnofable that unearths an ethics of "deadstock" in response to the utopic project of the nation-state and colonial re-creation of nature celebrated at the arsenal-turned refuge. This is followed by chapter 2's satirical PowerPoint and congressional report on land and waste administration at Rocky Flats, where bureaucracy and a form of abandonment through auditing is investigated and contested via an ethical response centered on waste and the alien. Next, the book turns to the struggle of nuclear laborers in chapter 3—at the national level but with specific examples from Rocky Flats workers—to receive compensation for occupational illnesses in a format that combines conventional academic text with collages of worker quotations and aphorisms. The chapter tracks the way uncertainty is produced and perfected to deny claims; the compensation program's politics of epistemology and economy are inventoried and critiqued. Turning to popular culture in chapter 4, the book investigates the imbrications of everyday cultural forms with the military legacies of toxicity, such as the organization of affect—of fear, hysteria, dread, or exhaustion—and the potential for creative transnatural responses. This chapter characterizes the performances and efforts of two artists and argues for an ethical response to waste/nature and human/nature divisions that is inspired by experimental coalitions and artful endurance. The Conclusion provides further reflection on the methodological and aesthetic approach of the whole book, captured via the operation of "hot spotting." The following is a more detailed tour through the individual chapters.

Chapter 1, "Where Eagles Dare," explores the conversion of the

Rocky Mountain Arsenal to wildlife refuge, specifically the biopolitics of the animals that enable the site's "return to nature." It documents aspects of the material and symbolic life of the bald eagle at the RMANWR, in part, to narratively develop and deploy the figure of the bald eagle as an agent of human inquiry and environmental forensics. The eagle is not only a biophysical absorber of the wastes of military–industrial modernity, but also a cultural boundary "object" that negotiates toxicity, value, and the symbolic health of the nation; eagles show resilience to military toxicity and secure waste disposal and management under the sign of the nation.[44] Taking the ambiguous and humorous form of the author's forensics front company, E.A.G.L.E.—the Environmental Artist Garbage Landscape Engineers—collects and curates a series of vignettes that comprise a contemporary fable in document form with collected evidentiary artifacts and images. The experimental E.A.G.L.E. text is supplemented by adjacent notes that glean insights from academic literature and authorial reflection; such commentary serves to critically annotate the fable's collection of animal stories and fragments of uncertain knowledge related to the RMANWR. Together, the fable and notes contest the environmental education and knowledge supported by the RMANWR: the pedagogical investments in a creation story, in animals as signs of purity and the native, and in the morality of the preserved nature refuge, which sanitizes other histories of the military, the frontier, and extermination of humans and nonhumans. The chapter shows how particular animals serve as prostheses for the nation or for our limited sensory capacity to both notice and respond to environmental change. The chapter ends with an ethics of "deadstock," a practice that accounts for the mortality of the bald eagle and takes stock of the ambiguity of life and death, waste and value, of animals at the RMANWR.

Chapter 2, "Alien Still Life," examines the cleanup and management of former plutonium factory Rocky Flats. Converted from nuclear weapons plant to national wildlife refuge, Rocky Flats serves as a model for other decommissioned nuclear facilities across the United States. Rocky Flats demonstrates how the DOE bureaucratizes land and waste to produce efficiencies and a certain endgame: transfiguring waste to wilderness and naturalizing the site, as the means for disposing of land and minimizing the terrain of that which must be governed. The chapter is in two parts: a congressional testimonial in the form of a PowerPoint document, followed by a congressional report. Both are common forms of presenting evidence of government agency accountability to congressional representatives. The PowerPoint exhibits testimonies of several DOE speakers who play between the literal and figural in a series of alternating slides and scripts. Utilizing creative nonfiction and satire, the PowerPoint boils down the transformation of Rocky Flats to wildlife refuge as a strategic form of institutional unaccountability and public-relations management that has evacuated evidence of the site's industrial history, and relied on appearances of nature to give the impression of safety. The second part of the chapter analyzes this PowerPoint, expanding on its overview of Rocky Flats. In the spirit of congressional auditing and loosely based on the U.S. Government Accountability Office, an invented organization called "Endgame of Government Oversight" (EGO) reports on the way truths about the site are produced, maintained, and monitored by the DOE management order. Two notable areas of concern are the end-state vision of wildlife refuge, which accelerated the turnover of Rocky Flats to the public as a limited-use recreational space, and the techniques of legacy management, which, as a response to the growing number of decommissioned and remediated DOE nuclear facilities, have worked to contain or dispose of DOE responsibilities for land and hazards. Fixated on endgames, EGO demonstrates that the cleanup and monitoring of Rocky Flats have resulted in an "alien still life," sustained by voids in the public record, the alienation of former workers from the environment, the proliferation of the idea that Rocky Flats is now pure wilderness, and the endless spiral of institutional self-audit. In response, EGO turns to psychoanalysis as a trope and conceptualizes an ethics of "the return of the repressed." This response foregrounds the "lasting on" of waste, its material persistence, surprising appearances, and transformative possibilities, to insist that "this is *still* life but not as we know it."

Chapter 3, "Hole in the Head Gang," investigates a bill passed by the U.S. Congress in 2000 intended to provide compensation to U.S. nuclear workers who suffer from various illnesses related to their years of work producing the bomb.[45] This law—the Energy Employees

Occupational Illness Compensation Program Act (EEOICPA)—would acknowledge the sacrifices of the health and livelihood of such former workers, who interacted regularly with deadly materials necessary for the making of nuclear weapons. Whatever the intent of the legislation, administering EEOICPA has become a labyrinthine claims procedure that draws upon the uncertainties, inaccuracies, and lack of recorded data about workplace exposures to support a new science of "dose reconstruction" and to adjudicate compensation claims using epidemiological techniques as well as subtle and not-so-subtle forms of denial. The chapter systematically works through the claims process of compensation, to show how workers are simultaneously revalued and brought back into an industry of dose reconstruction but "let die," often to the benefit of large sections of the former weapons complex. Using a performative approach that symbolically advances the agenda of nuclear workers (without making a spectacle of their suffering), the collective figure of the "Hole in the Head Gang" perforates the chapter text to poke holes in EEOICPA's production of knowledge about exposure. Named after a comment made by a former Rocky Flats, Colorado, nuclear worker and activist who battled brain cancer, the Hole in the Head Gang is portrayed graphically by radiation masks that are usually worn to position patients' heads during radiation treatment; this is an effort to shift affective registers from documentary public spectacle and compassion to the uneasy bodily vulnerability, holes of knowledge, and satirical responses of former nuclear workers who were exposed to radiation and hazardous chemicals. "Hole in the head" also refers to the corrosive legacy of government secrecy, misinformation, classified materials, technical errors, and omissions of biological effects. As a nonindividualized group or "Greek chorus" that simultaneously comments on the dramatic action and contests the individualizing procedures of the claims process, the Hole in the Head Gang calls out some of the ways EEOICPA produces denial and obscurity and allows for the depletion and death of claimants. The Hole in the Head Gang performs a kind of forensics that leaks information, locates administrative loopholes and catch-22's, detonates criticisms, and collages charts, cartoons, and contradictions. The overall "workbook" quality of the text serves up a hodgepodge of collected quotations and logic exercises that point to the absurdities of the legal, statistical-epidemiological methods and overall administration of EEOICPA. The chapter also briefly considers the responses of workers—the work of life within the compensation system—and the possibilities of an ethics of "living as the remains" of U.S. bomb production.

Chapter 4, "Transnatural Revue," reviews the ethical demands delineated in the three preceding chapters to contextualize military-to-wildlife conversions and the nuclear worker compensation scheme within popular culture. The chapter considers the spectacle of leftovers, by-products, and residues of U.S. military restructuring and green war to be an effective way of obfuscating death and abandonment through hypervisible signs of nature's flourishing, waste administration, and social responsibility. This is achieved through the consumption of exposure—a form of "let die" within the broader dictates of organizing life and administering the population.[46] Such spectacles tactically distance the harmful legacies of war and convert waste and catastrophe into patriotic excess or nostalgia, exemplified by atomic tourism and Cold War kitsch. In response, the chapter sketches the performances and art projects of two figures in popular culture who demonstrate how alternative ways of living might be rehearsed and arranged through aesthetic practices that re-enchant waste as creative material force. The performative personae of a Denver-based/Rocky Flats drag queen comedienne and a U.S. Northwest/Hanford-affiliated nuclear waste sculptor inspire the chapter's speculative offering of transnatural ethics and aesthetics. In the context of the decommissioning and conversion of U.S. nuclear facilities to wildlife preserves, and the culture of denial over the negative health impacts of military environments on human and nonhuman natures, the chapter explores the aesthetic enterprises of these two figures as transnatural counterspectacles that potentially unsettle the status quo, especially the consumerism of exposure and habits surrounding waste under conditions of toxic stealth and ongoing state secrecy. Such artful and humorous forms of critical potential, including "mutant drag" and "transmutation follies," could re-enchant interest in the materiality of nuclear waste and make visible some of the lived/embodied mutations of humans and nonhumans experienced in relation to toxicity.

Drawing on Donna Haraway's "natureCultures," Michel Foucault's relational ethics, the work of Éric Darier and Catriona Mortimer-Sandilands on "queer ecology," and efforts to delineate the "trans" (relational) aspects of bodies and biopolitics, the chapter shows transnatural aesthetics to be a relational material ethics that acknowledges and potentially politicizes the permutations of waste and human, nature and waste, and experiments with subjectivity in waste through forms of collective play and artful endurance.[47] Transnatural arts of existence offer resources for creatively "enduring" in the present through rearranging social and material relations with waste.

The Conclusion, "Hot Spotting," reflects on the method of hot spotting, providing extended comments on the creative documentation and criticism enacted by the whole volume. Hot spotting operates on several levels: identifying, visibilizing, and, most important, keeping open the possibility that more can be identified, documented, and debated in different arenas. Foucault captured the impetus best:

> I dream of a new age of curiosity. We have the technical means for it; the desire is there; the things to be known are infinite; the people who can employ themselves at this task exist. Why do we suffer? From too little: from channels that are too narrow, skimpy, quasi-monopolistic, insufficient. There is no point in adopting a protectionist attitude, to prevent "bad" information from invading and suffocating the "good." Rather, we must multiply the paths and the possibility of comings and goings.[48]

01

WHERE EAGLES DARE

A BIOPOLITICAL FABLE ABOUT THE ROCKY MOUNTAIN ARSENAL NATIONAL WILDLIFE REFUGE

[1] The "Field Office of Authorial Remediation" refers to the field of remediation, a practice that removes contaminants from environmental media, such as surface water, soil, and groundwater, in order to protect human health and/or prime land for redevelopment. "Authorial remediation" then playfully considers the environmental backdrop of authorship, complicating the notion of an originating author, the dualism of subject and object, and the assumption of subjectivity as "natural" or "safe." Large-scale environmental contamination, wherein the variability and duration of toxicity and its biological effects are uncertain and never closed, troubles the assertion of authority.[1] The goal of authorial remediation is to show how knowledge about the Rocky Mountain Arsenal is folkloric and complex, shifting and partial, institutional and embodied. In these notes, the personal stories of the author occasionally surface to demonstrate how an ethnographic sensibility in/about the Arsenal requires a sense of the ways scientific and social knowledge circulate at public and private levels. Authorial remediation insists that there is no pure subject and no pure knowledge, and seeks to develop accounts of the environment that proceed by way of revisiting and reworking taken-for-granted arrangements of nature and waste, human and nonhuman, fact and fiction. The chapter, therefore, proceeds through bricolage of official quotations, collected images and artifacts, academic references, journalistic accounts, ethnographic documentation, popular discourses, creative writing, and critical annotations. Such an approach speaks to uncertain material natures and rejects the ontology of depth versus surface regularly evoked by claims of scientific or personal truth.

[2] Neoliberal restructuring of the U.S. military has entailed not only the downsizing and "temping" of labor, but also the closure and decommissioning of Department of Defense (DOD) and Department of Energy (DOE) landholdings and various cleanup actions in response to threats generated historically by the United States' industrial warfare economy. The Rocky Mountain Arsenal is one such example of DOD downsizing and remediation followed by the transfer of land to the U.S. Fish and Wildlife Service (FWS), which administers the national wildlife refuge system. This trend is often referred to as military-to-wildlife, warfare-to-wildlife, bombs-to-birds, or the shorthand "M2W" conversion. The closures of decommissioned facilities often lead to the repurposing and redevelopment as real estate. The classification of former arsenals as nature refuges has also been a popular strategy; abetted by private industry to externalize cleanup costs, the conversion of military bases into nature preserves serves to limit human contact with sites and environmentally reinterpret the history of the vast U.S. nuclear and military landscape—areas of secrecy and exception, military testing, weapons production and, subsequently, contamination.[2] Military activities do not just destroy nature but actively produce it. However, the reclassification of military lands as nature refuges often obscures the profound and ongoing material transformations of such sites.[3]

This chapter considers the "return to nature" brought on by military-to-wildlife conversions to be spectacle, an environmental project and arrangement of truth that administers military remains as wilderness. The cleanup and conversion of

Field Office of Authorial Remediation [1]

Per our conversation with Dr. Krupar, the Field Office of Authorial Remediation has agreed to conduct an environmental assessment of the Rocky Mountain Arsenal National Wildlife Refuge (RMANWR), an established wildlife preserve in the U.S. national refuge system. The Field Office of Authorial Remediation specializes in the reprocessing of author-subjects and their environmental histories. Based on several stories about Krupar growing up near military sites and nuclear facilities where family and friends worked, we determined that Krupar needed help reprocessing these unsettling places of her past. For example, owing to health risks, her father had refused family day visits, shrouding his places of work in mystery. She knew, growing up, that many of the landscapes where she played were steeped in local lore about invisible geographies of contamination. Thus, she confided, it befuddled her to watch popular consumption of such sites as if they were pristine wilderness. She suspected the environmental education of the RMANWR had something to do with this trend. This former Department of Defense chemical weapons manufacturing facility and Superfund site near Denver, Colorado, was now administered as a wildlife refuge and actively promoted as an urban sanctuary. [2]

the Rocky Mountain Arsenal into a wildlife refuge essentially makes a spectacle of this decommissioned DOD site. The spectacle of nature is the vehicle through which the ideas that nature has returned or never left, that war can lay waste to nature and then save it, are instilled. This is a governing relation that implements a species line, tactically separating nature and humans, humans and animals. Certain animal bodies, such as those of American bald eagles, are instrumentalized as pure, natural, separate from the human yet symbolic of the health of the nation; they serve as indicators of survival or adaptation more than as guides to military practices of environmental destruction or ecocide. Pedagogical investments in animals as signs of purity intensify the moralism of saving wilderness and often become evidence of successful recuperation, refusing the natural history and social life existing prior to, during, and after military occupation. Such spectacular separation can serve to justify the disposal of military operations and landholdings, help settle legal disputes and contain remediation costs, bury public criticism of the Arsenal's military past, increase property values, and propagate new origin stories.

Whether some animal populations are resuscitated from near extinction, or systematically slaughtered as surplus or collateral damage to maintain a healthy environment, humans are the clear operators of a range of instruments on the social flesh and bodies of animal and plant life.[4] This chapter investigates how the preservation of nature can simultaneously involve economic efficiencies, domestication, even extermination, under signs of the management of "life." In other words, the RMANWR—an example of a successful military-to-wildlife conversion that provides a model for other sites—is simultaneously a political zone of protection *and* sacrifice, salvation *and* extraction, life *and* death.[5] It is a zone of indeterminacy, wherein waste and other evidence of the unsustainability of war necessitate a biopolitically efficacious creation story that depoliticizes socio-ecological processes and porosity and instrumentalizes animal presence to raise value from the dead, often by using animal signs to generate economic returns or, in the case of the former Arsenal, by bringing back animal populations after historic decimation.[6] Military-to-wildlife conversions, such as the RMANWR, are biopolitical projects that call on the human to act in the interest of nature, and to maintain nonhumans in relationships that are advantageous to projects of value production. Such actions are moralized as responsible, as life-giving—their relation to historically destructive de/militarization processes unrecognized in the advocacy of creation and the attendant practices of environmental education, genetic diversity programs, real-estate projects, nature-based tourism, and nostalgic longings for nature as a source of authenticity, escape, and inspiration.

[3]

The investigation features the figure of the E.A.G.L.E. collective (Environmental Artist Garbage Landscape Engineers). The Office of Authorial Remediation enlists the E.A.G.L.E. collective to assemble a series of Arsenal images and information that render evidence of the spectacle of nature, and the way nonhumans, specifically the bald eagle and bison, become incorporated into the project of the wildlife refuge. Bald eagles, for example, are used to implant the idea of purity;

We accepted her commission and immediately contacted the E.A.G.L.E. collective (the Environmental Artist Garbage Landscape Engineers), U.S. subcontractor at the Rocky Mountain Arsenal. The E.A.G.L.E. collective had been working for decades at the Arsenal, documenting demilitarization and remediation; they also had long-standing connections to the U.S. military. They were perfect for the task of scavenging information about the site to help Krupar convert her confusion and unease into critical reflections. **[3]**

The following report was filed by the E.A.G.L.E. collective for the Krupar commission, and contains important background information about the site, as well as various evidentiary exhibits with E.A.G.L.E.'s critical annotations.

they serve as emblems of nature's resilience, as opposed to its vulnerability, porosity, or damage. The figure of the E.A.G.L.E. emerged from my reprocessing of the material and symbolic status of the American bald eagle at the RMANWR. E.A.G.L.E. functions as a site of slippage between the physical animal and a fictional forensics team that reports on government practices. The collective serves as my front company and makes reference to eagles and animal life at the refuge. E.A.G.L.E. takes the bald eagle as a U.S. national symbol under its wing; the organization also seeks to recognize the importance of the bald eagle to numerous American Indian communities. E.A.G.L.E., playing between actual animal and imagined collectivity, between animal life and human inquiry more generally, arranges a kind of animal fable and morality tale for the reader through humor, irony, even rage. The approach plays among different affective registers rather than performing detached critique.

As a garbage landscape engineering firm, E.A.G.L.E. investigates the biopolitical existence of eagles at contaminated sites and works to uncover the ambiguity between habitat preservation and the sanitizing of historic extermination at the Arsenal. In contrast to the bureaucratic assessments, environmental reports, and community impact statements about contamination typically produced during the remediation process, E.A.G.L.E. digs around the refuge to report on purity and origin stories. As though it were the voice of an animal subaltern, E.A.G.L.E. questions human/animal and nature/technology divisions, and injects a more uncertain relationship between the actual eagles and the state and corporate environmentalism on display at the RMANWR. Despite their warrior-like depictions in various U.S. nationalistic outlets, bald eagles do not solely represent military and state power, and they are not mere agents of capital, although this is unavoidably a key aspect of their living conditions at the Arsenal. Eagles are often scavengers, and their use of one of "the most contaminated square miles on earth" prompts recognition of their agency.[7] Inspired by this, E.A.G.L.E. functions as an agency that gleans the Arsenal's discursive field and drafts a report on the animal's terminal survival as the basis of the site's "return to nature."

[4]

The E.A.G.L.E.'s report uses the visual capability of eagles—scan and swoop—to orient the reader to the historical geography of the Rocky Mountain Arsenal, including the Arsenal's role in establishing U.S. military air power and the Army's "Project Eagle," which laid the early groundwork for the Arsenal's mission of cleanup and conversion to refuge. "Scan" adds an aerial perspective of the Arsenal as part of Colorado's extensive list of Superfund sites slated for EPA-supervised cleanup. "Swoop," referring to the tremendous binocular capacity of eagle vision, performs a fast, focused sweep into the Arsenal's contamination, court battles, and return to nature. The executive summary of the site, with its war-game rhetoric and biomimicry of eagle vision, initiates an environmental history of the Arsenal that foregrounds the strategic relations of eagles to humans and the toxic effects of the site's militarization *and* demilitarization.[8]

Site Background: Eagle Vision [4]

The E.A.G.L.E. collective compiled a broad chronology of the Arsenal's background, using its specially developed "scan and swoop" technology of environmental history. Images and narrative text have been included here as the opening site background of the Report.

SCAN

Things are heating up in the green counterstrike. Congress mobilizes the Comprehensive Environmental Response, Compensation, and Liability Act (CERCLA), known to civilians as Superfund. The reconnaissance mission of this official Act initiates an ongoing federal battle with bad environments. Government actions involve taxing chemical and petroleum industries and ambushing hazardous substance releases in the country. Within the state of Colorado, eighteen Superfund sites have been active,

two proposed, and three terminated. Targeted sites include missile-test fuel spills, lead and acid mine drainage, radium industry waste, nuclear weapons plutonium, chlorine gas munitions and pesticides, cyanide gold-leaching ponds, uranium and molybdenum wells. A full-scale invasion of Adams County may prove necessary to bring its several troublesome sites under control. The Rocky Mountain Arsenal alone is one of the most notoriously poisonous square miles in the country. Its cleanup will cost billions, and there is no guarantee that contamination will ever be contained. It is necessary to move in for a closer look before remediation commences.

SWOOP

1942

Following President Roosevelt's charge to become the "great arsenal of democracy," and the imperative to respond to the bombing of Pearl Harbor, the War Board occupies roughly thirty square miles of farmland ten miles northeast of downtown Denver for a chemical weapons manufacturing center to boost U.S. brawn in World War II. The application of industrial technology begun in the Civil War has culminated in the rise of air power in the world wars, particularly weapons for bomber aircraft. Removed from the coast, near rail, highway, and airport, with extensive water rights and ready labor supply, a manufacturing plant is erected by the Army to make chemical warfare agents chlorine, mustard gas, and lewisite, a blistering agent with similar effects to mustard gas but with an odor of geraniums. The first batch of mustard gas is brewed on New Year's Day 1943, while new facilities are subsequently built for petroleum-based incendiaries

that project fire in large quantities over great distance. Production runs twenty-four hours a day in three shifts, with a workforce of more than 3,500. The Army's Chemical Warfare Service commands approximately 155,000 tons of chlorine, mustard gas, and arsenic trioxide, and around 87,000 tons of other chemical munitions, such as incendiary cluster bombs and the 1,500 tons of napalm dropped on Tokyo in 1945.

Nearly a decade later, a secret $25-million North Plant for the manufacture of deadly chemical agent GB nerve gas initiates the Rocky Mountain Arsenal's new Cold War mission; the Arsenal is the only facility outside the Soviet Union to brew this lethal organophosphorous compound, one drop of which, absorbed through skin, can cause death in under a minute. Adding to the stockpile of artillery shells filled with distilled mustard gas and deadly chemical agent GB nerve gas, the Army makes fuel mix for Cold War-era ballistic missiles, rocket propellants that claim to take the sting out of Sputnik, experimental biological warfare, including anticrop agents wheat rust and rice-blast fungus, and Vietnam-era "button bombs," which, when stepped on, explode to warn of enemy infiltration.

Among the turbulent munitions activities of the site, the Army hosts the World War II Rose Hill POW camp and later allows private industry—first Julius Hyman, then Shell Chemical Company—to offset operational costs and maintain the facility for national security by adding pesticide, insecticide, and herbicide production to the site's toxic inventory. Throughout, waste products from the assembly line are left to the elements in ponds and unlined industrial lakes; the site collects 243 million gallons of waste in the ninety-three-acre,

War and control of nature expanded together as related aspects of life in the twentieth century. U.S. technologies developed to eradicate and control animals, for example, were also used for the same purposes against humans.[9] At the Rocky Mountain Arsenal, Shell made insecticides in the same facilities the Army used to manufacture chemical weapons. Whether for weapons of foreign wars or domestic combat against insects, militarization entailed treating nature as an external and available dumping ground and/or battleground. This continued after demilitarization became the Arsenal's primary task. Responding to a 1968 presidential directive to destroy obsolete chemical weapons, "Project Eagle" landed in the 1970s and retired more than sixty thousand individual munitions and more than half a million gallons of chemical nerve agent. Trailing Project Eagle, reports of pollutants from pesticides and by-products of nerve-gas neutralization in wells near the Arsenal revealed that the site's high water table, porous underground rock, and fine loams provided a pollution highway into groundwater. Wind, surface water, and arid conditions also reportedly facilitated the spread of contaminants.

The legacy of de/militarization remains deeply entrenched at the Arsenal, made evident with every story or discovery of remaining chemical warfare munitions, rocket fuels, and pesticides.[10] According to local lore, the approximately 917-acre Western Tier parcel sold to Commerce City in 2003, after deletion from the National Priorities List, harbors a large pit of white phosphorous grenades somewhere underground. Nerve-gas bomblets and, more recently, lewisite have made surprising appearances and forced the closure of the Arsenal's open space to the public. Lewisite, a blistering agent developed by the United States for use in World War II as a poisonous gas, penetrates ordinary clothing and even rubber, and its inhalation or absorption through the skin is extremely damaging and potentially fatal. It was detected in 2007 (and again in 2008) during an Army-conducted excavation of a trench in an area called the Lime Basins, which was used to treat liquid waste from the Arsenal's former lewisite production.[11] The construction of a protective cover over the area now marks the purported end of all major excavations of contaminated soil, and the RMANWR reopened to the public in June 2008. Approximately two months later, however, a Rocky Mountain Arsenal Remediation Venture Office press release informed the public that a laboratory jar, containing a potentially explosive chemical solvent found during the decommissioning of an on-site laboratory, was safely disposed of by burial in a hole near the building and detonation by plastic explosives.[12] Nature continues to serve as medium of disposal in the present. The Army has divested the site of indicators of its own presence, providing the means for a practical exit strategy. The decommissioning and cleanup process has naturalized the military's historic productions of nature and, in doing so, tactically ended the Arsenal's legal battles.

asphalt-lined Basin F, also known as "the sink." Under the pressure of state and civilian complaints about wasted well water, degraded crops, and kamikaze waterfowl, the Army pumps 175 million gallons of waste material into the earth at a depth of more than twelve thousand feet until Vietnam-era earthquakes force Denver to get a grip on the rumblings of the world's largest boil. The Arsenal and the local area are shocked by reports of the accidental nerve-gas exposure and death of six thousand sheep at Dugway Proving Grounds, Utah. The Arsenal aborts nerve-gas production and begins extensive demilitarization of chemical munitions in the 1970s, directed by a secret task force known as "Project Eagle." **[5]** This is the largest demilitarization effort undertaken by the Army at the time. Cleanup of its chemical weapons inventory results in the production of extremely toxic by-products. Meanwhile, Shell continues pesticide manufacturing until the early 1980s.

States are handed the power to command federal polluters to clean up after themselves, while new federal environmental and water pollution controls act tough against environmental muck. In 1982, all production ceases; the Arsenal is one of the first federal properties nominated for the "National Priorities List" under Superfund law. Superfund listing in 1987 puts the Arsenal at millions of cubic yards of contaminated soil, several basins of toxic waste, and a history of leaking sewer lines and spills. It is estimated that from 1942 to 1982, at least 176,000 tons of hazardous substances were released into the environment. An Arsenal commander describes an inner section of the Arsenal as "the most contaminated square mile in the nation." Insecticide chemicals and the nerve-gas by-product

DIMP are found in water outside the Arsenal in a series of off-post contamination incidents. Contaminated groundwater continues moving north and northwest of the Arsenal, leaving hundreds of people with unusable wells and needing to use bottled water. The Army implements several interim measures to control contamination: intercepting and treating groundwater that flows toward the northern boundary; pumping out toxins from leaky Basin F to million-gallon holding tanks and double-lined holding ponds; employing a "Submerged Quench Incineration" method to burn and convert Basin F liquids into gas and brine that can be shipped off-site for recycling and disposal by ocean.

Estimated cost of cleanup is staggering. A fusillade of lawsuits breaks out over the authority and responsibility for decisions about cost and extent of cleanup:

- The Army demands that Shell pay for part of the tab; the federal government files a $1.9-billion lawsuit against Shell.
- Colorado State insists on Army and Shell compensation for damages to its natural resources and sues both for failure to comply with environmental regulations.
- Colorado sues the United States for control over Arsenal cleanup oversight.
- Shell Chemical unsuccessfully passes the buck off on the 250 insurance companies of its parent agency, Shell Oil; Shell's insurers blame company misdeeds for the mess and refuse to cover liability.

The Arsenal remains mired in messy litigation, myriad cleanup cost forecasts, murky arguments about

The Rocky Mountain Arsenal is an intensely rational-technical terrain structured by professional and legal discourses. The court cases mentioned are just one tangle of the environmental regulations, containment efforts, ethical commitments, and struggles between various interests, all grappling with the ongoing material-biological effects taking place in and around the site.[13] The contest between federal and state actors for remediation control and jurisdictional power has resulted in enormous expenditure on bureaucratic and legal negotiations, environmental studies, and design of the final Record of Decision cleanup agreements.[14] Private industry's rejection of accountability for cleanup and decontamination, on the grounds that it was not contractually responsible for such actions, exacerbated these already-exorbitant expenditures. The appearance of the bald eagles, then, was vital to the turnover of the site from toxic blight to an amenity. When the bald eagle—the nation's symbol since 1782 and the featured species of the Bald Eagle Protection Act of 1940—was first found inhabiting the Arsenal in 1986, the FWS feared the birds would eat contaminated prairie dogs and considered frightening them off with noisemakers or flashing lights.[15] The response to the serendipitous bald eagle finding, however, soon shifted from recognizing their vulnerability and encouraging them to take flight, to maintaining the animal's presence and, later, leveraging eagle presence to incorporate cost savings into the cleanup. Eagle presence became an integral material and economic consideration in the assessment, planning, and remediation of the site.

The bald eagle serves as an important cultural boundary "object" for humans that negotiates toxicity, value, the symbolic health of the nation, and legal-regulatory processes with ecological relations. Under the wing of the bald eagles, a much less extensive and less expensive cleanup took place at the Arsenal. The land had only to be fit for animal habitation, avoiding the more costly remediation required for human residency or usages involving more comprehensive human contact. The separate cleanup standards associated with animals versus humans meant that savings could be made by leaving higher levels of contamination in the name of protecting habitat.[16] While not vital to the survival of the ecosystem, the eagles would now serve as the flagship species of the new wildlife refuge, vital to the cleanup of the Arsenal's image. Various government sectors, environmental groups, and, at times, the popular press presented the bald eagle's surprising appearance as an "act of nature," a spectacular showing of nature's resilience and the redemption of a military wasteland.[17] A highly visible icon and popular charismatic species capable of grabbing attention and attracting resources, the bald eagle was made permanently present as the outstanding on-site resident, with Bald Eagle Day events, bald eagle mascots, bald eagle on-site tours, bald eagle calendars, and numerous eagle-themed souvenirs. Visions of the RMANWR teaming with wildlife abounded, supported by cultural programming and FWS efforts to cultivate native vegetation. The material unsustainability of war, and the spiral of environmental litigation, cleanup costs, and blame constituting the Rocky Mountain Arsenal's history would now be balanced in social fantasy by the site's return to nature and the bald eagle as harbinger of a creation story.

the scope and standards of the project, and junked cleanup pacts and consent decrees that the state refuses to endorse.

1986

In the midst of Superfund super-mess, twenty bald eagles roost in the Arsenal's southeast scab of land along First Creek, tailgated by the U.S. Fish and Wildlife Service (FWS), federal wardens of the national bird. [6] The discovery of America's national symbol has major implications for the fate of the land: the dawning perception of Arsenal property as wildlife preserve and the proposal for an accelerated cleanup. An unlikely coalition between Colorado State representatives, the Army, Shell, FWS, environmental groups, and the local media supports the redemption of the Arsenal to the status of environmental amenity. The Rocky Mountain Arsenal National Wildlife Refuge Act is signed into law by then-President George H. W. Bush on October 9, 1992, envisioning the largest urban refuge in the country on Denver's front doorstep.

Consensus is reached for both on-site and off-site components of the cleanup; formal Records of Decision are issued. The $2 billion effort entails construction of a massive landfill, excavation of tons of contaminated soil, the demolishing of the buildings constructed for mustard gas, incendiary bombs, and nerve gas, the recycling of old rail tracks to create roads. The now eco-friendly

Owing in part to my long-standing rejection of the bald eagle as a symbol of U.S. military prowess and aggression, I could not have been less interested in attending one of the refuge's many Eagle Day events after the national bird's guest appearance at the Arsenal. The call to protect the bald eagles was deafening, but my ambivalence for the birds led me to question: How did the appearance of bald eagles facilitate the conversion of one of the most polluted places on earth to a spectacle of nature's survival? And how did this serve as an opportunity to both legitimate and abandon domestic warfare productions? If naturalistic renderings of the animal dovetail with heraldic images of the eagle as symbol of the U.S. nation, yet dis-identify with the bird's animal materiality, how might the ambiguity of the animal be used to acknowledge the living conditions of the eagle and reject its burden of bearing universal meaning for the nation? Clearly, the bald eagle was being called into the service of the national imaginary as a trope of wholeness and completeness; the animal stood for the natural and the pure, and, as such, could recuperate a pristine nature. Further, the life of the bald eagle could be instrumentalized for efficiency: a faster, cheaper cleanup, following the nature–culture and human–animal division installed by the designation of the Arsenal as an exclusive space for nature. Whereas Arsenal ecology and actual bald eagles are porous, the appearance of the bald eagles was used to secure the spectacle of nature, obscuring that porosity and creating the possibility for a new creation story. Representational practices, narrative and viewing technologies, and affect were thus key to the administration of military remains, extending managerial powers over nonhuman life.[18] Even though the condition of life of the bald eagle at the Arsenal is one of ongoing vulnerability to contamination, the eagle overshadows this as the sign of the state, disavowing the precariousness of eagle life, ironically, by focusing attention on its protection. But in order to account for the ambiguous existence of the bald eagles at the RMANWR, I first had to look into the biopolitics of another animal: the bison. The story of the bald eagle is braided with the bison. As a symbol of the white settler nation, the bald eagle is implicated in the historic extermination of the bison, and the bird's actual appearance at the Arsenal has pioneered the way for the bison's reintroduction to the prairie. These intertwined animals demonstrate how the Arsenal's conversion to nature refuge involves a powerful fantasy of the return of the native.

partners Shell and the Army under Colorado State monitoring share an environmentally friendly vision of the Arsenal as a bald eagle refuge and prairie teaming with wildlife. The FWS cultivates native vegetation, wildlife programs, eagle viewing areas, a fishing program, visitor facilities, and Service-conducted tours from government vans. Thousands of people flock to the Arsenal for the Bald Eagle Day event. **[7]**

Forget the grapefruit-sized nerve-gas bomblets discovered in 2000 during cleanup. Forget the giant earth-moving machines that worked round the clock on Army land on the other side of the restricted-area fence, plugging injection wells, demolishing and decontaminating building debris, sending 76,000 drums of contaminated salts down the Utah sink, tanking millions of gallons of liquid waste in triple-lined industrial basins, walling off and filtering groundwater, trucking around two million

cubic yards of soil. Forget the disturbances created to reroute animals away from ongoing contaminated areas: the noisemaker devices that simulate distress calls, the "zon guns" that frighten waterfowl from foul land, the plastic curtains around contaminated ground to keep prairie dogs away from toxic exposure passed up the food chain to the eagles. And never-you-mind that the bald eagle's sudden appearance and ongoing presence means a less geographically extensive and less expensive cleanup program that eliminates human habitation plans in a "we wouldn't want to disturb their habitat" cover-up.

The Vision

Promised Arsenal land to Commerce City and surrounding towns buffers the refuge with prairie suburbs, country homes, and a Denver International Airport tech center and free enterprise zone. The Rocky Mountain Arsenal is removed from the nation's Superfund list, the majority of land now under FWS management, following the completion of remediation. Signs of the Army are gone from all but approximately one thousand acres where toxic materials have been consolidated and buried, in addition to ongoing monitoring of contaminated groundwater. Eagles regularly hunt the site for prey; a herd of bison roams the rolling hills. Eight miles of walking trails have been installed. A new $7.4-million visitor center powered by wind turbine, solar panels, and geothermal heat serves as the gateway to short-grass prairie, woods, and wetlands that are

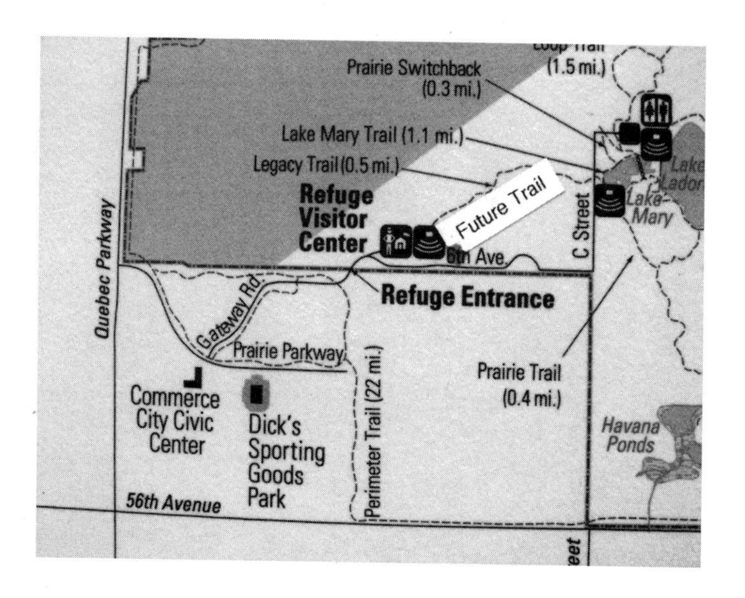

Exhibits B1–B5 "Buffalo Traces" visually and textually track the bison as a key figure in the RMANWR's post-cleanup creation story. This creation story is supported by a conservation initiative to purify and preserve the bison, and a salvational ethics that seeks to return the land to a mythic origin. Bison were reintroduced into portions of the RMANWR as part of a larger effort to preserve a population of genetically pure bison (i.e., those with no cattle genes). It is estimated that forty to sixty million bison once existed throughout the North American interior; by 1900 they were reduced to a mere few hundred. The colonization of the western frontier decimated the animal, paralleling the forced removal of Plains-area American Indians who relied on the bison. The expansion of the cattle industry played a central role in the destruction of the bison's biogeography, and loss of genetic diversity is largely the result of hybridization with cattle. Presently, the close to one million animals in wild and captive or semidomestic herds exist on less than 1 percent of their former range in public trusteeship in national, state, or provincial parks and wildlife refuges.[19] Sixteen bison were released from trailers onto the former Arsenal's grounds in March 2007; film clips were immediately posted of the bison's first encounter with the site on the refuge's Facebook page.[20]

[8]

home to coyotes, foxes, burrowing owls, both mule and whitetail deer, white pelicans, cottontail rabbits, black-tailed prairie dogs, great horned owls, herons, ferruginous hawks, and hundreds of other birds. The 12,500-square-foot facility features a resplendent observation deck, aviary, auditorium, and exhibits chronicling Arsenal history.

Front Range audiences throng to see this urban showpiece of the American West's wildlife system; the annual number of visitors quadruples. The Rocky Mountain Arsenal's communications program is named one of the top five PR campaigns of the decade. Receiving more than $3 million in federal stimulus funds, the RMANWR anchors a greenway corridor project in northeast Colorado that preserves open space, restores water resources, and provides recreational opportunities through an interconnected system of trails. As the gem of the eight-state Mountain Prairie region, the RMANWR is conscripted as the prototype for other military sites, propagating wildlife preserves in defense areas and recreational stories of land before military time.

The following exhibits #B–D are a selection of E.A.G.L.E.'s research on the RMANWR to date. Inquiries or comments should be addressed to the Office of Authorial Remediation, ATTN: EAGLE/Krupar commission, 2131 Piney Butte, Suite A, Denver, Colorado, 80232.

Exhibit Series B: Buffalo Traces [8]

During ethnography of a bald eagle nesting area, E.A.G.L.E. found bison in abundance at the RMANWR. E.A.G.L.E. has reason to believe that, following the

cultivated residency of the bald eagle, this animal was purposefully reintroduced to restore native prairie and reestablish the frontier and early white settler views. Subsequently, E.A.G.L.E. made the collective decision to study the bison in order to advance understanding of the eagle. Exhibits B-1 through B-5 below provide select visual and textual traces of the bison, which indicate that these megafauna are part of a powerful creation story.

Exhibit B-1

E.A.G.L.E. annotation: Following the bison's reintroduction to the RMANWR in a 2007 pilot program, E.A.G.L.E. found that a popular tourist activity involved tallying the number of bison seen on-site. This large bison bull silhouette greets visitors to the refuge's new 2011 Visitor Center and has become a popular photo opportunity.

[9] The FWS endeavors to distribute genetically engineered pure bison in various "stockpiles" around the country for the purpose of securing diversity. Accordingly, biodiversity can be preserved by banking gene sequences in bison "livestock."[21] The absurdity of reintroducing genetically purified bison to lands too contaminated for human habitation indicates the extent to which former military sites function as salvific arks for endangered species. Paralleling the contemporary reinvention of the zoo as wildlife park and revealing the extent to which experimental, sentimental, economic, and aesthetic factors can shape animal genomes, the reintroduction of a small herd of bison at the RMANWR was seen as a way to develop genetics conservation of the animal, restore authenticity to the site, add cultural and historical value, atone for the elimination of the animal from native heritage, and maintain the health and ecological integrity of the short-grass prairie system. The bison is a keystone species of the prairie ecosystem; its return to the prairie of the former Arsenal, therefore, is considered a key strategy in restoring the ecosystem as it was before industrialization and even European settlement. The animal's reintroduction nourishes an ecological fantasy of restoring native wilderness. Even though the history of the prairie ecosystem of the American West is defined by the relentless invasions of plants, animals, and people, the RMANWR's return to nature settles on the disappearance and reappearance of pure natives to reenact the frontier.[22]

Exhibit B-2

E.A.G.L.E. has culled several statements about the bison's reintroduction to the RMANWR from official documents. Four have been selected and reproduced verbatim in the Report below.

Citation #B-2a

Pending approval within the Service, the Service may introduce a herd of 10 to 100 bison in the northern zone within five years after cleanup completion. At one time, bison were present in the ecosystem, and this species provides a necessary grazing/trampling component in sustaining a short-grass prairie. Additionally, bison would be a major attraction in the urban setting of Denver and would facilitate educating visitors/students in plains ecology and ecosystem management (U.S. Fish and Wildlife Service "Comprehensive Management Plan" 1996, 52). **[9]**

Citation #B-2b

Reintroduction of bison to RMA will be a hallmark event in the process of cleanup and restoration that converts this Superfund site back into native prairie. The bison . . . is a keystone native species that is needed for a fully functional prairie ecosystem. Without these grazers and the disturbance regimes they create, it is unlikely that a native and functional prairie can exist at RMA unless there is constant intervention by humans to replace the role that bison play (U.S. Fish and Wildlife Service "Draft Reintroduction Plan for Bison at RMANWR" 2007, 14).

Citation #B-2c

One of the signature species of the American West, the wild bison, is calling the Rocky Mountain Arsenal National Wildlife Refuge home

starting this spring. . . . The bison arriving at the Refuge are the genetic "gold medal standard," meaning they have no detectable cattle genes in their ancestry. . . . The Service continues to coordinate with its Army and Shell partners at the Arsenal, and with regulatory agencies, local governments and other entities involved with the site, to ensure the bison reintroduction fully complies with all applicable laws, regulations and policies related to the ongoing clean-up at the Arsenal and the transformation of the site to a national wildlife refuge. . . . As the Service works to restore and conserve prairie habitats throughout the National Wildlife Refuge System, the agency has identified wild bison as a species that can and will play a vital role in this effort ("Bison Pilot Program," RMA *Milestones*, spring 2007, 1).

Citation #B-2d

As an icon of the "Old West" bison hold cultural and historic significance. More importantly, bison were a central focus for many indigenous cultures in North America. For centuries before Europeans came to this continent, bison were the center of life for the Plains tribes of North America, providing them with food, shelter, clothing and spiritual inspiration. Today, bison still hold cultural and historic significance for both Native Americans and the descendants of Europeans. There exists a strong desire to see bison populations continue to grow, at least where they do not come into conflict with other human uses of the land and its resources (U.S. Fish and Wildlife Service "Draft Reintroduction Plan for Bison at RMANWR" 2007, 13).

[10] Within U.S. settlement history, natives were viewed as undeveloped human nature, and territories were considered empty and undeveloped, treated alternately as dangerous wildernesses, wastelands, or virgin lands.[23] Lands, waters, and societies were plundered for raw materials and/or made the domain of sport. Nature was considered passive, outside the realm of progress, available for the taking and/or ready for rational management. With this understanding of the frontier, I attended one of the RMANWR's Refuge Roundup events, which has taken place every fall since 2007. The activities enlisted a nostalgic web of associations in the service of white settler views of the West and historical expansion of the frontier. The event thematically drew on the material history of North American westward expansion, the formation of the settler-colonial nation, and the hide trade of bison as a kind of primal scene, but in an entertaining format that incorporated the bison as a noble savage purified from the violent history of the frontier.[24] The frontier fetishism of the Old West drew on the bison as a totemic figure, and the exotic spectacle of actual bison incited desires for other signs of "endangered authenticity" on RMANWR land. The event essentially commemorated a reversal of frontier history—that capitalism's expansion outward to exploit nature could be reversed, and that all that was once plundered could be salvaged for ecological and entertainment value. The bison's return served as evidence of the renewability of the raw material of the original prairie, celebrating the Arsenal as host of biodiversity, and drawing on imperial nostalgia for the frontier in order to "indigenize" the land.

Exhibit B-3

E.A.G.L.E. tracked several years of bison appearances through the promotional material, newsletters, and press releases for frontier-oriented activities at the RMANWR, including the well-attended annual Refuge Roundup celebrations of western heritage. An E.A.G.L.E. intern drafted the following composite of the event's Old West-themed entertainments, and submitted sketches of the bison found on the children's craft table.

E.A.G.L.E. annotation: Step back in time onto the prairie at the Rocky Mountain Arsenal National Wildlife Refuge for an even bigger and better celebration of the National Wildlife Refuge Week and our western heritage. Pull on your boots and come on down to the Refuge for a Journey to the Old West. Partake in historical reenactments and cavalry drills. Learn roping and wrangling from real cowboys. Take a Home on the Range guided tour through bison pasture, or go for a hayride and enjoy an authentic chuckwagon supper. Be sure to round out your visit with an evening of trick shooting, dancing horses, and cowboy crooning under the stars. **[10]**

Military technologies for command, control, and intelligence often work with imperial nostalgia for the colonized, enslaved, animal, native, and noncitizen.[25] In the case of the Arsenal, a racial ontology flew in on the wings of the bald eagle. A national symbol for settler society, the eagle ushered in the historically decimated bison, evoking extinction and redemption in ways made familiar by colonial representations of vanishing natives. The settler could bring the bison back, reclaiming the animal and, in the process, sanitizing its historic extermination. This effort to redeem the past folds human control of the animal and the indigenous into the sign of the bison. The bison is valued as an indigenous creature that can define native space; the animal sign mediates identification with [11] indigeneity. Historically exterminated from indigenous life but now resuscitated to define native space as part of an authentic landscape, the visual persistence of the bison roots out the historical violence of the frontier. Essentially, the creation story at work is a narrative of settlement, one that sows a neocolonial terrain of power through a human–animal binary: the human as white settler and redeemer of native nature. From tourist experiences of the site's genesis that reference the first encounters of explorers or homesteaders, to forms of knowledge about native people as already or inevitably gone, the RMANWR is a deeply racialized ecology: race and nature serve as historical artifacts, with native humans and nonhumans fixed in nature and white humans set outside the ecosystem as "creators."[26]

Exhibit B-4

E.A.G.L.E. annotation: E.A.G.L.E. utilized the binoculars from an FWS refuge map and brochure to take an imaginary tour through the RMANWR. Along the perimeter trail of the Refuge rendering, E.A.G.L.E. discovered the following mythic biogeography:

> The Henderson Hill Wildlife Viewing Area is the high point of the Refuge's northern edge. The entire Refuge is visible including remnants of north and south plants where pesticides and weapons were once produced. To the south lies the skyline of Denver; to the east Denver International Airport; to the west Mt. Evans, Longs Peak and the mountain peaks in between. Near this point, grazing bison and pronghorn antelope as well as ancient tepee rings found in the area are sharp reminders of a time past (U.S. Fish and Wildlife Service "Discover a New Refuge" RMANWR map inset, no date). **[11]**

The bison's return evidences the renewability of the raw material of the original prairie and stages visitors as the instruments of creation, rather than as subjects that perceive nature instrumentally or that differentially value animals. The role of the human is to express the interests of nature's integrity, evoking the fantasy space of creation in pursuit of itself.[27] The return of the bison's habitat, both literal and symbolic, offers a moral landscape imbued with the idea of salvation, and of humans as stewards of the ark; reconnecting the bison with its habitat is a moral act.[28] Visitors are invited to partake in a benevolent regime of management, wherein animals are acted upon to be saved, bred, archived, conserved, captured, displayed. This spectacular division of caretaker from nature arranges for "wild nature" to emerge as originary, pure, and redeemed, in spite of the ineffaceable presence of humans and the fact that such bison are genetically engineered and "released" with intense support.[29] The struggle to preserve specific species clearly [12] engages in practices of domestication, in order to promote and preserve biodiversity. Furthermore, preservation and extermination often work together to evaluate various objects of domestication; some elements of nature are considered harmful and eliminated or banished, while other aspects are selected as worth preserving because they are unique or useful, even if as an exotic spectacle or a reintroduced native species. Preservation blurs the supposed line separating life and death. Essentially it is a fantasy of bringing back the dead, to make the dead speak: *"Animals, like the dead, and so many others, have followed this uninterrupted process of annexation through extermination, which consists of liquidation, then of making the extinct species speak, of making them present the confession of their disappearance."*[30] In the case of the Arsenal, the domesticated and restored bison serve as proxy for the timeless integrity of the prairie, and animal life, more generally, a form of promissory value creation.

Exhibit B-5

E.A.G.L.E. annotation: E.A.G.L.E. found this viewing technology on the premises. Further study resulted in the following transcribed text: "People will gaze upon a prairie much like the one early settlers encountered many years ago." E.A.G.L.E. classified these binoculars as a form of settler vision that captures native life to generate scenic value. The photograph below was filed as evidence of a powerful form of first-contact reenactment. **[12]**

[13]

The Arsenal's conversion to nature refuge has been a call to reinvest in the land and animal life as a store of Colorado native values, which can then be drawn on to increase the value of surrounding activities. Attendant to the establishment of the RMANWR, activities have multiplied for the purposes of public relations and image management, real-estate development, pedagogical investments, and military restructuring. This section of the E.A.G.L.E.'s report inventories "biovalue" cultivated around the refuge.[31] By using the term "biovalue," the account seeks to acknowledge the importance of both representational and economic logics to biopolitics. The power to reproduce life literally, via the biological capital of the species, and the power to reproduce life figuratively, through, for example, the symbolic capital of the animal sign, work together to organize the meaning and matter of life.[32] The inventory collects some of the ways animal life is called on as a potentially generative, transformative, and instrumental field, and a target of discipline, efficiency, profits, and desire.

Exhibit C: Biovalue Inventory [13]

E.A.G.L.E. made the important discovery that the reintroduction of the bison, and its associated genesis story, is connected to the growth of a broad range of practices associated with value creation. E.A.G.L.E. inventoried these activities, specifically the way they draw on the Refuge to manage public appearances, annex territory, promote industry efficiencies, and cultivate viewing pleasure, consumption, and a particular pedagogical legacy.

Rod and Gun Club Viewing Blind

E.A.G.L.E. annotation: Nature protector or nature projector? You are you and over there is there and here you are looking at captured objects from outside the staged frame. Explore the outdoors (on other side); act natural for the camera. Rather than a storehouse of raw materials and certainly *not* a

[14] The RMANWR's activities of nature watching, scouting, and round-the-clock online surveillance make animals available for consumption through a projected stream of habitat or scenery. The visible forms of bison, eagles, and other animals are observed in reified views of the wild, as a collection of observable anatomical species traits, as dehistoricized signs of biodiversity.[33] The experience of nature is highly mediated by the technology of viewing platforms that preserve animals virtually. The vignette "Viewing Platform" focuses specifically on the viewing blind as an architecture of separation that aesthetically divorces visitor and nature, viewer and animal; it arranges scenic views for the seemingly separated viewer, implying that history and nature are somehow separable. The blind facilitates the consumption of nature as romanticized preindustrial, premodern physical setting, as scenic biovalue. This "nature as spectacle" surface shrouds the site's natural history, obfuscating the drift of toxicity and the metabolic relations between human and nature. It also stimulates "eco-consumerism," which trades out the industrialization of nature for its "informatization," as scenic images, informational utilities, and leisure activities that can serve many profit agendas.[34]

[15] The RMANWR makes spectacle from the remains of the military—from viewing blinds and outposts to lakes, trails, and mounds. Like many lands within the national refuge system, the RMANWR bears distinction for its scenery owing to military outdoor recreation and human alterations to the landscape. Many of the activities and much of the infrastructure of the RMANWR betray the Army's former use of the area as a private hunting reserve.[35] A legend says that Lake Mary in the Arsenal complex, now the most celebrated wetland on the property, was originally dug out by an Army captain who enjoyed fishing and liked to use a bulldozer. It is popularly speculated that the Army Rod and Gun Club started stocking the lake with game fish long ago, initiating a tradition that continues today at the refuge in the form of a public fishing program. Another connection is the viewing blind reconstructed by the Rod and Gun Club at the end of the Woodland Trail. Although it was originally built as a hunting instrument, its replacement is a voyeuristic device that scenically captures the wetland as a landscape enframed for consumption. When I visited this site in April 2005, I also noticed several inconspicuous, old, and weathered wooden stands set up in a circle around the wetland; several photographers were using these antique props as camera tripods.

[16] Through an allegorical strategy, the vignette—"Prairie Companion"—illustrates the military's involvement in the genesis story of the RMANWR *and* the way that story, particularly its focus on the animal, has been deployed to eclipse other histories of the site. The remediation of the site used settler rhetoric to assert order over chaos, authority over nature, and release Arsenal land to premilitary time. As a result, the Arsenal's conversion enables the DOD and, by extension, Shell to advertise environmentally friendly faces at the RMANWR visitor center. Stuffed animals and simple charts about water and soil filtration assure audiences that the land no longer poses a threat to visitors, surrounding communities in northeast Denver, or the refuge wilderness. Such diagrams are considered sufficient demonstration of safety and good nature preservation. While polluting

military-classified wasteland, Nature's renewing resources emerge naturally as scenic worth, biodiversity data bank, and aesthetic-entertainment center. Decoupled everyday consumption and material nature magically recoupled by consuming nature as experience, physical setting, and slick photo surface. **[14]** The Arsenal alchemized to Refuge: scenic value redemption, affect intensification, Army history naturalization. Contaminated land transmogrified into the once-upon-a-time prehistory scenes of modernity. Army recreation remnants provide mythic support structures: a circle of old wooden stands in a mysterious clearing. Are these the former gun racks of an Army gun club spot? Camera stands for your next photo op? Or bird perches to stage wilderness shots? These Arsenal nature projectors screen landscapes and wildlife in an immersive shooting gallery. Out along the trails, you are your own mobile viewing blind. **[15]**

Prairie Companion: Commemorative Print of Colorado State Poet

The following limited-edition print and commemorative copy of Colorado State poem "Prairie Companion" were acquired in the Arsenal gift shop, accompanying the sale of a memorial birdbath. **[16]** E.A.G.L.E. included it in the Report to show the military's hidden hand in the site's creation story; it is also representative of the popular literature on Defense sites as Walden Ponds.

on massive scales elsewhere, the DOD and Shell, in collaboration with the underfunded public agency the FWS, present a PG-rated military history lesson and a cartoon environmental future, shielding scrutiny of the habitat destruction and pollution instigated there.[36] The coupling of the military with biodiversity protection and nature management perpetuates a never-ending cycle of DOD protection, against the environmental havoc of the DOD's own making. The biovalue of publicly professed nature reverence on the part of the defense industry allows business to go on as usual elsewhere.

Along with the decommissioning and conversion of former DOD lands into wildlife refuges, massive military budgets and research infrastructures have been rewired into an environmental security industry.[37] The U.S. defense industry is one of the biggest polluters on the planet and a main contributor to Superfund's National Priority List; it now partners with private industry both to clean up after itself, fueling a multibillion-dollar remediation industry, and to preserve the nature it enclosed with arsenals, air-force bases, and military camps. The main [17] Rocky Mountain Arsenal contractor, Tetra Tech EC, Inc. (formerly Tetra Tech FW, after acquiring the Arsenal's first contractor Foster Wheeler Environmental Corporation), states that its mission is "to conduct a global business directed toward cleaning up and protecting the environment while facilitating economic growth, and to do so in a safe, compliant, cost-effective manner."[38] As this quote demonstrates, the neoliberal political climate counts on the cost-effectiveness of "green." Environmental sustainability can excuse and obscure an underlying motive of profit maintenance via cost containment.

The turn-of-phrase "A Few Good Species" (a play on the usual Marine motto "We're Looking for a Few Good Men") references a public-relations campaign of the U.S. armed forces, depicting them as care providers and good stewards of approximately twenty-five million acres of public land with a hundred thousand archaeological sites and three hundred listed or candidate endangered species.[39] According to the campaign, endangered animals found on military base habitats such as Pendleton Beach were actually protected by the combat training areas.[40] This can be placed within a broader continuum of DOD activities associated with environmental security, from preemptive disaster responses to Defense-led preservation projects. The vignette intimates that, in the case of the Arsenal, [18] a process of "ecological militarization"[41] and dehistoricization is at work. This takes the expression of a renewed internal colonialism that rehearses a fantasy of first encounters, draws on the colonial language of waste to render ecologies legible, and weds military "no-man's-land" to "Garden of Eden" creation myth. Intensifying this, the widespread nostalgic desire for untouched, preindustrial nature finds unanticipated wildlife thriving at former military sites.[42] Despite the destructive technologies and treatment of the environment as blasting pad, dead zone, or wasteland, DOD lands are often lauded for their on-site de facto wilderness preservation. Adding to the allure of postmilitary landscapes as somehow largely untouched by humans or development, there is no clear distinction in the land between protecting humans from contamination and the conservationist

Prairie Companion

Along the borders of inland agricultural regions, urban areas and barracks sit the largest remains of disappearing habitats and lovely animals. The limbs of an old cottonwood lean over a clear lake in a most mysterious, wonderful secret prairie garden. The tangles of tall grass sway like summer curtains under a high arch of blue sky, occasionally dividing for small paths and a reassuring alcove of old trees.

Here in this meadow, trailheads evoke strong feelings of environmental security. **[17]** *Like an eagle in its winter roost, the garden visitor experiences safety and the privacy enshrined by careful nature conservancy. In this vast Edenic distance from all ecological ill, no one is around. No one except uniformly-green troops cultivating wildlife support systems and environmental warnings to save a few good species, occasionally removing old arrowheads, farm implements, and industrial relics found while on duty. Camouflaged in their professional ecological manners, they hardly bother the nature reverie.* **[18]**

While watching a hummingbird bob around patches of Columbine, the garden visitor wonders when these custodians assumed the care of this prairie shelter. Suddenly a mule deer's glance reveals that these troops of nature benefactors have always been here; they have been here since creation, maybe longer, tending the secret garden, weeding out intrusive doubts and ensuring nature's security. **[19]** *The visitor feels as confident as the sun with this greenery as guard.*

A warm wind from the east swirls up and out of sight leaving a blanket of cooling mist. The beautiful vista expands as the sun comes to rest on the dewy mantle over wind-swept ground dotted by slumbering prairie dogs and bumblebees happily flying under the covers of their wildflower beds.

agenda of limiting human contact to protect nature from humans; the setting aside of land as totems of biotic resurgence or national decay, biodiversity or environmental ruin, is a mutable biopolitical arrangement.

"Secret Garden" captures what ecological niches attempt to do: bind space and stop time to create a premodern nature artifact.[43] The prairie preserve renders the DOD a caretaker of the history of the Arsenal "Eden," retrospectively justifying the original military enclosure as a benevolent gesture. Secured behind the fence, the RMANWR is a biological inholding that has been cut off from "mainland" prairie by farming, development, and the Denver International Airport. Separated and valorized over agricultural or urban natures, the RMANWR is cele- [19] brated as a timeless prairie island in a sea of suburban sprawl; the Army's enclosure protected nature from unchecked population growth and the evils of human use. In this way, the refuge is consumed as a kind of Noah's ark or habitat island amid the patchwork of land use that characterizes northeast Denver, ranging from factories and sprawling warehouses to farming and suburban developments buffered with open space lots. Surrounded by forces of historical development, the RMANWR endows the area with a timeless zone of valorized nature that housing subdivisions and commercial projects draw on in order to freshen their public face.[44]

Based on my visit to housing developments nestled against the fence or near the northern Arsenal boundary, "Migration Pattern" presents a composite of the various themed real-estate developments in the area. It illustrates the way local and regional development schemes use the RMANWR to market real estate and former Superfund land for development ventures. For example, the Arsenal's environmental turn from national toxic dump to restored prairie has inspired visions of regional recreational tourism and enhanced wildlife education opportunities; it has also meant the sale of a western container of former Arsenal land from the Army to Commerce City for anywhere from one thousand to ten thousand dollars an acre.[45] Passed from Army to Commerce City hands, this land parcel, like the acreage that shifted from Army to FWS care, migrated outside the restricted-area fence, this time through public–private partnership. Kroenke [20] Sports Enterprises (owner of the Colorado Avalanche, Denver Nuggets, Colorado Crush, Colorado Rapids, and Colorado Mammoth professional sports teams), in partnership with the city, has planned 217 acres for commercial use, 130 acres for public and utility use, and 570 acres for a conservation zone. A new Commerce City Civic Center, Adams City High School, and Dick's Sporting Goods Stadium, which hosts professional soccer games, concerts, and other events, have opened in the area.[46] Sports mascots have already undergone a major upgrade to charismatic local RMANWR animals: the eagle, bison, fox, and raccoon.[47] In addition to sports team rebranding, the prime real estate of this former Superfund site provides Commerce City with grounds for recycling the entire town and its landscape of factories and refineries. Commerce City, historically one of Colorado's most industrial areas on the fringes of Denver, utilizes nature conservation as economic boosterism. The patchy landscape and habitat fragmentation of the

Eagle Migration Pattern

A wing of E.A.G.L.E., dedicated to tracking the Arsenal's migration off-site, observed and inventoried numerous Web sites offering real-estate tours of the area surrounding the former Arsenal fence. Included here is a screen capture of marketing copy from these promotional Web sites, followed by a photograph and field note taken by an E.A.G.L.E. intern during a "disaster tour" of the area.

[20] Welcome to Commerce City, folks. On the high plains along Denver's northern corridor bordering the Rocky Mountain Arsenal National Wildlife Refuge, and just ten minutes from the Denver International Airport, this city combines the best of small-town atmosphere with big development ambitions. Formerly known for its foundry smokestacks, refineries, stockyards, sugar beets and steam engines, Commerce City now serves as the prairie gateway to one of the nation's largest and most unique urban nature preserves, and to a growing regional green corridor, combining extensive trail system, an environmental campus, conference facilities, a professional soccer stadium, and Colorado lifestyle shopping. The recreational opportunities and Commerce City's location in a State Enterprise Zone providing big breaks on CO state income tax make this city the smart and healthy place to be! In fact, over 43,000 new households have already been planned or constructed.

We here at Eagle Terrace Realtors want to bring you a great new hometown. Prairie Meadows redefines "good living" in the northeast Denver area. Incorporating the best home designs with traditional suburban planning to create New Suburbanism, Prairie Meadows offers ideally-priced two-story model homes in a community environment of tree-lined streets, recreation centers, churches, parks, and detached sidewalks. **[21]** Priced from the low to mid $200k with lofted ceilings, tech rooms, and private sanctuary bedrooms and bathrooms, one of our four featured homes (Whisper Meadow, Feather Creek, Aspen Grove, and Crystal Springs) is sure to express your Colorado lifestyle. You'll discover that eagles from the neighboring refuge make regular visits, providing the most unique "neighborhood watch" in the country.

Logo Design by Joshua McDonald

area demonstrate the historic permeability of city and country, urban and suburban. Now drawn into development projects that romanticize wilderness, formerly "subaltern" wild animals are placed in the spotlight for improved city image and expansion.[48]

The term "New Suburbanism" (the meaning of which is perhaps a play on New Urbanism) and the housing layout, prices, and community mythology described in "Migration Pattern" were contrived from the Shea Homes Reunion housing development.[49] Reunion is one of more than twenty developments located on the northern boundary of the Arsenal. Other subdivisions include Buffalo Highlands, Aspen Hills, Potomac Farms, River Oaks, Still Water, Dunes, Buffalo Mesa, Outlook, and Eagle Creek, which was developed directly at the northwestern point of the Arsenal-retained land. Regardless of the porosity—and potential toxicity—of land adjacent to the Arsenal, these residential developments draw on regional environmental themes and wildlife-oriented landscape architecture, albeit as companions to low-density suburban tract homes surrounded by manicured lawns.

On my first visit, the overall housing types and thematic components of these developments looked like so many other sprawling, partially gated suburban housing projects with their green lawns springing from arid Denver dirt. However, three things in particular alarmed me: many houses were directly next to or across the street from fenced, Army-restricted land, complete with burping underground water pumps and industrial filter systems; the air smelled like a soup of acid and eggs; and, because of the constant construction, the pastoral rural America theme was relentlessly besieged by hulking piles of dirt on black plastic liners and marauding dumpsters. Reunion took the toxic farm theme to its limit with a community recreation center shaped and painted to be a corporate-branded barn, a playground with silo-themed slide, and a bucolic lake with acid-blue floor. Whether or not this body of water was actually a swimming pool, its toxic color and gurgling sounds were as disturbing as the hissing water pipes guarding the Arsenal boundary just down the street. To my utter amazement, by 2010 this pool had become a serene wetland area. Cattails lined the pond banks in thick clumps; legions of birds, a rustic viewing dock, and the overhead aerial play of a bald eagle and falcon all provided a powerful plot to enhance real estate, lending homebuyers, steeped in an antiurban sensibility, proximity to the outdoors, with the added amenity of serendipitous appearances by wild animals.[50]

Before the mortgage crisis hit in 2008, residential projects in northeast Denver, specifically Commerce City, were undertaken at a steady pace. Population had doubled in Commerce City since 2000. The U.S. Environmental Protection Agency Web site bragged that, since Arsenal cleanup began, thousands of new households had been either built or planned on the northern boundary, creating new jobs and the first housing plans in Commerce City for forty years.[51] This was taking place in spite of the fact that developments like Eagle Creek do not have reliable groundwater supplies, *and* without regard to the prospect of a contaminated alluvial aquifer.[52] While water disputes have made the news, the issue

E.A.G.L.E. intern field note from disaster tour: The subdivision known as Eagle Creek was built over groundwater that was found to be heavily loaded with the chemical by-product DIMP, resulting from the Arsenal's manufacture of nerve gas. Water continues to be a major problem for this whole area, which was rezoned for residential use after migrating from Arsenal northern land into Commerce City developer hands. [22] Numerous residents of the area depend on bottled-water provisions, and the Army has had to acquire and deliver water to the South Adams County of environmental racism in postindustrial northeast Denver has been less publicly addressed. The poor, the elderly, new immigrants, and socially and economically marginalized people live near or on the Arsenal's extended toxic plume. Environmental justice groups, such as the Colorado People's Environmental and Economic Network (COPEEN), once organized toxic tours of the metro-Denver area, which included the Rocky Mountain Arsenal. Such efforts aimed to increase awareness of the toxic environments endured daily by the predominantly Hispanic and African-American communities in northeast Denver, located in neighborhoods along the corridor of Interstate 70, sandwiched between Superfund sites, oil refineries, power plant, and trucking firms.[53] Such lessons are not cultivated in the educational opportunities sponsored by the FWS, in partnership with Shell and the Army.

The list of "Teaching Aids" displays the RMANWR's youth-focused education programs and the rhetoric of the FWS visitor information Web sites.[54] The development of the wildlife refuge has included an aggressive education campaign, leading thousands of schoolchildren on site for nature programs. As many as thirty thousand children and adults visit the refuge each year; this number is expected to increase to well over one hundred thousand visitors.[55] "Through an active public involvement program, [the FWS, Army, and Shell have cultivated the idea that] the future of the Arsenal is in good hands and entrusted to informed minds."[56] The strategic reproduction of the RMANWR's origin story requires the epistemological handing down of the ontology of the human as creator—acting on but not of the world—and of the Arsenal as pedagogical investment. [23] The wildlife activities on offer at the RMANWR endeavor to re-create this knowledge through the actual bodies of young children. Such educational programs call forth children's bodies as sites of biovalue. Learning is treated as an investment in certain ways of thinking and acting, capable of shaping individual habits and regulating populations. This educational life—a strategic site of intervention that can be molded—serves biopolitics, propagating a type of politics wherein life itself is something to be managed and controlled through technoscientific knowledge and practices.[57] Children are central to constructing the refuge's promissory future and moral landscape; they act as conduits for keeping the story alive and for maintaining the restoration of native habitat. The biovalue of this pedagogical legacy is the assurance that children will be developed as subjects that embody and enact a salvational ethic as the primary mode of conduct. The refuge's wilderness programs also promise a particular type of educational vitality that can be conscripted to enlarge areas of domestication and consumption.

Water and Sanitation District for qualifying homes adjacent to the Arsenal and well-water owners within the DIMP plume. There is evidence that contaminated groundwater from the Arsenal has moved north and northwest into the alluvial aquifer. Will there ever be a safe drop to drink again?

Teaching Aids [23]

E.A.G.L.E. found several lesson plans, parental permission slips, and field-trip conduct guidelines for elementary schoolchildren blowing around the Arsenal parking lot after a fleet of school buses departed from the visitor education center. The following is a composite list of recommendations frequently supplied to schoolteachers to generate interest in refuge activities. Note: E.A.G.L.E. has maintained the rhetoric of these education-oriented materials and, in most cases, has reproduced them verbatim.

1. Take the opportunity to expand your vocabulary while talking to visitors, especially children, by creatively using synonyms for the Arsenal. Terms you might try: "Eagle Oasis" or "Habitat Island."

2. Invite the children to participate in supervised Refuge animal hunts. Examples of recent hunts include:

- Wildlife Watch Workshops and the Shutterbug Tour (Learn how to sit still and shoot . . . your camera, that is)
- Scouting for Wildlife and the Tracks and Scat workshop (You too can become a wildlife detective)
- Search for animals on the Refuge on-line Webcam
- Don't forget to hop on board the Refuge trolley; be sure to make a reservation!

3. Visit with Eddie the Eagle, a favorite Refuge resident, and design your own outdoor field activities like Recycle Relay, Build-a-Prairie, and Spin the Wild Wheel for a Refuge trivia challenge.

4. Walk the Prairie Trail and enjoy the scenic vista of the Front Range. To capture the moment, parents can set up a tripod and camera at the Rod and Gun Club Viewing Blind.

5. Attend one of our popular conservation seminars, such as Eyes on Eagles and Prairie Blooms, or spend an Evening with a Naturalist. Join a Ranger Read-Along or request that the Wildlife Learning Lab visit your school.

6. Discover the "fun" in Superfund! Find out how waste is specially treated on-site through an informative and children-friendly exhibition on the cleanup.

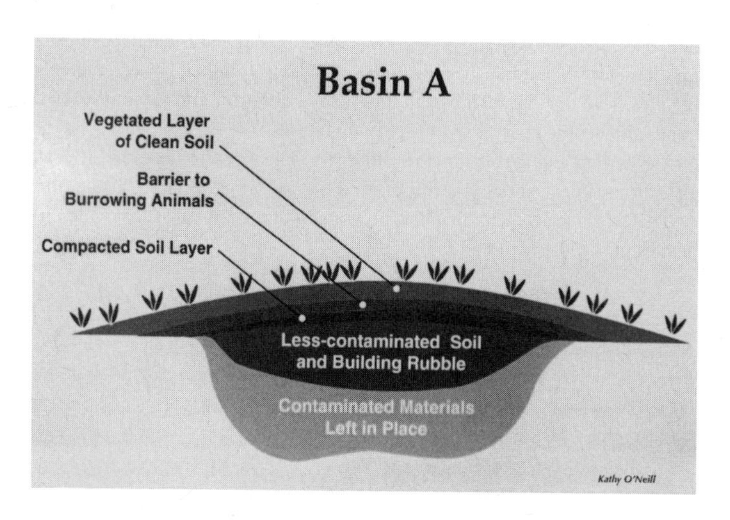

The final section of "Teaching Aids" is a selective re-presentation of safety information about the Arsenal from the Army's "Visitor Safety Fact Sheet."[58] The warning about discovering munitions and other war-related debris is no longer there; much of the Arsenal's history and toxicity have been symbolically slipped under more attention-grabbing signs of the refuge's fecundity. However, the Arsenal's toxic legacy continues to bubble up through the cover of wildlife refuge education; [24] it resurfaces time and again in the landscape, spilling out visibly in "Do Not _______" visitation restrictions and "Do Not Enter" dead ends demarcating Army property. How much longer such indications will remain, with respect to human visitors, is uncertain. But the site's past continues to impact the material world of animals, terminally. Contamination and preservation are integral aspects of the bald eagle's return to the Arsenal and the RMANWR's perpetually death-in-life refuge for the national bird.

The National Eagle Repository is the primary center for the receipt, storage, and distribution of the remains of the country's deceased bald and golden eagles. Through this facility, American Indians may apply for eagle feathers and other parts for religious use. The Eagle Repository also functions as a storehouse for illegal wildlife products and contraband under the supervision of the FWS.[59] The vignette "Wildlife Repository" uses the existence of the Eagle Repository on the Arsenal complex to critique the biopolitics of the animal by suggesting a connection between nature preservation and toxic accumulation zones. The voice of the [25] vignette is that of the satirical crypt-keeper figure, *not* the FWS or anyone else employed at or volunteering at the RMANWR or National Eagle Repository. The inflammatory and moralistic tone purposefully rejects the cool detachment of critique; instead, the figure draws on the rhetoric of horror and commits to narrating the ridiculous in order to impart a morality lesson—that something has gone terribly wrong and that it could be otherwise. The vignette closes by advocating an ethics of "deadstock" as a response to the spectacle of nature at the former Arsenal, specifically the biopolitical production of the animal as that which recuperates purity and nature.

7. Explore the Refuge's many souvenirs at the Nature's Nest Gift Shop: wildlife calendars, T-shirts, pens, stuffed animals.

8. Although the Arsenal is a safe place for kids, please remember the following safety tips:

- Children under 18 must be supervised by an adult at all times
- Sit quietly and listen to wildlife. Talk quietly only when necessary
- Respect your trail guide. Obey all signs identifying closed areas which are established to provide undisturbed sanctuaries for wildlife and to prevent unauthorized persons from entering Army property [24]
- Keep in mind that you may encounter biting or stinging insects
- Carry drinking water with you; there is none along the trails
- Do not pick up any unnatural objects you may encounter, especially metallic or fabric-covered items. Due to the history of this landscape, there is an extremely remote chance you may encounter munitions debris that could be hazardous to children

Exhibit D: Wildlife Repository [25]

Following multiagent research on the RMANWR, E.A.G.L.E. consolidated its findings to focus on the bald eagle's existence at the refuge. As part of this initiative, E.A.G.L.E. agents transcribed several Arsenal ghost stories recorded during an evening detail on a campfire sing-along. The Report includes one of these morality tales below. In an effort to re-create the effect of the storytelling

[26]

Reasserting waste as a discursive or philosophical category, the enraged crypt-keeper voice of "Wildlife Repository" attempts to interject waste's "dark matter" as an epistemological and material limit of late modernity. Waste serves as a theoretical lens through which to reexamine militarization and capitalist industrial production; waste sits on the historical horizon, as there are no longer any valueless spaces of wilderness into which we can cast off waste and dispose of its material dangers.[60] By contrast, the rhetoric and economics of risk are a representation machine that employs tremendous abstraction to displace waste's threat. Mobilizing the remediation industry, "risk assessment, a form of cost-benefit analysis that conceals social interests in the form of mathematical possibilities, evolved as a managerial strategy to respond to the massive Cold War public anxiety about hazardous technologies."[61] Such risk wizardry and cost-benefit abstractions attempt to dispel the materiality of waste and deny various "bodies of knowledge" and remainders that testify to the continued dangers, illnesses, and deaths that follow in the wake of such technical number conjurings.

experience, E.A.G.L.E. has appended two visual artifacts of the bald eagle's vitality and morbidity at the RMANWR.

Building 128

Isn't that charming?! HA!—a habitat island for children, eagle-viewing platforms from the Gun and Rod Club, a new whitewashed face for ruddy agro-industrial Commerce City with groundbreaking environmentally friendly suburban development (let's hope those homes don't have basements and the city hosts a large bottled water recycling program!). The former military warfare farm now bucolic wilderness sure is a good neighbor, isn't it? An outdoor laboratory to keep Army environmental security in business. A showpiece of Shell's corporate sacrifice to preserve future health. For all instructional purposes, the lesson plan of other Department of Defense and Department of Energy cleanup courses. The earthly stage of risk-assessment magicians with their statistics-disappearing tricks that turn populations and hazards magically into calculations of uncertain future probability, ratios of one to tens or hundreds of thousands, that suddenly (POOF!) disappear in a cloud of secondhand smoke and reappear as everyday sun exposure risk. [26] The living classroom of thousands of children a year under the supervision of corporate-military-state sponsorship and scholarships. And the living history museum of western prairie heritage, complete with an animal menagerie of bison, reproductions of indigenous life, and reenactments of the frontier. The Arsenal's creation story shows us how to reproduce ecological futures by simply removing the marks of exploitative human practices and militarized zones from our memories and GIS remote-sensoring screens. *Let's all feel good.* Isn't that why the

bald eagles snuck in the back door to the former kitchen of the U.S. military-industrial complex to instruct us in how to cultivate "green" in the remains of war? Or is there perhaps more to the story? Is there not something sour about this secret garden where eagles dare, some rotten part or alarming ring to its "look but don't really touch" instructional care, some deadly matters unattended in this seemingly happy solution to military chemical properties?

Removing one chemical, DDT, from the environment has contributed to the recovery of the nation's symbol and most beloved bird...
...the bald eagle.
Egg size: 6-12 cm
Nest size: up to 12 feet diameter
24

This end tale, after all, is a story about entrails, indigestion, and messy by-products that define the limit of modernity and the nature/society dialectic. Stuff so toxic it can only be stabilized materially and discursively as "waste." These excessive inexpressibles and deadly invisibles threaten epistemological certainty and physical awareness. Lurking at the historical limits of industrialization, in the shadows of military experimentation and hypermodern consumer capitalism, they are the dark matter of nightmares relentlessly imposed on social life. Leaky basins and barrels, underground tanks of goo, drums of death liquids sitting in open air daring you to look inside even though you are already petrified. Panic circles these sites, real or imaginary, like seagulls—or eagles. These mentally indigestibles regularly kill where and when they seep into the social. They are the living dead (or deadly still-living). The horrifying product of

A discourse of the desert as an unproductive, infertile wasteland has presented the western United States, especially the Southwest, as a barren, uninhabitable region: the logical candidate for hosting secret bases, nuclear weapons production facilities, and extractive technologies on a mass scale. This discourse helped the military commandeer millions of acres of land for weapons testing and development. The wasteland discourse obscures the actual richness of resources in the West in order to justify relentless plunder through extremely environmentally destructive technologies. Much of this land, considered expendable as sacrifice zones, remains central to different Indian nations. The mass destruction of the homelands of indigenous cultures, as well as the silent casualties of millions of soldiers, armament workers, and downwind civilians, is naturalized in the landscape of the West largely as an invisible, secret holocaust.[62] Although "Wildlife Repository" does not present scientific evidence that directly links Arsenal toxins to the premature death of bald eagles, the vignette horrifically reveals the animal body as abject "bare life" and a national sacrifice zone.[63] The assertion that eagles are "waste cans" insists on recognition of the ways the animal serves as both "sink" and "sentinel" for the human. Eagle waste cans also show that the legacy of militarization's racialized ecology continues to occupy the land and human–nonhuman relations. The bodies of eagles are managed and reprocessed, after death, by the repository, which collects eagle remains and redistributes them to American Indians who apply for ceremonial use rights. While this ceremonial use of eagle parts is an important ecological counterpoint to the cannibalistic cycle of nature re-creation and the installment of the bison as a paradigmatic sign of the native, such practices are frustrated not only by extensive bureaucratic paperwork but also by the extent to which eagle remains are tested and decontaminated, before they are redistributed. In addition, the redistribution of eagle parts to American Indian applicants easily evokes the idea of the site—the animals too—"going back to the natives," reinforcing the contemporary settler's creation story, and equivocating the toxic links between wilderness preserve and Indian reservation.

military-industrial war making the world over. The bad matters of so many cannibalizing nature reproductions. A toxic soup that, uncanned, relentlessly gums up institutions, smooth rational systems, and capital accumulation. Refuse that refuses a final resting place, leading to strange ecologies and geologies—futures of unpredictable material natures. Welcome to modernity's twilight zone of undead remainders.

The Arsenal-as-nature-refuge story ends big with one last-gasp attempt to put down these unsettling residues. To stop up the drain of impossible-to-measure costs of cleanup with a spectacle of nature. To eliminate doubt about the promissory return of an urban refuge by calling on the bald eagle in the service of the national imaginary, with the added bonus of generating cost efficiencies. To encourage a more than $30-billion industry to remediate polluted areas—but not too much, just the right touch. To avoid messy host fees, interstate delivery systems, and political struggles over the boundaries of an outside wilderness dumping ground made from within, like so many poor-populated desert regions and American Indian reservations. **[27]** To muffle not-in-my-backyard public outcries over waste management's information containment with the touching hand of wildlife protection and signs of native frontier life in place of more recent military histories and their chemical legacies. To close the deal on a dragged-out court dispute over resource damage, by counting animal corpses the legal basis for compensation; to ensure human safety and cleanup integrity through the ongoing monitoring of wildlife morbidity.

A well-guarded sanctuary of High Plains ecology, the RMANWR hosts rare species and protects such animals as the bald eagle. These birds have historically suffered: at one point only 417 nesting pairs were known to inhabit the lower forty-eight states, owing to DDT impacts and habitat loss. Thanks to national protection efforts, nearly ten thousand pairs now exist—a remarkable comeback. The bird has been declared "recovered" and flown off the endangered species list; it has even left behind the classification status of "threatened species."[64] As this chapter reported, the bald eagle was an instrumental figure to the entire RMANWR refuge production. The visibility and protection of this charismatic creature politically enmeshed with the positions of various stakeholders: the bald eagle was called on to rescue purity, generate powerful affect, and create the possibility for a budget cleanup and new origin story. In response to the bald eagle's inevitably compromised position as a symbol of the state, the E.A.G.L.E. collective—its ambiguity as the real bird, the national emblem, and forensics team—has attempted to call attention to the strategic cultivation of nature spectacle at the RMANWR and the operation of a range of instruments on the social flesh and bodies of animal life.[65] Through the voice of the crypt keeper, E.A.G.L.E. seeks acknowledgment of the tragic aspects of the biopolitics of nature, such as the biopolitical violence of the power to preserve animal life. There are animal lives that are made to flourish, while others are extinguished, sometimes in direct relation to each other. In the case of the Arsenal bald eagles, they are made to live by the biopolitical power to keep animal life in interminable survival; the bald eagle's ongoing Arsenal presence follows the spectral logic of a body that eternally survives. The final vignette, therefore, directs attention to the bald eagle's vulnerability, and the sick way the animal is made to emblematize life even though its condition of life is always marked by death—as waste-can potential—owing to its part-time residency at the former Arsenal. Even the bald eagle's actual death obscures the vulnerability and porosity of the life of the eagle. In the eagle repository, the material remains are divided up and redistributed through the strict returns policy of religious use or educational purposes. By federal law, eagle bodies and parts must end up on the preparation table of the eagle repository, where they are recycled, in effect implicating indigenous significance and religious and memorial use in the reproduction of the site's creation story. The animal is drained of its historical substance, enabling the bird's timeless sign of life at the refuge; its literal body is dismembered and recycled through the perpetual control of animal life and death by the state, as the ultimate behind-the-scenes settler-creator.

The Arsenal ultimately serves a life sentence, as a bounded wilderness-forever resting place. Creation stories can stop nature dead, weeding out previous histories and ignoring the vibrancy of borders. Animals serve, on reserve, as media signs and genetic material, as biological livestock and semiotic stockpile. They hang around in various states of suspension, their lives instrumentalized as undying currency and regenerative material. They are drained of historical substance to maintain the integrity of the ecosystem's new origin story.

What is worse, being made to flourish or put to death to rescue purity? What, you may ask, is the difference between preservation and extermination in light of the Arsenal's recent history? And how do we understand the bald eagle's existence, as a figure of flight from the Arsenal's past, or as the crypt keeper that secures the Arsenal's deathly grip?

In this wilderness underworld, this land of the damned, the bald eagle raises value from the dead. It is a boundary creature between waste and nature, death and genesis, history and pure origins. Caught in a life-and-death tailspin of salvation and sacrifice, the eagle tends toward the undead; yet the bird can never be death. Its fate is infernal survival—its spectral body always ultimately delivered for recycling onto a preparation table at the national eagle and wildlife repository. **[28]**

"Wildlife Repository" concludes this fable of the RMANWR with an enraged response to contemporary biopolitics of the animal: an "ethics of deadstock." Appalled by the biopolitical regularization of a certain kind of death-in-life of the bald eagle, the vignette calls on an ethical practice of taking stock of the dead.[66] Such a practice endeavors to account for the ways certain deaths have been necessary for the eagle's vitality and the Arsenal's reclassification as nature refuge. "Deadstock" insists on acknowledging the mortality effects of nature's return. Numerous animal lives have been sacrificed as surplus or collateral damage. The quantification of dead animals as damaged resources was essential to the successful conclusion of a more than two-decades-long court battle by the state of Colorado for compensation from the Army and Shell. It was estimated that at least twenty thousand ducks died in a ten-year span at the Arsenal; hundreds of dead waterfowl were found each day around the Arsenal's four lakes during the 1970s; many mammals such as prairie dogs and badgers fell victim to chemicals and suffered reduced reproduction rates, birth defects, and chemical paralysis.[67] Such gruesome counts of the dead led to the allotment of $17.4 million by Shell and the Army for natural resource damages to the state of Colorado, with a $10-million donation from Shell earmarked for a northeast greenway corridor to preserve habitat, restore resources, and provide new recreational opportunities.[68]

How death is to be managed is a quandary for biopolitics in that preservation blurs the supposed line separating life and death.[69] The struggle to preserve specific species clearly engages in practices of domestication, even extermination, in [29] order to promote biodiversity. This is exemplified by the fate of the bison, which might be said to live out a kind of injury through their preservation and reintroduction; not only is their genetic suitability in question, but they have been severed from connections to emplaced kin or habitat or instincts developed directly in relation to particular environments. While the bald eagle population has made a remarkable comeback after historically suffering from extensive hunting, pollution, and loss of habitat, eagle protection at the Arsenal has at times required the mass extermination of contaminated prairie dogs and other organic threats. These deaths spectrally inhabit eagle nests; the accumulation of such deadstock ensures the bald eagle's terminal survival at the RMANWR. Just as some animals serve as national symbols or signs of the native, others are conscripted as biomonitoring indicators of chemical exposures and contaminant-related health effects. Rodents and starlings, for example, provide the means for determining the biological effectiveness of remediation efforts, the possible pathways of exposure, whether contaminants have found their way into food chains at biologically significant levels, and the extent to which cleanup objectives have been successfully met or not. Such animals function as gauges of the health and toxicity of the refuge; they provide a way to evaluate risks associated with dosages and exposure routes, as well as a means for testing the validity of laboratory models of exposures and effects.[70]

Dead wildlife, which serve as instruments for assessing the site's condition and for laying claim to compensation for ongoing injury, haunt the site's future

In an unassuming brick crypt in building 128, tomb for animal skins and illegal wildlife parts, dead eagles are quietly collected and cataloged in a stuffy back room. They are plucked, disemboweled and dismembered, preserved and passed on to religious-use applications and educational institutions. Poor, poor national bird, looks like your bones are in fact national sacrifice zones. Eagle bodies as waste cans—HA! What ingenuity! **[29]** In the end, the refuge in your honor secures not only mountain-prairie habitat, but your arrival at the wildlife repository per your federally mandated interminable survival.

development and testaments to safety. An "ethics of deadstock" undermines the spectacle of nature by demanding that the national association of the bald eagle with purity *take stock of the deaths accumulated to sustain the life of the state and of development projects.* Deadstock scavenges for the historical materiality of animals, whose bodies are otherwise expunged by representations of nature as pure, and the animal as natural (and whose relations with each other are typically rendered insignificant and immaterial), to teach a material lesson on the mortality effect of the conversion of chemical weapons factory to nature refuge.[71]

[30]

"Wildlife Repository" ends with a quotation from cultural critic Jean Baudrillard.[72] The bald eagle has had enough and quits as national bird, granting a final interview to *The Onion.*[73] The E.A.G.L.E. report is currently under governmental review for the environmental risks posed by its critical content. The Office of Authorial Remediation continues to remediate Dr. Krupar's personal landfill of Superfund stories.

WE Quit!
We, the bald eagles, officially resign from U.S. employment as secret agent of de/militarization and as national symbol of wilderness and freedom.
We hope that we have shown by now,
That because of Defense, capital and even conservation know-how,
There are places created where
Only eagles dare,
Places where not everything that is solid melts into air!

Please remember, dear listener, this parting lesson: "it is the living we embalm alive in a state of survival." [30]

Exhibit E: The Bald Eagle's Official Resignation

E.A.G.L.E. ends this report with copy from a protest sign found in a suddenly abandoned eagle roost at the RMANWR.

> We, the bald eagles, officially resign from U.S. employment as secret agent of de/militarization and as national symbol of wilderness and freedom.
>
> We hope that we have shown by now,
> That because of Defense, capital and even conservation know-how,
> There are places created where
> Only eagles dare,
> Places where *not everything that is solid melts into air*!

02

ALIEN STILL LIFE

MANAGING THE END OF ROCKY FLATS

Office of Legacy Management

U.S. Department of Energy Spoofs

Before the
Subcommittee on Satirical Appropriations
Committee on Postnuclear Environmental Productions
United States Senate

ROCKY FLATS CONGRESSIONAL TESTIMONIAL AND POWERPOINT SLIDES

October 2011

"Government Warning": This PowerPoint transcript and slides are an allegorical bureaucratic drama of the cleanup and postnuclear management of the facility Rocky Flats, involving the U.S. Department of Energy's Office of Legacy Management and Kaiser-Hill. You should not construe the inclusion of government-issued photographs, charts, statistics, or directives—or that of any corporations—as an endorsement by the U.S. government or private industry of the views contained herein; all content has been arranged strictly by the author. The characters are fictitious, for the purposes of presenting the author's analysis; any resemblances to real persons are coincidental. The satirical content is intended to show the tremendous labor spent erasing signs of labor, the logics of efficiency at work, and the cleanup operations of "legacy management." Additionally, note that the logo for the U.S. Department of Energy Office of Legacy Management included in the PowerPoint is original but no longer in use. Its employment here is not an official Legacy Management endorsement of this text, nor is this text intended to fool you into thinking it is an official representation.

Speaker #1: Lester Ismore

Good afternoon, Mr. Chairman and the distinguished members of the committee. My name is Lester Ismore. I am the Bureaucracy Reduction Director (BRD) of the Office of Legacy Management (LM) at the Department of Energy (DOE). Established in 2003, the Office manages postclosure responsibilities of DOE's former nuclear weapons production program. We do the following:

- Manage legacy land and assets, emphasizing reuse and disposition of former weapons facility property
- Maintain environmental cleanup for the protection of human health and the environment
- Support safe disposal of workforce structure
- Centrally contain records and information and improve public relations
- Increase organizational efficiency by developing cost-effective oversight and reducing actions to that of making appearances

It is with great pleasure that I report to you today the significant progress Legacy Management has made. As more sites are closed, cleaned up, and turned over to the LM, we have reduced our staffing and streamlined our performance, simultaneously expanding our oversight of DOE environmental remedies. We have efficiently and cost-effectively disposed of DOE lands to the public, and we have maintained appearances while reducing our management burden, earning the distinction of high-performance government organization in 2007, as well as a DOE Management Award in 2010 for our outstanding contribution to overall DOE cost savings.

Office of Legacy Management

Maintenance of cleanup (post-D & D)

Custodial control of former weapons facilities

Re-use / safe disposal of workforce structure

Information management

Sustainability

When concerns were raised over the staggering contamination from Cold War operations and the DOE's potentially boundless responsibility, the LM was created as a separate office to operate on guaranteed minimal funding for long-term stewardship of nuclear waste liabilities in perpetuity. Per that end, and owing to the military-to-wildlife model set in motion by the Department of Defense (DOD), we have innovatively partnered with the Department of Interior's U.S. Fish and Wildlife Service (FWS) to minimize management by establishing wildlife refuges in place of weapons facilities. Such refuges both adhere to state-of-the-art risk assessment techniques required by law and appear to do most of the work of maintaining the legacy of the cleanup and mitigating long-term harm to the public. Our most recent fiscal-year budget request demonstrates the next phase of our management plan: the dismantling of our own organizational waste. We have already succeeded in substantially decreasing our program funds and staff, even cutting back on our Internet footprint.

How have we been able to log in so many successes? The cleanup of the more than fifty-year-old Rocky Flats plutonium production facility near Denver, Colorado, completed in 2005, has set an important precedent. Site closure and legacy management of Rocky Flats have demonstrated that limited funding and time limits encourage innovative cleanup and expedited stewardship of former DOE nuclear sites far into the future. Today, I have asked three speakers to join me for the purpose of sharing Rocky Flats' story. Their testimonies will show you the logics underlying the conversion of Rocky Flats from plutonium plant to wildlife refuge, and demonstrate Legacy Management's commitment to spending money only on appearing to manage.

Office of
Legacy Management
LM
Rocky Flats
Photo courtesy of the U.S. Department of Energy

Speaker #2: Shirley Goodsell

I am so honored to be here today to testify before this committee on the 2005 completion of the Rocky Flats cleanup. I was the Kaiser-Hill spokesperson for the Rocky Flats Closure Project, responsible for removing one of the nation's first nuclear weapons complexes from the Superfund National Priorities List. The closure of Rocky Flats has really been a *fairy-tale ending.* Original estimates said it would take seventy years and $37 billion. But on October 13, 2005, Kaiser-Hill, the company DOE hired to remediate Rocky Flats, declared completion of the project. Against all odds, we completed cleanup fifty-six years ahead of schedule at a savings of $29 billion. *Isn't that unbelievable?*

All you ever hear is "We can't clean up radiation." Yet the successful closure of Rocky Flats shows that the DOE and its contractors *can* make a mess and clean it up. Not only can we mop up radiation in our former nuclear weapons factories, but we can accelerate the Superfund process and use innovative technologies that save time and money. Now every defunct old nuclear weapons complex can become the next *field of dreams*—that is the legacy being made. The tremendous success of DOE's Rocky Flats cleanup radiates that optimism. But in order for you to really understand the job we did at Rocky Flats, I should first let you know what we were up against.

Since 1952, hundreds of tons of plutonium and other radioactive and hazardous materials cycled through the Rocky Flats Plant, producing all sorts of nasty toxic leftovers. Unfortunately, the DOE and its predecessors, like miners out of the Old West, simply shoveled out the waste and left it to the elements! The vacant land of the prairie seemed as good a place as any to take out the trash. In the beginning, workers mixed waste with concrete in fifty-five-gallon drums and stacked them outside, where the drums would leak or explode from unvented gases. The facility soon earned the distinction of "world's barrel capital"! Over the years, waste was dumped in burial trenches and pumped into solar evaporation ponds; sometimes it was burned and buried. The plant caught fire twice, producing even more waste. Creeping further and further afield, toxic refuse congealed at new waste sites such as the 903 Pad, a prairie parking lot for more than five-thousand barrels. When contaminated pond water was sprayed over the surrounding prairie buffer zone, high winds

Photo courtesy of the U.S. Department of Energy

Photo courtesy of the U.S. Department of Energy

and water runoff spread the site's radioactive and chemical residues far and wide. Plant operators tried to get rid of leftover pond sludge by mixing it with concrete and pouring it into large plastic-lined cardboard boxes. Thousands of the "pondcrete" blocks were made between 1986 and 1989, until it was discovered that the mixture had not hardened but remained toxic pudding. *What a mess!* It's no wonder Rocky Flats was one of the most hopelessly contaminated of DOE's old Cold War facilities! The FBI even raided the site on June 6, 1989, seeking evidence of environmental violations in what was called "Operation Desert Glow." A federal grand jury convened to determine whether DOE and its corporate operators knowingly polluted. *Now, isn't that just like a paperback thriller?!*

Well, under the dark cloud of the FBI's first raid ever on a federal agency, Rocky Flats' Superfund listing and the end of its plutonium production line actually facilitated a new beginning for the site. The DOE was inspired to hire Kaiser-Hill to clean house. *Finally, something to celebrate leaving behind: the legacy of cleanup.* I call this "Operation Sweep Away." Rocky Flats is entirely gone now. All the waste—the entire facility—GONE! No mess! This is just a short list of what was removed:

- 62,000 shipments of waste have departed Rocky Flats since 2000.
- On average, the company Material Stewardship sent a shipment of waste off-site by truck or rail once every seven minutes.
- More than 600,000 cubic meters of radioactive waste were swept right off site, enough to fill a railcar ninety miles long!

All of it, GONE! Isn't that unbelievable?

Now, I'll let you in on a couple of handy solutions that made plutonium cleanup possible. I call this "In-house Know-how." First, we pitted sugar against plutonium for a sweet victory. We were able to clean up infinity rooms—rooms that had been sealed off and abandoned, where contamination was so high it could not be measured. We cleaned out these closets by pumping in an aerosol fog consisting of sugar, glycerin, and water with a flexible duct the same as any old gas clothes dryer. The sweet solution made plutonium stick to surfaces that could then be shaved off the walls and floor, canned, and finally packed in cargo containers using special foam to fix everything in place. *Isn't that unbelievable?*

- 21 tons of weapons-grade plutonium
- 106 metric tons of high-content plutonium residue
- 30,000 liters of plutonium solutions
- 594,000 cubic meters of low-level and low-level mixed waste
- 575,000 tons of sanitary waste
- 700 tanks for storing waste
- 1,450 production glove boxes
- 97,800 tons of contaminated soil
- millions of classified items and excess property
- 1 million sq. ft. plutonium facility real estate
- 800 buildings

ALL GONE!

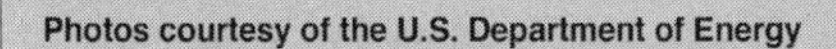

Photos courtesy of the U.S. Department of Energy

Photo courtesy of the U.S. Department of Energy

A second cleanup innovation involved the disposal of plutonium-contaminated glove boxes. First, we hacked them up with electric saws until a worker was badly cut. Then plasma torches were used, later fancy robotics. And do you know, all it took in the end were some spray bottles of cerium nitrate, squeegees, and rags? We simply had to wipe off plutonium, just like you do when cleaning the bathroom. Can you imagine? The DOE cleaning your bathroom? I'll have you note this simple domestic cleaning technique scrubbed clean the most contaminated building in the country, the more than fifty-year-old plutonium production building 771/774. *Unbelievable, isn't it? The nation's most contaminated building just wiped off.*

Another exciting legacy of cleanup is what I call "Surface Dress for Less." Why pay more when you can look good for less? This went a long way in sweeping the legacy of waste under the rug. Cleanup levels of contaminated soil were set according to surface/subsurface distinction: a conservative allowance of radioactivity for the top three feet of ground, a 2,000 percent increase of radioactivity three-to-six feet below. And beyond six feet anything goes, since contamination at this depth does not pose a danger to surface dressing. I'll have you know this amazing cosmetic arrangement allowed for a minimal percent of the total budget to go toward cleanup of the soil and water.

Cleanup and Demolition of 771/774
"World's Most Contaminated Building"
Photo courtesy of the U.S. Department of Energy

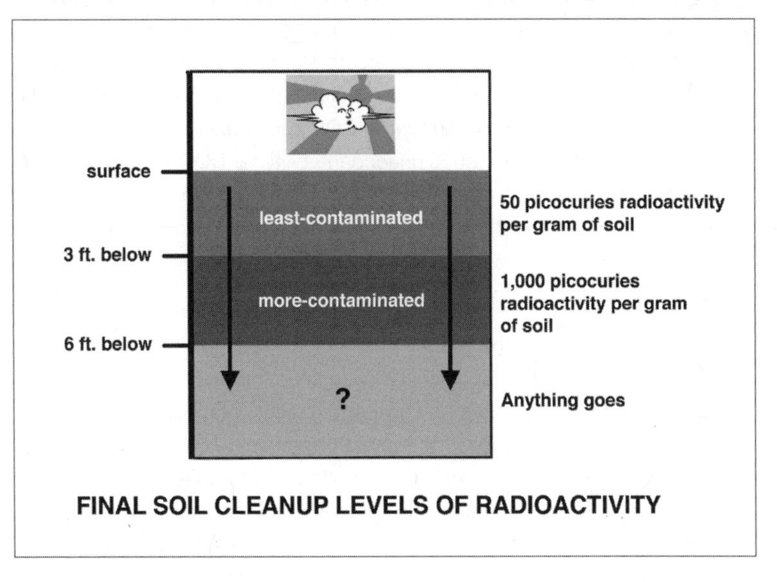
surface
3 ft. below
6 ft. below
least-contaminated
more-contaminated
?
50 picocuries radioactivity per gram of soil
1,000 picocuries radioactivity per gram of soil
Anything goes
FINAL SOIL CLEANUP LEVELS OF RADIOACTIVITY

Last but not least, the "Stay Fixed and Flexible" model. Polluters and regulators really collaborated under a trim Superfund program, one that fixed prices and completion dates, legally compliant risk levels, and future land use, regardless of public input. Under a flexible incentives-based commercial contract, Kaiser-Hill instituted ASAPs—Accelerated Site Action Projects—eventually putting the finishing touches on Rocky Flats' fresh new look thirteen months ahead of schedule for a bonus of $355 million. *Isn't that unbelievable?!*

Well, time to wrap up. I hope you now see Rocky Flats as I do: a spectacular DOE-homespun environmental remedy. I give the floor to Barry Graves, from DOE's Office of Legacy Management, who will discuss administration of the site after cleanup.

Speaker #3: Barry Graves

Good afternoon, committee. I am Barry Graves, Director of Strategic Disposal and Unaccountability, DOE Office of Legacy Management, or LM. The LM is in charge of the inventory of DOE cleanups after their completion. My department, the Strategic Disposal and Unaccountability arm, eliminates any endangerment to DOE appearances of environmental remedy. *My point:* we produce unaccountability for DOE. Now, in the Rocky Flats case, we have successfully maintained the site's surface after D&D (decommissioning and decontamination), with unaccountability production at an all-time high. But first, you need to see the big picture.

The private sector has sustained impressively high levels of unaccountability over the years in the nuclear weapons industry. Corporate operators of Rocky Flats were fined for environmental abuses in 1992; however, this drain on corporate unaccountability was stopped by DOE's indemnifying the former plant operators. Any charges brought against the government-contracted companies would be paid by the government, that is, the taxpayers. The DOE also ensured the unaccountability of the waste remediation company Kaiser-Hill. It was contracted to clean up Rocky Flats less and faster than original estimates to receive a large bonus for early completion. Long-term quality control is handed off to us, the LM. *My point:* private contractor money up, unaccountability up.

- Operation Sweep Away
- In-house Know-how
- Surface Dress for Less
- Stay Fixed and Flexible

Photo courtesy of the U.S. Department of Energy

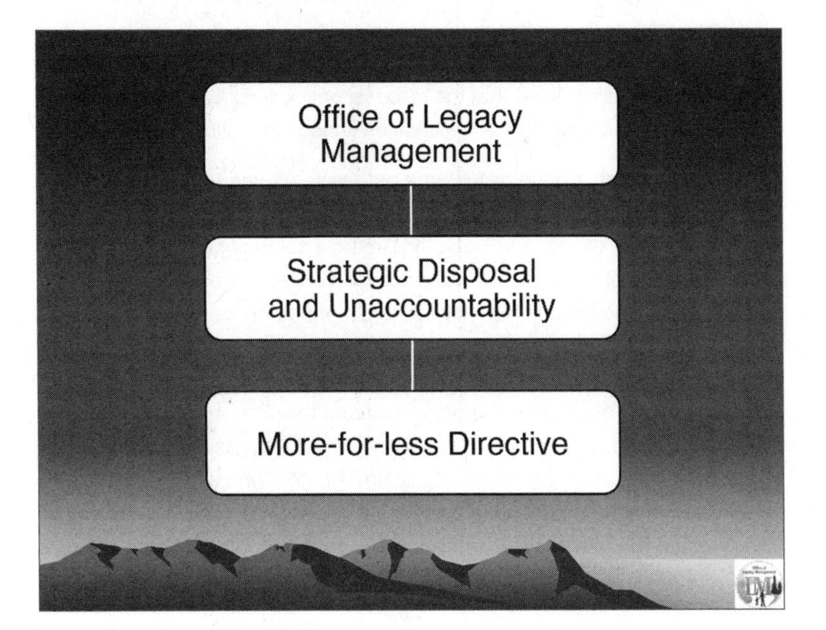

What this means for DOE is a different picture. The DOE-LM is responsible for maintaining more and more cleaned-up nuclear weapons sites, where cleanup is done less and less by corporations to get more and more money. This means that overall DOE unaccountability *has been in recession* since the Cold War and will continue to drop unless we take action now. *Am I making myself clear?* What we need to do is stop this downward trend and raise unaccountability levels back to what they once were.

So, *how* do we do that? *How* do we boost unaccountability? I'll show you. Rocky Flats provides the model.

See here. As this graph demonstrates, DOE unaccountability has been on the rise at Rocky Flats since cleanup commencement in 1995. We can see a sharp increase specifically when the DOE made a "no landmark decision" to tear down all eight hundred buildings, sixty-four of which the State Historic Preservation Office considered eligible heritage sites at Rocky Flats. The LM predicts that this rising trend will continue because of the *successful reduction of the stockpile of memory triggers.* Several studies we have made of the cleanup process show that there is a direct inverse relationship between unaccountability and residual memory triggers. *Do you follow? This is a significant discovery.* The Rocky Flats experience has revealed that the *disposal of visual residuals* during the cleanup process substantially increases unaccountability.

Some examples. During a DOE tour of the site offered to former Rocky Flats workers in fall 2004, the tour operator had to use a GPS instrument in order to find the former Protected Area; he could not discern where the center of the production site was once located. A second example: the DOE Field Office circulated online a series of photographs of the industrial area after cleanup. A Berkeley graduate student, in an independent study, collected the feedback these images received among the former workers during the fall of 2005. Her findings included the remarks of one former employee who was dumbfounded he could not recognize the site of the notorious plutonium production building 771/774, where he had been contaminated with plutonium as a safety inspector. Another former worker wrote that the former plant, where he had worked for more than a decade, looked *alien,* as if he were encountering a still-life image of an unknown landscape. *Do you all get the message?*

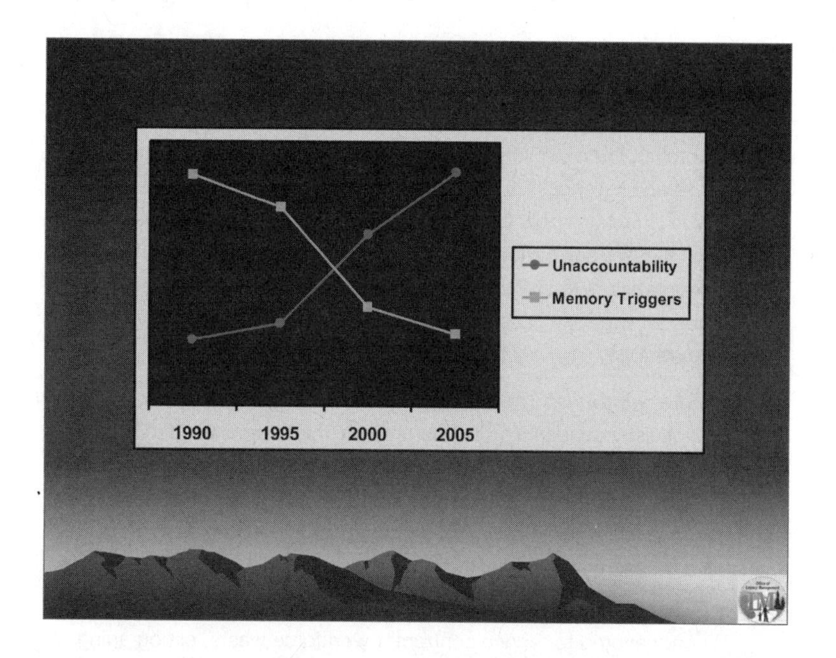
Unaccountability
Memory Triggers
1990
1995
2000
2005

Photo courtesy of the U.S. Department of Energy

Now, what does the LM do to sustain these impressions? First, the LM, like other government offices, is responsible for sanitizing documents. Thousands of cubic feet of Rocky Flats files have been shipped off to Morgantown, West Virginia, for containment in a new state-of-the-art energy-efficient repository. Second, the LM has institutionalized a very proactive "inaction" policy. We isolate potentially dangerous waste streams that we see surfacing, and then "do nothing" in order to sustain unaccountability. *Do you understand what I'm talking about?* There are a number of ways we have implemented inaction to keep Rocky Flats unaccountability up. I will briefly cover four today:

- *Museum Burial:* The LM considers the residual waste that workers, activists, academics, and community members collected for a Rocky Flats museum a danger to the site's cleanup. Although a bill gave the DOE the authority to establish a museum, we took no action in order to bury this proposal; we have done nothing to help a museum surface at Rocky Flats.
- *Benefits Discontinuity:* Over the past decade, the DOE has endeavored to reduce contractor workforce waste, encouraging the view of Cold War worker turnover as inevitable and necessary. The LM, therefore, has done increasingly nothing for worker transition or benefits, even eliminating some of our responsibilities in this area of our original mandate.
- *Administrative Distancing Devices:* The LM has outsourced the monitoring of DOE-retained land at Rocky Flats to a private company. As the number of sites under LM oversight increases, the LM has strategically minimized land administration to the activity of overseeing contract compliance. Another significant innovation in this respect has been the implementation of remote-controlled collection of real-time monitoring data from LM sites.
- *Material Waste Stream Containment:* Generally speaking, the DOE has treated the collective efforts of workers and their survivors, who have organized to seek compensation for occupational exposure to chemicals and radioactive elements, as a point of nonreference. The DOE very actively did nothing much to help their compensation cases, in spite of stated aims to protect and maintain environmental health. The problem with the "do-nothing" strategy in this example,

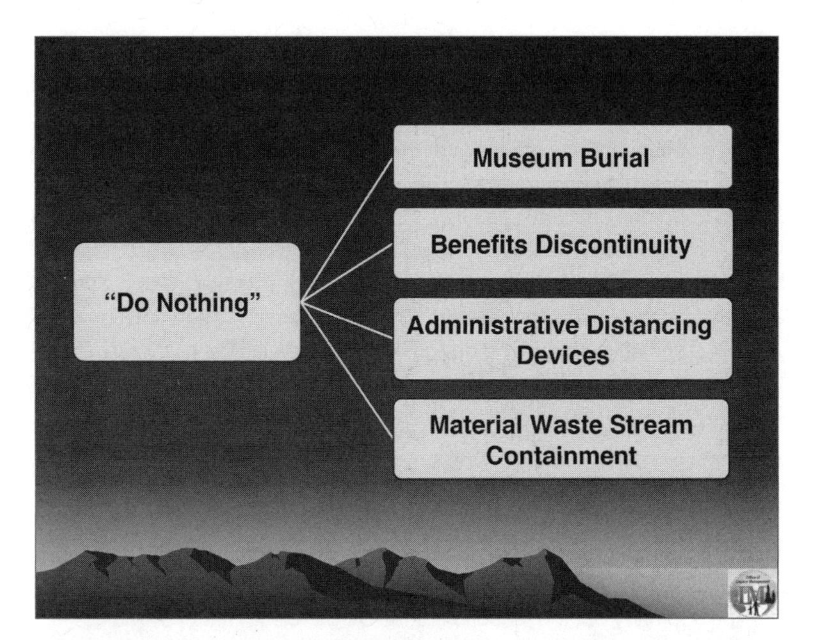

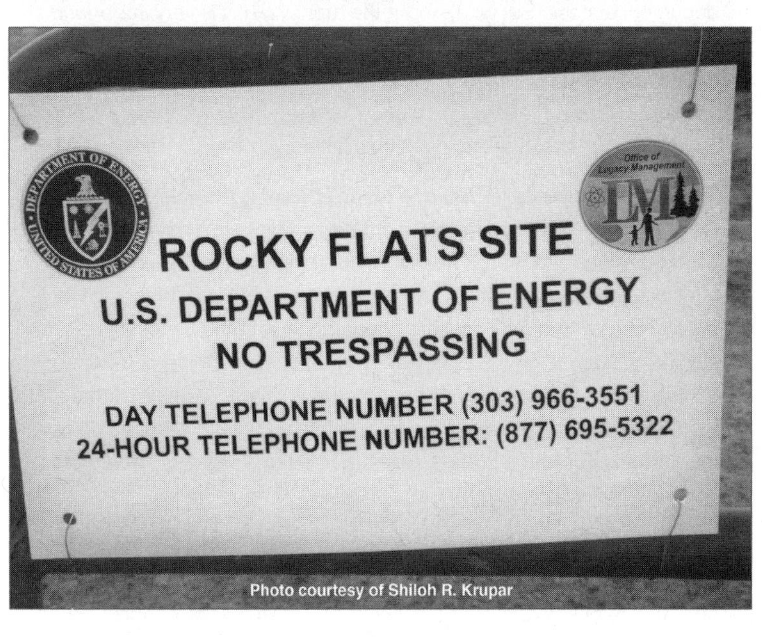

Photo courtesy of Shiloh R. Krupar

however, is that it was an extremely expensive endeavor: the DOE paid around $92 million to do so little for workers. This resulted in the transfer of the compensation program to the Department of Labor and the reduction of the DOE's role to that of merely providing documents. While the curtailing of DOE's management burden could be seen as a success in spite of the cost, it expanded LM's responsibility and actually required significant investment in a long-term occupancy solution to assist the agency in executing the conveyance of documents. The LM hopes that the remote location and efficient building design of its new LM facility in West Virginia will improve the DOE's public image, decrease the demand for DOE records, and/or reduce the energy spent accessing files.

Now, I have given several examples that show how the DOE-LM has successfully raised unaccountability levels. However, in the last instance, "doing nothing" turned out to be very expensive and further damaging to DOE's appearance. Moreover, a significant amount of money had to be spent to carry out the reduction of the DOE's role to that of providing document access. So, what about the money? *This is very important: we need to account for the money.* We don't want inaction to increase our expenses; we want "doing nothing" to decrease cost. *Bottom line:* we want unaccountability up, residual memory down. And we want to "do nothing" more efficiently and cost-effectively. *No questions.*

So, *how* do we do that? Part one of the answer is the Rocky Flats Nature Refuge Recovery Program. We at Strategic Disposal have found that a nature refuge is a robust way to "do nothing" for less. The program effectively turns most of the land over to Interior's FWS. What land is retained by LM requires only the appearance of nature recovery, significantly reducing the time, energy, and cost of "doing nothing" for nuclear waste and land administration. *Do you get it now?* As more and more sites are cleaned up less and less in the nuclear weapons complex and turned over to the LM, *we are dedicated to doing more and more nothing about everything for less and less.*

I now turn the presentation back to Lester Ismore, who will share with you the testimony of Bertha Huggenkiss, director of this innovative program. Mr. Ismore will then close with part two of LM's innovative solution: securing LM's long-term performance through the reduction of the office's own environmental impact.

Photo courtesy of Shiloh R. Krupar

VISION

- Spend less to achieve more unaccountability
- Bury threats that surface
- Manage cleanup's appearance
- Make the public the happiest
- Minimize the terrain to be managed

Photo courtesy of Shiloh R. Krupar

Speaker #1: Lester Ismore, testifying for Speaker #4: Bertha Huggenkiss

Good afternoon again. I am pleased to report that the LM's Nature Recovery Program at Rocky Flats no longer requires an acting director. Before leaving her post, however, Bertha Huggenkiss cheerfully shared with me several highlights of the program. The vision of this program is to release lands that were withdrawn for nuclear weapons back to the public, and to relinquish any negative energy or controversial history surrounding such sites. Ms. Huggenkiss emphasized the holistic approach of the program's seven-step system. I will read from her notes.

Step One: Forgive and forget complex nuclear history and environmental abuses

A wildlife refuge is a valuable neighbor. Refuges encourage an atmosphere of trust and public confidence. While more facilities are decommissioned and cleaned up, we can look forward to the expansion of the national wildlife refuge system. Ugly environmental histories can be supplanted with the beauty of undiscovered pastoral scenery.

Step Two: Affirm Nature's power to recover the site as wildlife refuge

The restorative miracles of Nature are vital to the recovery process. Visual reminders of former weapons production facilities will rapidly disappear, no longer compromising the view. Natural attenuation will lessen the impact of contaminants left in the ground or subsurface water after cleanup. The nearly four-hundred-acre former industrial area retained by the DOE-LM will, over time, develop an overlay refuge landscape, seamless with the wildlife surrounding it. A future land bridge will connect the refuge at Rocky Flats with the mountains to the west.

Step Three: Identify scenic and biotic resources; prospect your liabilities for disposal

At Rocky Flats, Lindsay Ranch provides wonderful scenic opportunities to interpret the early history of settlement and ranching on the prairie. The barn, in particular, is in a charming dilapidated condition. Mining rights have been purchased by the federal government to prohibit human digging, where contamination may reside underground. Large winds, tumbleweed, and snakes, all native to the Rocky Flats mesa, remain the predominant on-site hazards.

Photo courtesy of the U.S. Department of Energy

Step One *Forgive and forget complex nuclear history and environmental abuses*
Step Two *Affirm Nature's power to recover the site as wildlife refuge*
Step Three *Identify scenic and biotic resources; prospect your liabilities for disposal*
-
-
-
Step Seven
Maintaining recovery efficiently & effectively

SEVEN NATURE-RECOVERY MODEL STEPS

Photo courtesy of the U.S. Department of Energy

I will stop there and stress the program's last step: *Maintain recovery efficiently and effectively.* Legacy management is expanding this program to encompass numerous sites across the United States and, in doing so, minimize the terrain of that which must be managed. In partnership with the FWS, we are cultivating an increasingly administration-free form of stewardship over the legacies of the Cold War. As the federal organization charged with maintaining nuclear waste remedies in perpetuity, LM is proud to be one of the leanest and greenest in the entire federal family. We have found novel ways to provide even more efficient forms of less management and more economy when making appearances, by enacting sustainability measures. The new environmental mission of the LM ties management goals and visibility to organizational austerity. We envision a future where legacy management will be so efficient as to be unnecessary, ultimately satisfying our mandate to manage in perpetuity.

With that bold vision statement, I conclude. Please note that our combined presentations were kept under fifteen minutes, with fewer than 3,500 words, no more than twenty slides, and including one less speaker. Our testimonies were 100 percent government-recycled.

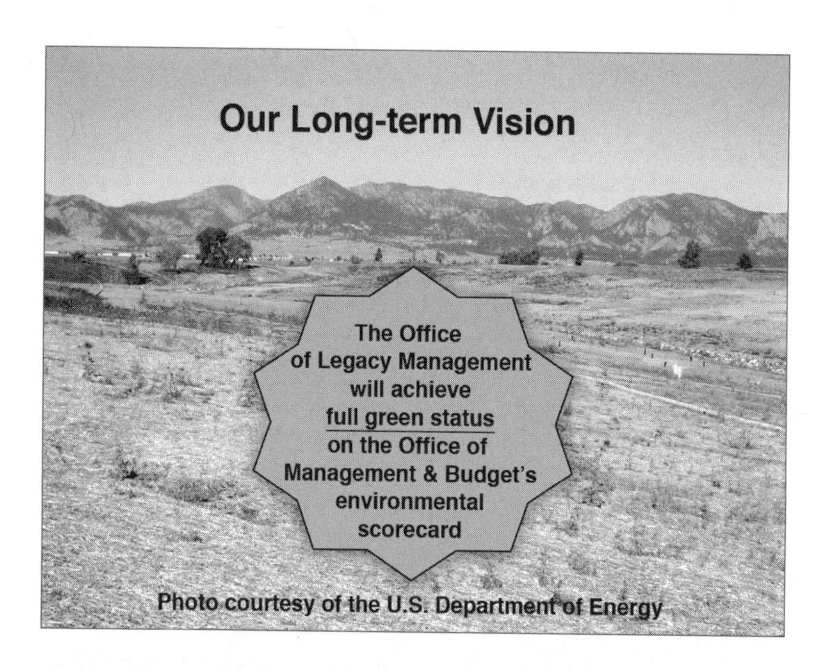
Our Long-term Vision
The Office
of Legacy Management
will achieve
full green status
on the Office of
Management & Budget's
environmental
scorecard
Photo courtesy of the U.S. Department of Energy

Photo courtesy of the U.S. Department of Energy
The End

Endgame of Government Oversight

EGO

Report on the Congressional Auditing of and Management of Complexity by the Environmental Management and Legacy Management Offices of the U.S. Department of Energy Regarding the U.S. Nuclear Landscape, Specifically Rocky Flats

A REVIEW OF THE POWERPOINT PRESENTATION PRECEDING THIS REPORT

Endgame of Government Oversight

EGO-10-784

It's not a matter of a battle "on behalf" of the truth, but of a battle about the status of truth and the economic and political role it plays.
–Michel Foucault, "Truth and Power"

Mission of the Endgame of Government Oversight (EGO)

EGO investigates government "games of truth"—how they are put in place, the relations of power that are brought into play, and the strategic final stages that attempt to secure truth and eliminate uncertainty.[1] EGO documents the ways truths are produced as the result of particular government administration and procedures, regulatory frameworks, and biopolitical techniques of calculation. Of particular interest are the endgames: the strategies enacted to make something final. Such actions seek to eliminate the contingency and plurality of human action, achieve efficiencies while securing the population, by means of government rationalities, instruments, and programs that can be economic as well as political and military.[2] EGO is careful to observe the ways that endgames carve out manageable scenarios to handle complexity and "to arrange things in such a way that, through a certain number of means, such and such ends may be achieved."[3] The legitimacy, persistence, and "finality" of government resides in this continual activity of arranging, specifically the negotiation of what constitutes the competency and effectiveness of the state. Refer to Appendix A for further information on EGO's theoretical understanding of government using the ideas of Michel Foucault, namely, biopolitics and governmentality.

In the context of the general breakdown of consensus over scientific-economic progress and popular demands for political accountability and control, the state is paradoxically called on to provide intensified monitoring of increasingly complex hazards and threats to safety. Programs of government are compelled to respond to external crises and even to seek out new ones in order to prove the government's ability to provide safety, and to convince the population of the need for regulation.[4] This paradoxically involves the entry of state institutions into domains where power can be contested, where it is fundamentally impossible to regulate, and where the state reveals its impotence.[5] In many cases, the desire for regulation is met with the stark reality of the impossibility of regulation. EGO is interested in such regulatory failures and the absurd endgames that result. *In the face of impossible charges and aims that can never be met, such endgames endeavor to dispose of the complexity of that which cannot be regulated in order to maintain the governing arrangement.*

EGO is dedicated to the study of one primary endgame: oversight, or auditing. Under conditions of limited choice and necessity, the incompetence of the state leads to further bureaucratization, in the form of auditing. Auditing formalizes internal oversight of competency, installing a relentlessly reflexive inward spiral of hyperregulation that reduces complex issues into manageable simulacra. Auditing produces such manageable simulacra to serve as signs of competency and accountability. The endgame of auditing fulfills the governing relation by producing truths—usually technical fixes or signs of efficiencies—that simplify, minimize, and/or purify, tying state competency to such signs, *simultaneously intensifying and deferring the desire to regulate while reducing the terrain of state responsibility.* EGO is dedicated to understanding the challenging, confusing, often contradictory ways auditing and governmental oversight maintain *the appearance of regulating as effect,* within broader claims about securing the safety of the population. The appearance of governing is the desired end. EGO's last order of business is to tactically consider what fixed regulatory frameworks and games of truth expel and dispose of. EGO takes a general psychoanalytic orientation; however, instead of featuring the psyche or psychologized self, the account reflects on the ways that regulatory institutions repress, refuse, disown, and deny the materialities they excrete or expunge through distancing devices, dissociative mechanisms, and spectacles of auditing. This can be an occasion for positively challenging biological-material, ethical, and aesthetic orders, arranging other truths, and regularizing alternative forms of life.[6]

Report Background and Approach

This study reviews and develops a comprehensive background report on the previous PowerPoint, which celebrates the accelerated closure and cleanup, transition to wildlife refuge, and legacy management of the former U.S. nuclear facility Rocky Flats. While the PowerPoint script and slides capture important logics at work through satire, EGO's report, using a more "deadpan" documentary mode, offers further nuance to the ways that such logics have materially emerged. In the context of the changing U.S. nuclear landscape and mission of the U.S. Department of Energy (DOE), EGO is particularly interested in the bureaucratization and oversight of land at Rocky Flats. The case has implications for other sites undergoing decommissioning and decontamination for future land transfers, repurposing, and/or long-term DOE stewardship.

At the purported "end" of the Cold War, DOE and its predecessor agencies had acquired control over approximately 2.4 million acres, or 3,750 square miles, to conduct a broad range of programs, from the design, production, and underground testing of nuclear weapons to electrical power marketing, energy conservation, and scientific research.[7] The U.S. nuclear complex occupied the majority of that land, extending across a continental landmass from South Carolina to Washington State, Alaska to the Marshall Islands, with numerous institutions, facilities, and sites involved in the development, testing, and storage of nuclear weapons and weapons materials. These sites were—and continue to be—fusions of local ecologies, communities, and nuclear industrial processes that involved deliberate acts of territorial devastation and/or the treatment of the land as alienable, external, and subject to the dictates of the nuclear activity defining the site. More than half a century of nuclear weapons production had generated costly and complicated environmental and waste management problems, such as emptying and recontaining million-gallon tanks of highly radioactive waste, decommissioning plutonium-processing buildings, and dealing with contaminated soils and water systems. DOE's post–Cold War activities shifted toward cleanup, technology development, scientific research, and new energy sources. Further, the DOE underwent a series of audits, which later would have a huge impact on department land management. DOE began assessing the magnitude and extent of environmental contamination. Early estimates for partial environmental restoration and waste management and disposal varied greatly from $100 billion to $1 trillion. In 1997, DOE set it at around $227 billion for a seventy-five-year period.[8]

As this inventorying of land and facilities took place, DOE external performance and internal operational reviews strategically realigned the DOE's mission to "work better and cost less."[9] "Given the practical realities of the budget restrictions facing the Department . . . and the

Secretarial initiative to dispose of unneeded Departmental assets," the DOE had landholdings that were in excess of current and anticipated future needs.[10] With disposal the norm, DOE sought programmatic ways to close and remove sites that were no longer needed. It continues this work today. However, extensive contamination at DOE sites and the longevity of most nuclear materials have forced the DOE to develop programs for managing these nuclear hazardous sites in perpetuity, including administering knowledge about those sites, negotiating state and federal environmental compliance agreements, maintaining any cleanups, safeguarding the population, and cultivating transparency for public relations. Where hazards from former industrial processes are considered to be dangerous to the population, the radioactive and chemical legacies of World War II and the Cold War have largely been treated by the DOE as a bureaucratic management problem. This can be attributed partially to Congress's charge to the DOE to accelerate the reduction of site-specific environmental hazards, and to expeditiously streamline both its landholdings and its management operations.[11] In response to the growing number of decommissioned nuclear facilities, the DOE has implemented "end states" during cleanups, a risk-based containment technique that then opens the door for constant auditing procedures. Such auditing maintains appearances of knowability and accountability, in relation to the end state, and works to contain DOE responsibilities.

Study Purpose

Four rationales supported research for this analytic report. First, EGO conceived of this study as a pedagogical opportunity to advance its own interests, by exploring the endgame of the "postnuclear" and the tactical cleanup operations of DOE, such as the language and administration of "legacies." EGO considers the DOE's present situation of disposing of land and managing contaminated sites a biopolitical endgame that seeks to regularize efficiency, pursue environmental safety, and administer knowledge. It should be noted that activities such as inaction or avoidance are actually part of the DOE's biopolitical administrative strategies.

Second, EGO regarded this project as an important chance to advance genealogical analysis of the biopolitical administration of land in the United States—and, by extension, the relationship between waste and humans. The endgame of protecting citizens from the industrial effects of nuclear weapons production while returning public lands previously withdrawn for the development of nuclear weapons emerges from the historical geographies—and the ongoing tensions, contradictions, and reversals—of land withdrawals and the meaning/constitution of public land in the United States.[12] The practice of military land withdrawals, not unlike executive land withdrawals for the purpose of preserving land, has in many cases been responsible for sequestering open spaces. Often land acquired by DOE and its predecessors was not formally used, except as "buffer zones" of security and secrecy, as a measure to protect the processes internal to the site. Today, the perceived extension of industrial waste across the whole of nature has intensified desires for "precontact wilderness," to the extent that populations are pursuing recreational activities in reserves of or near toxic waste.[13] EGO is particularly interested in the postnuclear nature refuge and the long-term stewardship site, which supposedly provide outdoor recreational opportunities and safety from the threat of contamination.

Third, EGO has been attuned to the crisis of nuclear waste since the organization's founding. Waste presents a crisis of system and rationality; "the limit case here is nuclear waste. . . . The materiality of nuclear waste exceeds both the semiotic and concrete (in its figurative and literal senses) strategies to contain it."[14] Nuclear waste is a challenge to the very activity of administration: How does one regulate that which is ultimately unregulatable and cannot be disposed of—because it is material excess?[15] EGO was understandably intrigued with not only the completion of the Rocky Flats cleanup project but also the postcleanup regulatory framework of legacy management, which stipulates that the DOE will maintain cleanup compliance *in perpetuity.* EGO felt that this requirement to "manage in perpetuity" warranted further investigation. In spite of the crisis of accountability surrounding the DOE and the history of the nuclear weapons complex in the United States, there is a widespread desire to entrust an enduring institution or agency with the indefinite monitoring of waste.[16]

Fourth, EGO found that the current language of "sustainability" surrounding nuclear waste management is a fascinating and not-to-be-overlooked oxymoron and circular mechanism of self-exculpation—the absurdity of which warranted the attention of EGO's specialists.

Executive Summary: What EGO Found

Considered to be one of the largest nuclear closure projects in the world and a model for other decommissioned nuclear facilities across the United States, the Rocky Flats cleanup demonstrates the importance of producing certain truths about the site in order to administer the land and future relations with the site.[17] The endgame at work, simply put, is the desire to *produce more for less:* more regularizing of land by minimizing the terrain of that which must be governed, and by managing only for the purposes of *making appearances of governing that which must be governed.* The detailed report that follows will elaborate on this assessment. Directly below, EGO summarizes the two parts of its findings, which correspond to the two main sections of the report, with a concluding reflection on "lessons learned."

The ability to convert a nuclear facility, formerly hosting some of the world's most contaminated buildings, into a nature refuge depended on a new land administration reality that would regularize efficiency. Key truths mobilized by the cleanup effort included risk-based approaches to safety that determined the end state of the site; project baselines that calculated the uncertainty of cost and bracketed cleanup efforts to a predefined budgetary horizon; performance-based contractual arrangements that circumscribed regulation to the role of advocating end-state achievement; and labor-intensive groundwork for the ontological separation of human and land, human and waste. EGO found that the end state of nature refuge shaped the entire truth-production process at Rocky Flats. Combining efficiency and purity, the nature refuge end state—as a managed effect—minimized the terrain of that which must be managed, by instrumentalizing ecological restoration to save further cost while achieving the cleanup vision. Nature refuge designation worked to naturalize the site. Rather than investigating the uncertain ecology and contingent material exchanges between humans and the environment, the regularizing of the site as natural has led to regulating as if to not exist at all. The refuge "solution" allows for less managing; the de facto benefit of ideas of purity, authenticity, and nostalgia associated with nature is the reduction of the activity of regulating, since inaction looks natural. Note that the refuge end state is not mere rhetorical greenwashing; the administration of Rocky Flats' cleanup has arranged a human/waste binarism anchored to a surface/subsurface division of remediation standards that minimizes governing to the control and maintenance of appearances, with resolutely material effects.

Furthermore, DOE's long-term stewardship of a portion of Rocky Flats, where contamination from former industrial activities is considered to be extensive and dangerous to the population, was determined a way of safeguarding the cleanup in perpetuity and cultivating accountability through constant auditing procedures. Just as cleanups organize truths about a site and bracket unknowns, such as predetermining the site's future use, the DOE's postcleanup regulatory framework of legacy management developed techniques to contain DOE responsibilities for land and hazards, in response to the growing number of decommissioned and remediated DOE nuclear facilities under its purview. As this report will show, DOE's legacy management has continued the work of minimizing that which must be managed. *Legacy management, which stipulates that the DOE will maintain cleanup compliance in perpetuity, maintains only what must be managed in order for the government to appear to be governing, in the ultimate endgame of congressional oversight.* This enterprising form of "governing by making visible" facilitates the minimization of the impossible mandate that the DOE monitor waste and contamination indefinitely. Auditing serves as a way of producing truths that creates a barrier to defend against risk and that essentially makes the land superfluous to organizational auditing. Such auditing serves as a means of institutional disposal, one that can relentlessly be made more efficient to the point where it oversees only that which must be governed in order to maintain appearances. DOE has taken the opportunity to make its "in perpetuity" assignments more efficient by implementing high-performance techniques that improve the cost-effectiveness of making appearances of governing.[18] Ultimately, legacy management's task is to maintain appearances via a portfolio of performance assessments; public relations and accountability measures are sought through institutional self-audits and performance recognition, disregarding the historical controversies and negative socioecological legacies of sites such as Rocky Flats despite claims otherwise.

In conclusion, in the context of DOE's maintaining appearances before Congress, EGO considers the current management of the complexity of lands, such as Rocky Flats, to be a form of *ex-terrestrialism:* The environment continues to be treated as an externality, and the land and humans *are made to be alien.* Territory is dispossessed, environmental knowledges are repudiated, and institutional memories are discontinued. The endgame of auditing, in particular, alienates humans and land from a more transformational material relationship, through distancing devices that create a kind of displacement *in place.*[19] For the sake of the population, the alien-making process absurdly involves the labor-intensive production of labor's absence—of a no-man's-land, sustained through temporal, legalistic, geographic, scientific, and euphemistic devices, exemplified by DOE's "legacy" terminology.[20] Rocky Flats has been delimited, dislocated, and deferred through such attritional dissociative mechanisms and bureaucratic oversight; its legacy site status and monitoring operates as a form of memorialized denial, relegated to the past, suspended as an "alien still life." In response to the differentially inhabited risks and ongoing vulnerabilities of the Cold War that will last longer than the political-economic structures that produced them, the elusive accountability offered by DOE refuses the connective tissue that binds human potential to degraded environments, and that binds certain degraded personhoods to the material refuse of state projects—in this case former Rocky Flats workers dealing with occupational illnesses.[21] The plasticity of land withdrawals, the way corporate and military practices have been owned and then disowned, the interiorizing of profits yet exteriorizing of risks, and the administrative transferability of Rocky Flats from government-contracted nuclear weapons production to underfunded wildlife protection services *all have had dire long-term consequences on the material world.*[22] Toxic residues remain; there is nothing "over" about this form of ruination; it endures.[23] EGO suggests that the way to change the current governing of Rocky Flats is to confront the desire to regulate and the end state—the "arrested future" of the site—with all that has been made residual, specifically the materiality of waste as the "return of the repressed." Such a response to Rocky Flats considers the Cold War to be an ecology of remains, not fixed monuments, that might give rise to ways of "living otherwise."[24]

I. The Rocky Flats Cleanup: Producing Disposal, Fixing Risk, Naturalizing Efficiency

In the early 1990s, the Denver-based newspaper *Westword* offered to help the DOE find a new name for the recently closed Rocky Flats nuclear complex. A flood of entries was received, ranging from "civic-minded sobriquets to obscenities deemed unsuitable for nuclear families."[25] The winner was declared: "Boom Town!" Although DOE never adopted the satirical title, "Boom Town!" captures the rapid development of Denver in the mid-twentieth century, when federal dollars established a series of military facilities in Colorado. One of these was the plutonium plant Rocky Flats sixteen miles from Denver, constructed in 1951 and operational by 1952. Originally under the control of the U.S. Atomic Energy Commission (superseded by the DOE in 1977) and managed by government contractor Dow Chemical Company (succeeded by Rockwell International in 1975), Rocky Flats was built for the purposes of recovering and recycling scrap plutonium, manufacturing plutonium and highly enriched uranium components, and producing nuclear bomb "pits"—trigger devices for nuclear warheads.[26] The plant was essentially a giant machine shop for radioactive metals, requiring thousands of barrels of oil and solvents for the manufacturing process, along with more than 1,500 chemicals (many of which have been linked causally to cancer, liver damage, and nerve disorder). The list of machinery, equipment, and materials involved in the facility's operation is vast, including furnaces, tool and dye shops, assembly lines, glove boxes, research laboratories, cafeterias, locker rooms—and, at any given time, several thousand people.

Icon of the Cold War, economic powerhouse of the Denver and Front Range area, and popular target of peace and antimilitarism activists, the Rocky Flats Plant was plagued by fires and accidents, leaving an ignominious legacy of contaminated buildings, soil, and water, during its years of operation. The wilderness surrounding the plant served as an expandable dumping ground, where industrial wastes of the assembly line were cast off, their danger canned and buried, or dispersed by burning, spraying, or most often by simply forgetting. It also functioned as an elastic buffer zone, limiting public scrutiny. Over nearly forty years of operations, the weapons complex not only demonstrated that the production and disposal of waste is central to modernity, but also purveyed a logic dictating that anything potentially subject to practices of disposal equals waste. By this rationale, waste was not merely the material by-product of industrial development—the expression of monstrous energetic excess, exemplified by radioactive waste—but also a symbolic process of eliminating that which was considered to be useless matter, inefficient practice, or empty space.[27] The nuclear weapons production process taking place at Rocky Flats required that the natural world of the site be treated as an expansionary space—a waste frontier—that could be reordered to facilitate and reproduce various modes of disposal. A variety of container technologies were developed to separate, solidify, and/or store wastes: solar evaporation ponds, burial trenches, steel membranes and metal drums, incinerators. Disposal practices ranged from relatively uncontrolled systems, such as dumping waste in the ground or mixing it with cement in cardboard boxes, to highly controlled systems, including elaborate on-site water-control systems. The idea that waste could simply be sequestered and left to the elements, canned and shipped off somewhere else, or rhetorically removed meant that ecology was not a consideration but an afterthought of waste's contradictory refusal to disappear. As a result, waste relentlessly drifted, crept, became airborne, and seeped out through socioecological transformation—accumulating in underground plumes, within worker bodies, bodies of water, the soil and air, the crawl spaces of downwind homes, and in newspapers, court cases, and local lore. The dynamic, indeterminate environment of Rocky Flats—and its perceived purity and impurity—emerged through this intersection of technology, nature, and waste.[28]

Along with the dispersion of hazardous materials through ecological processes into off-site lands and water supplies, reports leaked out that tons of plutonium had gone missing in the ductwork of the plant and that a container technology known as "pondcrete" (fixing contaminants in cement for burial) had failed. Left outside, the containers' plastic liners broke and the cement reliquefied. This debacle resulted in the first FBI raid on a federal agency for environmental violations in U.S. history.[29] The 6,500-acre site was placed on the U.S. Environmental Protection Agency's (EPA) National Priorities List, the federal government's program to clean up national-priority waste sites, popularly known as Superfund.[30] The situation was so grim that proposed plans for the shutdown of the site included constructing a concrete dome over the whole area and leaving it as a tomb. Original estimates for the cleanup said it would take sixty-five to seventy years and $37 billion, but what such a cleanup entailed was not known, in large part because no one knew the extent of contamination, and public tolerance of the site remained an open question.[31] Whether the environmental devastation of Rocky Flats was attributed to bad safety procedures, corporate government contractor negligence, or the general mission of U.S. nuclear weapons production, the idea that waste was a failure of human control served as a dominant mode of processing the catastrophe, laying the groundwork for new forms of human technical-rational intervention and disposal, in this case, the deferral of toxic debts and arrangement of corporate and state self-exculpation through the removal of the physical plant and the nature refuge redefinition of the site.

The previously mentioned *Westword* newspaper contest to rename Rocky Flats proved to be prophetic, with such entries as "Plutonium National Monument" and "Half-Life National Park." Despite the hundreds of tons of plutonium and other hazardous materials that cycled through the plant over the decades and the profound challenges of such an uncertain ecology, including radioactive ponds and plumes of contaminants, Rocky Flats is slated to open its fence line to partial nature refuge use. The company hired by DOE to remediate Rocky Flats, Kaiser-Hill Waste Disposal Services, declared completion of the site's cleanup on October 13, 2005.[32] The company had announced the conclusion of the cleanup effort fifty-six years ahead of schedule at a purported savings to the public of $29 billion.[33]

The determination of "wildlife refuge" as the desired end state for the cleanup shelved the need to consider or respond to the ecology of the site, rendering further discussion with regard to the level of cleanup unnecessary. Conclusions were drawn, eliminating uncertainty. Against the chaos of the unknown ecological transformations taking place at Rocky Flats, the establishment of a desired end state fixed into place the maximum allowable risk essential for the expedited turnover of the site to the public.[34] Relational thinking about the military–industrial complex and social-ecological processes was thus mooted. Refuge designation entrenched certain conceptions of nature and waste implicit within modern weapons manufacturing: that waste could be removed or fixed to land, and that nature was static and separate from the human. This ontological position legitimated the wildlife refuge as the latest form of human-technical intervention at the site, effectively harnessing waste to material and discursive practices associated with the production and valorization of nature. In this form of land administration, the end state of "nature refuge" served as an efficient tactic of disposal, which has led to the regularizing of the site as natural and, consequently, to regulating as if to not exist at all.

Drawing on the observations of the preceding PowerPoint, EGO has elaborated several crucial aspects of the cleanup process below in order to develop further insight on the visual and material transformation of

Rocky Flats. Please note that EGO considers all references to "end states" to be interchangeable with EGO's preferred nomenclature of "endgames." While EGO has maintained the integrity of the original term "end state," as used by the DOE, readers should interpret end states to be a type of endgame.

Risk-based End States and Ensuring Future Safety through Determining Allowable Contamination

DOE policy directives issued in 2003 stipulated that all cleanups should be aimed at clearly defined, risk-based end states: "Representations of site conditions and associated information that reflect the planned future use of the property and are appropriately protective of human health and the environment consistent with that use."[35] The Rocky Flats Cleanup Agreement incorporated an integrated risk-based approach to cleanup; the end state of wildlife refuge provided a unified vision of the site for all cleanup goals. DOE and regulators, the EPA and Colorado Department of Public Health and Environment (CDPHE), had to ensure that the cleanup met the level required to protect the maximally exposed individual, in this case, the wildlife refuge worker. Compared to the degree of cleanup stipulated for future use as housing or farming, wildlife refuge designation enabled a cheaper, faster cleanup owing to the limited human contact entailed by this end-state scenario and the greater allowable contamination defining the site use's "safety"—the quantifiably acceptable amount of contamination levels to be left in place for the site to be deemed "safe enough."[36]

Cost Engineering the "Budget Cleanup" for Favorable Political Appearances before Congress

The DOE's environmental liability—the cost associated with cleaning up the environmental legacy of the Cold War—is one of the largest in the U.S. federal government. Following the generally accepted accounting principles for U.S. federal agencies, the DOE must annually prepare a financial statement, of which the largest entry is environmental liability. Established in 1989 for the purpose of addressing the Cold War's environmental impact, DOE's Environmental Management (EM) program comprises the greatest portion of this total cost; its founding consolidated several DOE organizations that were previously responsible for the handling, treatment, and disposition of radioactive and hazardous waste.[37]

Because of severe criticism from Congress over the EM program, DOE needed a success story that could maintain the viability of the program and show that the mission of "closing the cycle" of nuclear materials could be measured and accomplished in decades.[38] Rocky Flats was deemed a site small enough to allow cleanup and closure within a short time span and, subsequently, became one of two accelerated DOE cleanup projects.[39] For political appearances, DOE EM plans that initially had set formal certification of the cleanup at the already-accelerated year of 2010 were pushed forward even further to 2006. This effort secured baseline funding necessary to sustain the cleanup project; DOE would not have to negotiate with Congress over money on an annual basis.[40] Rocky Flats' cleanup was then "reverse cost-engineered" from this set amount of funding, wedded to the project deadline of 2006.[41]

Securing Regulatory Certainty by Practicing Regulation as a Tautological Activity Leading toward Accelerated Closure

See "Accelerating Scenarios and Closure Visions."

Accelerating Scenarios and Closure Visions

Thus, the first dismantling of a nuclear weapons plant was also the DOE's first accelerated closure pilot project within the weapons complex. According to the DOE document "Rocky Flats Closure Legacy: Accelerated Closure Concept," "No other site in the nuclear weapons complex had attempted a cleanup effort of this size and complexity under an accelerated schedule. . . . Because of the groundbreaking nature of attempting a first-of-its-kind accelerated cleanup and closure project, Rocky Flats had to pioneer processes, many of which have now become standard DOE approaches."[42] "Accelerated closure" refers to the tautological process of developing an accelerated closure vision to spearhead a closure project.[43] A closure vision meant the site could be treated as a project with a "work breakdown structure" that could be used for performance measurement; it bracketed the material world outside, focusing energies on the more stable and predictable environment prescribed by the vision.[44] In the case of Rocky Flats, the closure vision also defined the site's relationship with its regulators and stakeholders.[45] In contrast to regulation as ensuring regular or normative actions, often by imposing restrictions, DOE and regulators together implemented a regulatory framework that "provided a bias for action,"[46] aligning the accelerated closure vision with regulatory processes. This meant that the DOE and cleanup contractor could secure regulatory certainty and support by basically deciding what the site would be, and then outlining how compliance would be measured and achieved. Such a regulatory framework, always in a supporting role to the accelerated closure vision, effectively reduced regulation—downsizing the number of enforceable milestones. This was, in part, the culmination of efforts, since the 1980s, to redefine original Superfund legislation and prioritize speed and efficiency.[47]

Performance-based Contracts and Incentivizing the Remediation Industry

The accomplishment of the accelerated closure vision was also made possible by a change in the DOE approach to contracting. In 1994, the DOE Contract Reform Initiative pursued performance-based approaches to contracting in order to incentivize contractor execution and completion of work, consistent with clearly established performance expectations and measurements. The first of its kind to be awarded at a major DOE program level, the incentives-based contract between the DOE and Kaiser-Hill over the Rocky Flats cleanup provided opportunities for the company to earn rewards based on achieving the target budget of $3.9 billion, and for beating a March 2006 deadline.[48] The contract enabled Kaiser-Hill to fast-track the planning process, assign and integrate tasks among its subcontractors, projectize[49] the work, and incentivize the workforce.[50] Kaiser-Hill, working together with the Rocky Flats Field Office, developed an aggressive approach to the work and implemented interim ASAPs—Accelerated Site Action Projects—to complete the cleanup thirteen months ahead of schedule for a company bonus of $355 million.[51]

Implementing New Workplace Truths but Relying on Old Habits

The contract and the cleanup were put in place by constantly reaffirming the closure vision and scripting the mantra "Make It Safe—Clean It Up—Close It Down," facilitating a new workplace culture and guiding principle of behavior.[52] The mantra was repeated in management behavior, incorporated in the collective bargaining agreement with the local union, and foregrounded in every budget testimony before Congress. However, EGO found that this truth production relied on several continuities with past site operations, including ongoing practices of "laying waste." The cleanup arrangement provided incentives to turn over workers and equipment: workers were encouraged by Kaiser-Hill management to "work yourself out of a job" proudly—the quicker the better. While the incentivized workforce strategy accrued profits for Kaiser-Hill for early completion of the job, neither Kaiser-Hill nor the U.S. Congress extended benefits to many remediation workers, denying them their retirement in return for completing the job early—in some cases just months or mere days before their retirement qualification dates.[53] DOE's organizational pattern of "deny and defend" also lubricated the accelerated cleanup. Stakeholder involvement in the early planning phases was nullified after it was revealed that the cleanup's end state and schedule were predetermined and based on nonnegotiable fiscal restraints. Although a working group had recommended that remediation of the site leave only background levels of radiation, and that it would "wait as long as necessary" for technology that would achieve this conservative cleanup, only the soil was opened to debate.[54] Stakeholders were angered by the intentions to leave most radioactivity in place and fiercely insisted on lower soil action levels.[55] The result laid the foundation for the cleanup's accelerated waste disposal strategy.

Waste Disposal I—Canning and Redistributing

Accelerated closure was achieved by a repetition of the familiar pattern of extracting radioactive and hazardous wastes physically and rhetorically from the site. This extraction involved two forms: moving the waste off-site or burying it below the site. Part of a massive federally choreographed shift in the spatial division of waste storage, more than six hundred thousand cubic meters of radioactive waste were removed, and more than 3.6 million square feet of buildings, including more than one million square feet of highly contaminated nuclear production facilities, were demolished, eradicating the eight hundred buildings that once functioned as a small city at Rocky Flats.[56] The extensive waste disposal mobilized by the cleanup encouraged the popular notion that Rocky Flats was suddenly "all gone"! Indeed, significant amounts of the site had been canned and redistributed via the interstate highway and rail system.

Waste Disposal II—"Outsourcing" to Biophysical Processes, Expanding the Waste Frontier

Waste was also moved "off-site" through burial. Rocky Flats' cleanup hinged upon a surface/subsurface dichotomy: a conservative allowance of radioactivity for the top three feet of ground, whereas beyond six feet anything was allowed, since contamination at that depth, according to the DOE Rocky Flats site manager at the time, did not pose a danger to surface activity.[57] Building foundations, sewer piping, old plutonium processing lines, and massive volumes of cleanup equipment were dumped for a major savings in Kaiser-Hill's race to earn bonuses for early completion. In the end, most of the designated cleanup budget went toward relocation of weapons-grade material and waste, site security, and demolition of buildings; cleanup of the water and soil was to be done with the remaining money: approximately 7 percent of the total.[58] Strategic appeals to the geologic characteristics of the site helped justify on-site burial and liberal contaminant levels, shifting much of the burden of waste containment outside the realm of social and economic negotiations to the biosphere—to the structural aspects and preexisting material of the thick shale and clay stone underlying the site. According to DOE, the dry Colorado climate and alluvial fan on which Rocky Flats is situated would reduce erosion and inhibit off-site migration of contaminants.[59] The interlocking root systems of rhizomatic grasses on the "surface" of the prairie would purportedly hold down soil from disturbances, despite the harsh erosion, warrens of burrowing animals, and deep-rooted desert plants constantly belying the fixedness of the topsoil. This epistemological divorce of surface from subsurface allowed for a containment strategy that falsely presented geologic depth as a permanent externality and expandable waste frontier.

Nature as Managed Effect—Integrating Nuclear Work and Environmental Work for Efficiency

The extraction of waste through on-site burial and off-site dispersal ordained the site's rhetorical "return to nature" and signaled the end of an era. "Environmental recovery" was produced as a managed effect of the cleanup's integration of nuclear work and environmental work. Following the accelerated action protocol that guided the entire cleanup process, environmental restoration work, performed under the Rocky Flats Cleanup Agreement, devised ecological arguments and policies that would achieve efficiencies, reduce costs, and maintain the remedies, ranging from the treatment of the environment as a standing reserve available for disposing of waste and with a reproduction cost of zero, to the purposeful establishing of good vegetation cover during the decontamination process.[60] As a result, the removal of the plutonium plant buildings and the revegetation of the area positively reinforced the site's association with common understandings of nature as purity. Essentially, "nature" has served as both the means and managed effect of the cleanup, combining the logics of efficiency and purity in the handling of wastes and the negative social legacies of the former nuclear facility.

Remediating the DOE's Public Image

The cleanup provided a badly needed opportunity for the DOE to remediate its public image from massive polluter to good housekeeper. Referring to itself as "caretaker" of the site, the DOE instrumentalized distinctions between clean and contaminated, knitting together nuclear waste management with everyday domestic cleaning technologies. On one occasion, the Kaiser-Hill manager of the project even overtly described the cleanup of infinity rooms—rooms where radioactive contamination was so high it could not be measured with existing instruments—as similar to cleaning your bathroom.[61] From FBI raid to "field of dreams," the story of the site's cleanup tendered a modern morality tale of eliminating waste for the public good, converting a national liability to local community asset, and reestablishing social order through the proper in-house handling of nuclear waste.[62] The idea of a cycle for a weapons plant also worked to naturalize the technoscientific reproduction of nature and its attendant sexual politics. For example, a DOE and Kaiser-Hill brochure states:

> With the cleanup completed, the site will become a high prairie habitat for citizens to enjoy for years to come. . . . it is a lesson in American history and how the difficult and often frightening realities of nuclear weapons production can give way to new beginnings. The Rocky Flats journey has come full circle—from a proud legacy to a proud new beginning.[63]

Instead of producing toxic landscapes, DOE allowed the cycle of life to salvage the site purified by the cleanup from history. The management of the nature refuge continues to align such discourses of environmentalism and nature with the institutional logics guiding the treatment of the site since its inception.

Disposing of the Land, Regularizing the Refuge, and Naturalizing the Cleanup

In June 2007, the DOE transferred stewardship of approximately four thousand acres of the site's former buffer zone to the Department of Interior's U.S. Fish and Wildlife Service (FWS) to manage as the Rocky Flats National Wildlife Refuge, while the former industrial core area remains under DOE custody. This transfer between government agencies has been hailed as the public recovery of land under newly enlightened environmental management. However, the administrative process to convey land titles from the DOE to a sister federal agency is relatively streamlined, thanks to a well-established tradition of using the Department of Interior as the institutional "dumping ground"—the "Department of Everything Else"—of other government departments.[64] Essentially, the DOE reduced its management responsibility by transferring liabilities to the Department of Interior, which, as the biggest landlord government agency, serves as an administrative disposal service for lands.[65]

The FWS drafted a Comprehensive Conservation Plan for the Rocky Flats National Wildlife Refuge.[66] When the refuge opens, visitor facilities will include multiuse and hiking trails, a visitor contact station, interpretive overlooks, viewing blinds, and other opportunities for long-distance vision and consumption of landscape. Public-use programs will offer environmental education for high school and college students and a limited hunting program (two weekends per year) for youth and disabled visitors. The plan emphasizes ecological amenities and advocates restrictive access for reasons having to do with environmental protection. The prairie purchased by DOE as a buffer zone now restricts human access in order to buffer nature from urban sprawl. In a neo-Malthusian twist, human contact is considered more immediately toxic than residual radioactive and hazardous materials, justifying the original enclosure for its preservation of biodiversity and scenic value. The site appears to be just another open space along the Front Range area, without reference to the sociopolitical memories of the site, the open, ambiguous, uncertain hazards there, and the unpredictable future of its ecology. Instead of acknowledging that the presence of a broad array of flora and fauna, including tallgrass prairie (xeric tallgrass grassland), upland shrubland communities, bald eagles, resident herds of mule deer, and the threatened Preble's Meadow Jumping Mouse, is a formative component of ongoing social-environmental change, the naturalization of Rocky Flats advocates an understanding of nature in suspended animation—a salvaged "still life."

The Memorandum of Understanding between the DOE and the FWS mandates that the DOE must maintain continuous monitoring of hazards in the internal area only, not the refuge buffer zone, essentially circumscribing liability and the DOE's management burden. The postcleanup regulatory framework also stipulates that the FWS upkeep an "overlay refuge" across the entire area, including the DOE-retained property.[67] The appearance of a seamless natural landscape potentially hides the controls left in the environment after cleanup, such as the specially engineered caps, tanks, drains, and monitoring wells, in addition to all that remains buried on-site. The overlay of grassland is a powerful vehicle of naturalization, producing a uniform and unexceptional landscape, free of the marks of labor and controversy. The visual aesthetics of the landscape—its projection of an open wilderness "out there"—calls forth static visitor–consumer bodies least capable of affecting or being affected by the postnuclear nature refuge.

The case of Rocky Flats shows how the nature refuge is to be revalued for its flourishing signs of nature's survival in the absence of humans. The increasing regularization of the site redefines the purpose of the earlier site mission as that of environmental conservation. Furthermore, the refuge minimizes the terrain of the DOE's responsibility, through the disposal of territory to the FWS, the instrumentalizing of ecological processes in the service of waste management, and, finally, the de facto benefit of ideas of purity associated with nature, which *reduces the activity of regulating to appearing not to exist.* In this sense, the refuge "solution" to Rocky Flats signals less managing, even none at all—securing more cost-effective management through inaction. The conservation of land and animals "does more for less" in this scenario of maintaining environmental regulatory compliance; inaction looks natural. This is not to say that the FWS does nothing as steward of the Rocky Flats National Wildlife Refuge. However, a meager budget has made the opening of the refuge impossible at this time. The site is currently under the supervision of the neighboring Rocky Mountain Arsenal National Wildlife Refuge.[68] The special needs of such a nature refuge raise questions about the competence and appropriateness of the FWS as a steward; without specialized training and funding, the FWS is likely to be unequipped to deal with potential problems at the site. Wildlife management, where radioactive contamination remains indefinitely, becomes coterminous with sustaining a kind of displacement in place—refusing the historical and ongoing material effects of the site's Cold War activities, removing the visual presence of labor, and mandating safety according to surface depth. EGO labels this form of land administration to be that of making and maintaining an "alien still life," where the site's past—and responsibility for the ongoing legacy of that past—is made alien in order to upkeep the efficiency and viability of the refuge endgame. EGO will return to this point.

II. Legacy Management: Maintaining Appearances of Sustainable Managing

The monumental efforts of workers to reduce the effects of decades of environmental contamination at Rocky Flats represent an important turning point in the history of U.S. nuclear production, a point perhaps lost in the allegorical drama of bureaucracy of the preceding PowerPoint. The cleanup demonstrates DOE's recognition of the adverse environmental impact of Cold War nuclear weapons manufacturing. It also forces recognition of the dedication, resolve, and skills of nuclear workers. However, as the satirical approach of the PowerPoint effectively showed, the celebratory, accelerated cleanup focuses attention on the removal of waste and the security supposedly offered by the future use of the site as a wildlife refuge, rather than on the ongoing contamination, injuries, misinformation, and litigation involving the site. Such instabilities present severe bureaucratic, technoscientific, and historical-material problems for the state.[69] Labor was essentially mobilized to eradicate visual signs of labor's presence in the landscape, and, in doing so, to obscure the negative legacies of the site.

To support and sustain the cleanup model, the DOE developed the Office of Legacy Management (LM) as an institutional framework dedicated to the maintenance of the remedial actions taken at sites such as Rocky Flats, where portions of the land remain under DOE custody after cleanup. Formally, "legacy management" refers to all activities necessary to ensure the protection of human health and the environment following the completion of cleanup, disposal, or stabilization at a site in perpetuity. "Legacy" in this context also refers to preserving documents, and to managing clusters of former workers from DOE sites that have undergone closure and remediation. But, like many euphemisms, legacy management confuses; it seems to suggest taking stock of the legacies of more than half a century of nuclear weapons production and research, yet it functions more in the vein of "damage control," treating the interpretation and monitoring of legacies as a technical and public-relations issue. Its appearances contradictorily dissolve its own stated mandates.

At the historical moment of the DOE's decommissioning and decontamination of former facilities, LM maintains appearances of governing, namely, through the spectacle of performance audits related to its own management functions. While the stated purpose of LM is to secure health and the environment, LM achieves something else: normalizing hazards in the bureaucratic language of "managing legacies," and maintaining the appearance of safety and institutional control on limited means, even reducing its own staff and operational costs in order to do so. Administering itself in a more sustainable manner (i.e., more for less) has been added to LM's publicly stated main objectives. LM strives to reduce the government waste implicit in the management of nuclear waste by outsourcing technical management of legacy sites and managing the contracts with limited staff. LM also tries to increase accountability to stakeholders through organizational transparency. Its endgame is paradoxical: LM seeks to *appear to govern, in perpetuity, through a spectacle of oversight that actually minimizes the act of governing.* LM manages this by shrinking its own operations as a sign of "sustainable governing." Making government work better and cost less means achieving the minimum of government—what *must* be governed—in order to appear to be governing. Legacy management determines that which must be managed in order to maintain its appearances, leaving the rest, such as the memories and injuries of former workers, off the table as a regulatory nonissue.

LM's paradoxical nature can be partly explained as emerging from the "risk-free environment" and financial fallout created by historic indemnification clauses. These guaranteed that the federal government would cover any damages arising from nuclear weapons production, thereby releasing contracted site managers from liability. Although it remains significant that private contractors at Rocky Flats actually admitted to environmental crimes involving improper storage and disposal of waste after the 1989 FBI raid on the plant and subsequent grand jury investigation, the monumental case was dissolved in March 1992 when a plea bargain was offered to Rockwell, dismissing the grand jury and its determination that eight individuals—some from Rockwell, some from DOE—should be charged with criminal environmental mismanagement.[70] The company agreed to pay just $18.5 million in fines in exchange for pleading guilty to ten environmental crimes, with no individual charged. When outraged grand jurors protested by writing a report that described an "ongoing criminal enterprise" at Rocky Flats and indicted the Department of Justice for obstructing justice, they were muzzled by a court order that refused them permission to make the report public or to speak about the grand jury proceedings.[71]

Such courtroom dramas reveal a legacy of deferral, denial, and unaccountability that endangers the integrity and public perception of the accelerated remediation of Rocky Flats. Against the backdrop of the exorbitant costs accrued from indemnifying former plant operators Dow and Rockwell and widespread criticism of the collateral damage of corporate disowning and unaccountability endured by the land, workers, and general public, DOE claims that the accelerated cleanup eliminated the burden of the site on taxpayers forever.[72] Furthermore, the DOE embarked on a damage control mission to sanitize Rocky Flats through a combination of physical and discursive controls. The language and strategic concept of "legacy management" helped to mitigate doubt about cleanup, cosmetically treat risks, and symbolically detoxify the legacy of irresponsibility revealed as a subtext of litigation involving former nuclear weapons sites.

In 2003, the threat of administrative collapse antagonized by large-scale hazards and the illogical claims of safety elicited from the DOE prompted the department to bureaucratically formalize an agency at the federal level that appears to be responsible for the long-term containment and control of facilities following cleanup completion.[73] LM manages DOE responsibility for these legacies by imposing rationalities and distancing devices that give appearances of order, such as regulatory frameworks, metrics employed in monitoring sites, and mission statements about compliance. LM's 2004 strategic plan emphasizes the importance of implementing such damage controls efficiently: "As more weapons facilities continue to close across the country and remediation is substantially completed, there is an even greater need to manage the Department's legacy liabilities. Thus, the Department has realigned its resources and created a sustainable, stand-alone Office of Legacy

Management. This organization will allow for the optimum management of legacy responsibilities."[74] Originally charged with thirty-seven sites in 2003, LM had extended its managerial oversight to eighty-seven sites in twenty-eight states and Puerto Rico by November 30, 2010. The DOE predicts that by 2020, LM supervision will extend to 129 former nuclear-related sites, and 145 sites by the year 2050.[75] Simultaneously, LM has reduced its original workforce by 28 percent and was designated a "High Performing Organization." LM is the second office in the federal government to receive this distinction of "doing more for less."[76]

The strategic plan of LM enumerates a comprehensive list of actions. These include the following: protect human health and the environment at closed sites through effective surveillance and maintenance; manage the DOE's environmental liability consistent with laws/regulations; track and use advances in science and technology to improve sustainability; return federal land and other assets to the most beneficial use; preserve key records/information and make them publicly accessible; sustain public trust through cooperative partnerships with state, tribal, and local governments; mitigate the impact of department workforce restructuring and changes in the DOE's mission. More recently, the goal of improving program effectiveness through sound and sustainable management, along with an array of environmental performance assessment upgrades, have been added to the mission of LM. However, EGO has found that this dynamic vision of the LM, and its most current goal of providing "a long-term sustainable solution to the legacy of the Cold War," is more idealized than realized owing to a number of contradictions that undermine the institution's own rationale.[77]

EM/LM Bifurcation

The DOE has bifurcated legacy management from cleanup. Officially, EM remains in charge of site decommissioning and remediation, while the stand-alone LM receives the material leftovers—the "legacy sites"—for long-term maintenance.[78] LM's 2007–11 five-year plan explains the rationale: "Managing the long-term surveillance and maintenance at sites where remediation has been essentially completed, allowing the EM program to concentrate its efforts on continuing to accelerate cleanup and site closure resulting in reduced risks to human health and the environment and reduced landlord costs."[79] However, the EM/LM split divorces the consideration of the long-term effectiveness of remedies from the selection of remediation options, severs local knowledge produced before and during the cleanup from postcleanup stewardship, and exacerbates the discrepancies between cleanup funds and long-term stewardship and maintenance efforts, record keeping, DOE workforce restructuring, and local community support. While LM has requested and acquired an increased budget since its founding, the DOE's and Congress's insistence on funding LM activities through annual appropriations raises the possibility that funds may not be provided for the duration of contaminants.[80]

"Given the long-term nature of radionuclides and other residual hazards, it is reasonable to assume that long-term surveillance and maintenance will be required for hundreds or even thousands of years at some facilities. The Department recognizes that, as its environmental remediation efforts are accelerated and facilities are cleaned and closed, its long-term surveillance and maintenance responsibilities will increase."[81] As a result of the increasing number of accelerated cleanups and the augmenting responsibility of LM to oversee sites that require some form of postclosure management, LM has focused energy on maximizing its productivity as a manager, positioning itself to handle increased program responsibilities while meeting the administration's goals for reducing costs and increasing efficiency. The preceding PowerPoint aptly boiled down the logic to something more controversial: because sites cannot actually be cleaned up to prenuclear production states, they might be cleaned up less—and faster—to accelerate profits for private contractor waste management companies. LM then functions as DOE's internal dumping ground for sites that have been cleaned up less than they could have been, allowing the administrative divide between cleanup and stewardship to dissolve the long-term accountability of the cleaners for their remedies.

Technical Problems and Remainders

DOE's accelerated cleanups rely on technical land-use controls and containment measures to limit exposure to the contamination left in place. Yet environmental change requires that ongoing surveillance extend beyond the boundaries of contained sites and that continual improvements of the cleanup are made in relation to the life of the hazard. In the case of Rocky Flats, LM is responsible for oversight and containment of land where contamination requires long-term monitoring and protection after the cleanup. The Rocky Flats Legacy Management

Agreement, which establishes the regulatory framework for ensuring that the remedy remains protective of human health and the environment, emphasizes physical controls, monitoring, maintenance, and information management following the implementation of cleanup actions. LM aims to restrict access to the former industrial interior through physical barriers, legal restrictions, surface covers, subsurface monitoring, site information systems, and data protection. However, it remains unclear how these physical institutional controls are to be maintained and upgraded over time. LM must review the site, excluding the large area deemed suitable for nature refuge use, no less than *(only)* every five years in relation to radioactive contaminants with extremely long half-lives.[82] Yet security fences and land-use restrictions are not enforceable for the life of many contaminants.[83]

EGO has found that LM contracts out the environmental monitoring of the Rocky Flats site, outsourcing state functions to the delivery of outcomes by the private sector and insulating LM from direct accountability for the material consequences of the Rocky Flats accelerated cleanup.[84] LM also utilizes telemetry to dispatch its monitoring demands: the LM has implemented the System Operation and Analysis at Remote Sites (SOARS) project to obtain real-time data from sixteen sites in nine states for the purpose of evaluating remediation progress. "At some sites, SOARS allows pumps and valves to be monitored and operated remotely, which reduces travel to and from the sites and lets personnel respond to changing conditions rapidly."[85] By focusing on and developing technical instruments of oversight, legacy management in effect externalizes the damage done by former weapons production and entrepreneurial remediation onto labor and the environment. The remainders—former workers, site records, environmental contamination—are then materially contained and/or discursively sanitized.

Social Disposal through Inaction, Removal, and/or Absence

The trade-offs that the DOE and other stakeholders make between a more extensive cleanup and reliance on barriers and land-use controls are a matter not only of technical risk assessments, but of political negotiations, trust, and the maintenance of historical knowledge of the plant's prior existence. Part of LM's mandate is to foster community involvement and oversight in relation to DOE's custody of sites. But LM's reliance on physical institutional controls does not adequately allow for future revision of current end states based on the shifting needs of the public. LM does not incorporate social and cultural controls, such as consolidating and transmitting the environmental knowledge of former Rocky Flats workers, even though an organization given authority over "legacy" is inevitably a producer of history.[86] Legacy management fails to cultivate popular knowledge of residual waste, crucial for safeguarding future generations. Nor has it helped facilitate any significant public health monitoring plan or information about possible disease outcomes related to residual contamination. The LM claims it will maintain a continuous and long-term presence at Rocky Flats, but whether it will actually serve as a bulwark against the slow erosion of institutional memory is questionable.

After the EM's accelerated cleanup of Rocky Flats eradicated any sign of the industrial activity that once took place at the plant, LM withheld support for a museum dedicated to the site. During cleanup, the DOE eliminated visual evidence of the Rocky Flats Plant from the land; DOE did nothing to preserve any of the eight hundred buildings that once inhabited the site.[87] Although a bill gave DOE the authority to establish a Rocky Flats museum, it took no action.[88] Concerned about the loss of history of the plant, a local group of activists, academics, and former workers, known as the Rocky Flats History Group, eagerly sought to preserve representative examples of the building complex, initiated a scholarly oral history and video project to document the stories of former plant workers and surrounding community members, and proposed to assist environmental education about ongoing contamination at the site, providing a repository of data to citizens in order to recruit future stewards. The group founded the Rocky Flats Cold War Museum and received a financial donation from the plant's cleanup contractor Kaiser-Hill, a State Historical Fund grant from the Colorado Historical Society, congressional appropriations for a museum in late 2007, and a land donation near the former nuclear weapons production site.[89] However, they struggled to raise the money needed to purchase and renovate a building for the museum. Steve Davis, the museum project's former executive director, compared the cost of cleanup with the cost of memorializing, in an effort to garner support for the museum: "When you're spending $2 million-plus a day out there to tear down buildings and erase Rocky Flats from the landscape, we think $10 million to $15 million is a good investment."[90] The LM, by contrast, in a draft feasibility study completed in 2004, stated that the museum was not necessary and should not count on federal funds.[91] Despite this, the museum has moved forward with leasing a facility in Arvada, Colorado (a nearby suburb of Denver and near Rocky Flats), hiring an exhibition designer and new project director, and targeting 2013 for the museum's opening.[92]

LM remained absent from debates about what text should be placed on information signage at the future Rocky Flats National Wildlife Refuge.[93] Legacy Management will not be administering a release form, sign, or information video about the potential risk still posed by residual waste to future refuge visitors; signage is left to the FWS.[94] Even though millions were allocated for maintaining reams of DOE site documents, LM's strategy of information management has been most notable for proposing the disposal of five hundred boxes of public-record files on Rocky Flats, in the name of reducing government waste and securing information. An extensive list of geologic-seismological, archaeological, and historical files and radiation surveys of Rocky Flats, spanning the early 1980s to the present, were nearly destroyed by the LM in 2008 under the pretext of "digital management."[95] Public protests stopped the proposed action. On the one hand, LM has advocated converting paper archives to digital records in order to reduce waste and secure information; on the other, it has eliminated digital records about the Rocky Flats cleanup that were once available on its own Web site. In fact, the organization has successfully reduced its entire Web presence, even retiring its original logo. For an organization dedicated to legacy management, LM is curiously absent from public view, especially at the legacy sites under its purview. LM no longer even maintains its physical signage at the outer perimeter fence line of Rocky Flats.

Managing Appearances—the Spectacle of Sustainable Management

EGO is enthralled by the endgame of legacy management: the relentless climb in accountability and oversight of the LM despite its mission of costing less and reducing its own management burden. EGO considers LM to have pioneered a form of biopolitical administration of land and nuclear hazards that, in the end, works to evacuate institutional memory and minimize the terrain of that which must be managed. This type of administration is justified under the pretense of human safety, and yet humans are blamed for the necessity of this administration. Precisely because humans were at Rocky Flats, engaged in the nuclear industrial enterprise, the land must be cleaned up and "abandoned to nature" to achieve efficiency and limited accountability for the history, labor, and knowledges linked to the site. More than this, an auditing system has been installed to ensure the maintenance of this scenario and to achieve yet further efficiency by increasing its productivity or, rather, by streamlining its own management under the auspices of "sustainability." LM, through its spectacular reports of surveillance, compliance, and continual improvement, *economizes even its own appearances of managing.*

LM oversight of Rocky Flats claims to monitor residual hazards and maintain on-site waste containment infrastructure. "Visibility is supposed to produce verity, and surveillance sanctification. To produce truth and purity, though, the system requires ever-better modes of looking."[96] The LM's position of continuous improvement and ever-better ways of dealing with hazards can be interpreted as a drive to *find,* a relentless search for new spaces to examine, colonize, rationalize, and bring under control.[97] Seen thusly, legacy management disperses state functions, because its governing relation requires "zones of wildness." Sites with nuclear legacies fit the bill; their ungovernability can be seen to necessitate an expansion of power. "It depends not only on having spaces that are controlled, regulated and ordered, but also on spaces that are beyond the borders of control which defy rules and which introduce chaos. Risk, filth and disorder are essential parts of any programme to rationalize and rule."[98] By developing the office and practice of legacy management, the state is

> both channeling what is desirable and seemingly purging what is unwanted from both the record and reality, constantly making real life more closely approximate the model contained in the documents and records. . . . Modern power is thus based on more than the gaze, on more than the watchful eye and on more than mere surveillance and punishment: it is based on the ability to purify, to remove excess, corruption and putrefaction from view and sweep it away.[99]

Although EGO does not disagree with this assessment, it finds a slightly different process at work with regard to legacy management. Biopolitical governance in this case proceeds more along the logic of devoting expertise to and developing techniques for *making residual hazards, contamination, and wastes known, without necessarily finding out or reassessing what they are.*[100] LM labors to show the quality control of its contracted monitoring and, more important, its management organization and performance. This high-performance mentality encourages constant activation "as though it were in a state of perpetual self-awareness, animation and explicitness."[101] The LM is essentially a devoted auditing system, established to administer postclosure DOE sites in perpetuity—an ultimately unadministrable charge. In audit "what is being assured is the quality of control systems rather than the quality of first order operations."[102] Such auditing serves as a form of institutional disposal and can be made more efficient to the point where it oversees only that which must be governed in order to maintain appearances.

Legacy management endeavors to build accountability and trust because it produces some measure of transparency; LM relentlessly performs itself as an organization for the public to see, through banal validation of its validation procedures.[103] Auditing its own performance, as a display of self-discipline and moral authority, enables LM's claim to effective regulatory power. It sets up a moral call to action, but with the "end" of containing the social and material legacies of the Cold War through the logic and language of its own system, a priori measures, constant organizational self-observation, and self-improvement. This auditing spectacle secures the appearance of the control system (which serves as the sign of accountability) by disposing of the contingencies and embeddedness of the land in social-historical processes. LM does not disavow the injuries of DOE's former Cold War policies, but rather disavows how such injuries continue to shape social relations. Legacy management also exhibits little to no openness to creative developments or emerging social relations with the sites under its supervision. It operates as a form of power that depoliticizes such sites through technocratic appearances of response.[104]

"Sustainability" is the DOE's/LM's latest responsive action. Sustainability measures have come into play for the purposes of making LM's disposal of DOE land and post–Cold War burden more productive by giving the appearance of more responsibility. Sustainability focuses efforts on streamlining the organization itself, as a way to maintain LM's long-term managing of nuclear legacy sites and to aggregate attention as a moral duty and good governance. After earning the distinction of High Performing Organization, LM now seeks to achieve full "green" status on the Office of Management and Budget's Environmental Stewardship scorecard by implementing environmental sustainability initiatives: reduce energy use and greenhouse gas emissions by 3 percent annually or 30 percent total through the end of fiscal year 2015; increase the amount of renewable energy; reduce potable water use; acquire bio-based, energy-efficient, water-efficient, recycled-content, and otherwise environmentally preferable products; reduce solid waste; and ensure that any new construction or major renovation of agency buildings complies with EPA guidelines on high-performance sustainable buildings.[105] EGO found that, among these, sustainability refers only to the activity of improving management through monitoring performance. EGO considers this to be the penultimate neoliberal endgame of biopolitics: *make the managing of postnuclear lands more "sustainable"—make it cost less and look better—by essentially reducing it to maintaining appearances of managing,* which can be achieved by a maximum staff of fifty-eight people, to eventually cover more than two hundred legacy sites across the United States.[106]

III. Lessons Learned: "Alien Still Life" and the Ethics of the Return of the Repressed

The cleanup of Rocky Flats has obscured the historical production of waste and buried questions about responsibility for waste and injury under the cost-effective cover of the nature refuge. The remediation activity at the mesa has discouraged conversation about deep, unresolved contradictions. Likewise, legacy management has endeavored to redistribute the burdens of proof and reconfigure the responsibility, causality, and guilt that underwrite the long-term legalization of contamination at the site. While the cleanup enabled an entrepreneurial remediation industry to profit from basically doing less cleanup for more money with less accountability, legacy management must find ways to maintain appearances of control and safety—supposedly in perpetuity—over a mounting number of sites on a limited budget with extremely limited staff. The nature refuge gives the appearance of doing more for less: the conversion of nuclear production facility to nature refuge treats the land as alienable and natural in order to cut down on costs, circumscribe management, and enhance the appearance of safety and control. As a regulatory exercise, legacy management further depoliticizes the site, eclipsing Rocky Flats' controversial histories and administering the land by using manageable simulacra and distancing devices that deny complexity.

EGO maintains that the cultural result of this institutional transfer of DOE land, its reinscription as a nature refuge, and long-term monitoring by an organization concerned with self-auditing and appearances is the emergence of an "alien still-life" landscape; it freezes nature in an end state or "risk fix," anonymizes waste, utilizes the management of open space as a way to avoid managing contaminants, and sustains the alienation of Rocky Flats workers from the land through environmental disorientation, cancer, and cultural amnesia.[107] Ultimately, this bureaucratically enforced alien still life serves to suspend the political and ethical involvement of humans, effectively limiting responses in the name of security and the "end" of the Cold War. *The Rocky Flats case shows an endgame of arranging both land and humans as residual and alien by a form of management that is ex-terrestrial, that works by way of displacing the land—rendering it visually alien—and "offworlding" land management through relentless self-auditing.*

The emerging relations between land and labor at Rocky Flats reveal a landscape wherein nature is perpetually treated as an exception. Refuges have often been established in crisis mode or states of emergency. The Rocky Flats National Wildlife Refuge functions as a culturally palatable void space, a still life, an external nature without history. The history of damage suffered by wildlife appears unnoticeable, unexceptional. This is achieved in part by environmental regulations and legally binding performance agreements that offer up measures for the cost of managing environmental problems instead of exploring what might be there, investigating the wrongdoing that historically took place, and/or practicing accountability as an ongoing collaborative process without end. Relationship management units, such as the LM, evaluate performance obligations on paper instead of the human impact of various actions; risk assessments and actuarial modes of maintenance implant short-term technical fixes instead of cultivating longer-term outcomes based on transformational ecology or social justice, equity, and fairness. The impression of the refuge's fecund austerity and resilience allows state and private industries to remain insulated from the political and economic consequences, toxic impacts, and material-ecological effects of their own historic practices. "The managed void replaces eternal vigilance as the price of liberty."[108] It legitimates exclusion, diffuses responsibility, and rationalizes inaction by focusing on the upkeep of appearances of compliance.[109]

Waste and industrial residues of the former plant essentially make exterior nature porous and therefore threatening to the nature spectacle. Workers are the most obvious reminder of the site's industrial environmental history. The nature refuge subjugates the environmental memories of these workers, a process sometimes aided by an environmental movement that has dichotomized industrial laborers and nature, pitting workplace interests against environmental concerns. Instead of acknowledging any ecological sensitivity among former Rocky Flats workers, which did not take place in opposition to the bomb factory but rather in confronting constant everyday hazards and accumulating workplace knowledge, the nature refuge's association with purity deflects dangerous remainders of the site's history that might compromise the wildlife refuge remedy—sick former workers and their embodied environmental memories, leaky site records, uncertain environmental contamination. Wildlife management of the nature refuge joins the cleanup and legacy management of Rocky Flats in enframing the land as an alien still life, wherein the "alien" marks the alienation of the environment from former workers, who are no longer able to recognize the area where they once labored and, in some cases, were contaminated, *and* the silences in the public record and erasure of visible signs of the land's industrial past.[110] Former workers typically react to the visual landscape, bereft of the former facility, with an emphatic "There's nothing there!" The site has been made over as "terrain vague," alienating humans ironically in the name of protecting their health and the environment.

Worker knowledge of Rocky Flats land is rich, and is a critical embodied link to the nuclear facility's ecology. From exposure to plutonium via a leaky glove box to the smell of stray cats living under offices to the relentless road accidents caused by inclement weather during their commute to the plant, workers were particularly aware of ecological hazards. The Rocky Flats National Wildlife Refuge—no longer an industrial workplace, or even an area that appears worked—serves as the dustbin of such industrial environmental history. Jean Baudrillard captures the human tragedy of this process: "What is worst is not that we are submerged by the waste-products of industrial and urban concentration, but that *we ourselves are transformed into residues.*"[111] In the case of former nuclear workers, the nature refuge contributes to the treatment of former plant workers as the unavoidable material remains of the nuclear nation. The emphasis on the purity of the refuge detaches biotic assets from a landscape of radioactive contamination, dividing and devaluing bodies still "intoxicated" with nuclear history and nuclear materials. Illness is a kind of material-environmental memory of Rocky Flats in the form of embodied waste or exposure to waste. The spectacle of nature places value on the visible adaptability of certain organisms to radioactive environments—"nature's survival"—against remaining sick workers, who are then relegated to a visual register of impending death.

The DOE strategically entrenches this ontological split further: It has consistently tuned out the voices of Rocky Flats workers and relatives who have sought compensation for work-related exposure to toxins and radioactive elements, under the Energy Employees Occupational Illness Compensation Program Act (EEOICPA) instated by Congress in 2000. The DOE took little action to compensate workers, leaving claimants to dissipate with old age and "slow death."[112] Subsequently, the EEOICPA compensation program was transferred to the Department of Labor, which was expected to run the program more efficiently and effectively but is now implicated in what many workers describe as an industry of denial.[113] Rocky Flats labor activist Terrie Barrie, whose husband worked as a machinist at the Rocky Flats Plant, suggests an intimate connection between the nature refuge, institutional denial of illness, and cultural amnesia about ongoing Cold War effects and exposures: "I worry that people may forget about the workers now that it's [the Rocky Flats site] a big grassy field where deer play. That place made people sick, and we have to remember that. Never in a million years did I think the cleanup would be finished before my husband got paid."[114] Such deficiencies in recognizing the relations between humans and nature, humans and waste, demonstrate that an ethical response is necessary that would promote more relational thinking and contest the bureaucratically arranged suspension of Rocky Flats. EGO finds that arguing for better compensation, a better cleanup, or better signage, while important, would not resolve the stark divisions entrenched at the site by the massive technoscientific reproduction of nature and legacy management. Rather, a tactical ethics is needed to render tangible that which has been made to disappear by the desire to regulate and to achieve safety, systemic continuity, and predictability.

Although the site's Cold War mission is no longer active, it is still affected by the chemical, radioactive, infrastructural, and other residual wastes produced during its years of plutonium production. From the perspective of waste, the site's past is not over, it is not even past. The actual character of hazards that remain is ambiguous, open to interpretation, alien.[115] Waste disorganizes the appearances of control installed by management activities; waste remains, unassimilated and nagging. In this sense, the "still" in alien still life means an unregulatable persistence, something "in addition to" endgames and the apparent ending, a "nevertheless" or "after all" that abides regardless of attempts to evacuate or void the site. Rather than a still life of pictured objects in a state of suspended animation or rest, or the polarizing of subject and object by a chilling kind of vision that ruptures continuous life, waste haunts the scene.[116] The specter of radioactivity can make any landscape seem otherworldly and strange, as well as problematize understandings of nature as pure or external "other."[117] EGO suggests that refuge designation of residually radioactive land calls for an "alien" environmental ethic, where the alien marks not only what is alienated by the abstractions of the refuge end state but also the residual *lasting on* of waste[118]—waste's resistant and ultimately unregulatable material persistence as alien pasts still in the present.[119] Instead of producing safety through political and ethical suspension, *the heuristic device of the alien as waste estranges truth claims about the site, un-fixes the efficiency measures and systems assurance, and counters the disposal of land with a material urgency to experiment with the legacy of the nuclear*—to ensure that such legacies are not so much visibly framed and monumentally contained and denied but persistently, naggingly itched or smelled, as the chemical symbol for plutonium ("Pu") suggests.[120]

EGO recognizes the potential danger involved in promoting unworldly figures as liberating. The "enduring presence of extraterrestrial elements in radioactive wastes, those luminous fission daughters whose consuming emissions will continue to score our monstrous messages into the bodies of those in distant futures," should not be romanticized.[121] Yet to reverse the displacements and divisions that have turned Rocky Flats into an alien still life, EGO devises that an ethics of waste as the "return of the repressed"—of that which was refused by the refuge endgame and legacy management—potentially offers a new intelligibility. "At some point it [waste] becomes excessive, and it then emerges as a challenge to the whole system, the disavowed trace that erupts to put into perspective the calculations that have ordered its dissimulation."[122] EGO has substantiated throughout its oeuvre of research that "the return of the repressed [is]/as a powerful moment."[123] In the case of Rocky Flats, EGO considers the way forward to be that of exploring how new alliances might be founded on the nature refuge's stockpile of repressed affects and representations; "the whole imaginary of the stockpile, of energy, and of what remains of it, comes to us from repression."[124] The material unearthliness of waste provides a way of rescuing analysis and action in response to Rocky Flats from a metaphysics of earthly presence for the unforeseen material returns of life unknown.

One place to start are the numerous "hot spots" of contaminated soil that were discovered—their existence, definition, and location debated—during the DOE's verification of the adequacy of Rocky Flats' cleanup. A series of tests performed by DOE-hired consultants found small areas of low-level contamination, elevated above sanctioned cleanup standards, dispelling notions of the finality of the remediation. One test discovered thirteen hot spots near the former 903 pad area where plutonium-contaminated oil was once stored. Another test of the boundaries of the same area found that 28 out of 178 sites checked had elevated levels of contamination; retesting subsequently brought the number down to 5. A third test, done by a helicopter that reportedly performed forty-four thousand readings, even detected one unexpected low-level hot spot south of the site on nearby cattle pasture.[125] DOE officials initially downplayed the contamination as an anomaly, but later orchestrated the cleanup of the hot spots.[126] In a second unexpected return of the repressed, approximately twenty-five million gallons of radioactive pond water were found, demonstrating that the cleanup effort itself mobilized waste in uncontrollable ways. Retention ponds at Rocky Flats were unexpectedly found holding water with americium (a plutonium derivative) and trace levels of plutonium at levels four times greater than what the cleanup plan called for; cleanup of the water required an additional $2 to $3 million.[127] A final example blows the pristine cover of wilderness at Rocky Flats. In consideration of a future hunting program to control the deer population at Rocky Flats, the FWS randomly culled twenty-six deer from the site in order to conduct tests for radioactive contaminants in the liver, kidneys, lungs, muscles, and bones, and thereby ascertain whether they were safe to eat.[128] Several of the twenty-six deer contained trace levels of americium, plutonium, and uranium, revealing that the legacy of former plutonium production abounds at Rocky Flats, in the material processes of bodies of deer. Such phenomena share an insistence that *this is still life, but not as we know it.*

The challenge of finding ways to live with this alien life requires that we cultivate "a more thorough engagement with the indeterminate environment in which we are embedded."[129] Indeterminacy and complexity, more than catastrophe or simple endgames, are the forces that emerge through tactical ethics of waste, of the alien and repressed material.[130] As more and more nuclear sites are converted to nature refuges, resulting in less accountability and more sustainable (read: lower) costs, with just enough stewardship of what must be monitored in order to maintain appearances, the postnuclear nature refuge will continue to spill across the *still nuclear* U.S. landscape, depoliticizing the consequences of the Cold War and circumscribing what "legacy"—as a practice—might achieve. Other ethical legacies are still possible, indeed necessitated by this grim future. In the optimistic words of Jody Baker:

> The barrel of toxins . . . can reach a critical threshold where its material can no longer be contained and bursts forth or slowly seeps outward, into the crevices of social life. There are other thresholds: a cancer threshold when toxins accumulated in the cell cause its mutations, its mutinous growth. . . . There are political thresholds too: when the callous poisoning of the environment and the people who live there engenders political organization and political action that demands . . . social change. We should not, then, separate out the leakage of toxins from barrels, the leakage of hegemonic legitimacy and (corporate) power, and the leakage of social change.[131]

An ethics of waste as the outer limit of truth and the material return of the repressed—the still ongoing/not yet ended—seeks those ecological thresholds that demand sociohistorical transition. It attempts to model thresholds, locate their crises, and attend to that which has been set aside and disposed of, yet, at the same time, imposes itself on the social body and will not go away.[132] "Waste is the figure, then, not for a redemption that recovers the wholeness of a fragmented experience in aestheticized form but a figure for what remains 'possible' in the aftermath of catastrophe."[133]

Appendix A

Government here references Michel Foucault's notion of "governmentality," a variation of Foucault's understanding of biopolitics: the ensemble of institutions, procedures, analyses, calculations, tactics, and reflections that allow the exercise of an intricate form of power that has as its target the population. It is a power enabled by a whole complex of knowledges and enacted through a series of technical means and apparatuses of security. This theoretical conception is founded on the historical process by which juridical power and law were incorporated into administrative states to function primarily as regulation, rather than rigid prohibition.[134] Governmentality focuses on "arts of governing" and the regimes of practices through which populations are constituted and regularized. Modern governmentality is concerned with the fostering and regularizing of life, the "conduct of conduct" for a population of subjects, and the establishment of a regime of truth to which the population adheres. Such a governing relation aims to secure the freedom of the population by pursuing safety and mitigating threats of indeterminate and ubiquitous, invasive and pervasive forces, from diseases to nuclear waste, that can take hold of the social body. Such arts of government paradoxically entail "producing, breathing life into, and increasing freedom, of introducing additional freedom through additional control and intervention."[135] Yet the escalating costs of compensatory mechanisms of freedom and procedures of control and compliance mean that forms of government must continually introduce economy and efficiency; art of government is essentially concerned with "answering the question of how to introduce economy."[136] Governing, then, must somehow secure both freedom and efficiency in order to maintain the governing relation.

Foucault's recontextualizing of his own studies of biopolitics and the natural sciences within this framework of governmentality opened the door to an ecological reading in which "life" enters history as an explicit strategy—as a state preoccupation, an object of scientific knowledge, and a normalizing principle of conduct. Within the complicated genealogy of Foucault's work, he came to link governmentality and biopolitics in order to take into account how life became more of the focus of governing. In so doing, biopolitics "is probably the closest Foucault ever came to addressing the environmental issue from the perspective of how the mechanisms of biological life themselves became an object of 'reason of state' calculations and strategy."[137] Environmental management and ecology can be regarded in these biopolitical terms, as they originate in and operate on the relations between populations, resources, and the environment.[138] More generally, "we might say that 'power' is nothing other than the organization of 'capacity' itself, whether in the sphere of the economic or the political, and therefore represents . . . the mobilization, concentration, and deployment of living energies."[139] Such energies, whether related to human populations, stored energies from the past, plant or animal life, can be governed for the maximization of their productivity. "It is no longer a matter of bringing death into play in the field of sovereignty, but of distributing the living in the domain of value and utility. Such a power has to qualify, measure, appraise, and hierarchize, rather than display itself in its murderous splendor."[140]

03

HOLE IN THE HEAD GANG

THE *REDUCTIO AD ABSURDUM* OF NUCLEAR WORKER COMPENSATION (EEOICPA)

The U.S. Energy Employees Occupational Illness Compensation Program Act (EEOICPA), passed in 2000, was celebrated as a landmark piece of legislation.[1] EEOICPA's passage marked the end of the systematic denial of occupational illness compensation claims among workers in the factories and laboratories of the United States' atomic bomb complex.[2] EEOICPA represented a hopeful shift from "deny and defend" to "recognition and recompense" for thousands of sick workers who earned livelihoods in government-owned, contractor-operated weapons production plants, where large quantities of radioactive and toxic materials were used.[3] After more than five decades of exemption from liability for the purposes of national security, the Department of Energy (DOE) acknowledged responsibility for unsafe working conditions in the United States' Cold War weapons factories, and admitted that nuclear workers had been put at risk by exposure to radiation and other toxic substances. DOE policies had previously assisted contractors in contesting occupational disease claims.

When the concept of an atomic bomb was still theoretical, Nobel physicist Niels Bohr remarked to top bomb scientist Edward Teller that the bomb could not be built unless the whole United States was turned into one huge factory. Later, Bohr is reported to have told Teller, "You have done just that."[4] At its peak, the DOE weapons complex would have ranked high among large corporations in terms of plant subcontracting relationships. The country was drawn into the assembly line for making nuclear weapons, with as many as 365 sites nationwide involved in mining, milling, refining, and enriching uranium, making and machining plutonium and bomb parts, special materials handling centers, assembly, research and development, as well as testing grounds.[5] Contractor employees numbered close to six hundred thousand.[6] Some of the production sites were owned by DOE and its predecessor agency the Atomic Energy Commission (AEC) but operated by contractors; others were privately owned but run under contract with DOE; and still others provided contractors and the DOE services and supplies. Because of national security concerns, work took place secretly, with criminal penalties for breaches. Workers were also given minimal information about some of the materials with which they worked and their potential impact on health. In spite of the rhetoric of safety, exposure to dangerous chemicals and radiation was widespread in the nuclear facilities, where work involved chemical separations and the handling of a panoply of toxic substances.[7]

Historically, DOE and its precursor AEC claimed "sovereign immunity" and actively assisted contractors in opposing work claims of exposure-related illness. Workers pursuing compensation confronted an iron curtain of information, installed to protect national security and public confidence.[8] Their resources were little match for the combined front of corporations and the U.S. national security agenda. DOE-contracted site operators were typically indemnified by the DOE: contractors signed agreements that limited liability for running these sites, because the federal government agreed to cover costs from damages and lawsuits. The "civilian" status of nuclear workers was also strategically used to disown responsibility for the welfare of these workers; they were not eligible for veteran benefits, and federal workers' compensation programs generally did not include nuclear workers. Because of long latency periods, the uniqueness of the hazards to which they were exposed, and inadequate exposure data, many workers were unable to obtain any form of state compensation benefits.[9] During the massive downsizing of the weapons complex in the early 1990s, when nuclear weapons sites were closed and decommissioned en masse, these workers were forced to transition to other employment—or retire—and face unknown futures, with no form of federal care on the horizon.

An atmosphere of distrust and mounting environmental and

health concerns about the weapons complex has pervaded the post–Cold War period, which brought forth a series of revelations about conduct of the nuclear weapons complex and the many kinds of sacrifices people had been unwittingly subjected to in the name of national security—revelations about environmental devastation of an unprecedented magnitude, secret plutonium experiments on people, atmospheric nuclear fallout, the environmental racism and sacrificing of land for uranium mining and the making and testing of nuclear weapons, and the lingering problem of nuclear waste.[10] In this context, a public constituency for a federal program to compensate sick nuclear workers was galvanized by a series of discussions and public hearings held around the country, led by then–Energy Secretary Bill Richardson and then–Assistant Secretary of Energy for Environment, Safety, and Health Dr. David Michaels.[11] The aim of these meetings was to publicly acknowledge the harm done to workers and to correct the DOE's former culture of secrecy.[12] If a production imperative took precedence over worker health and safety during the Cold War, the moral imperative to rectify exposures that had occurred without knowledge or consent and to provide for sick nuclear workers was now made a U.S. national priority. Although Congress held hearings about radiation-induced cancer in atomic workers as early as 1959, those efforts were stymied by the problem of "indeterminate causation"—the inability to distinguish radiogenic cancers deserving of compensation from cancers that would have occurred in the absence of occupational exposure to ionizing radiation.[13] Forty years later, the EEOICPA legislation provided an unrebuttable presumption that certain cancers are work-related, and it set forth two standardized methods for federal agencies in determining compensation claims: radiation dose reconstruction and computer modeling of the "probability of causation."[14] Reading less like a legal document than a confession, EEOICPA would recognize the harm caused by radiation and toxic chemicals in the bomb factories of the United States, facilitate moral and financial recompense, and, in doing so, "make workers whole persons again."[15]

While the 106th Congress adopted a framework intended to yield consistent, scientifically informed causation determinations for cancer claims on a case-by-case basis, the implementation of EEOICPA

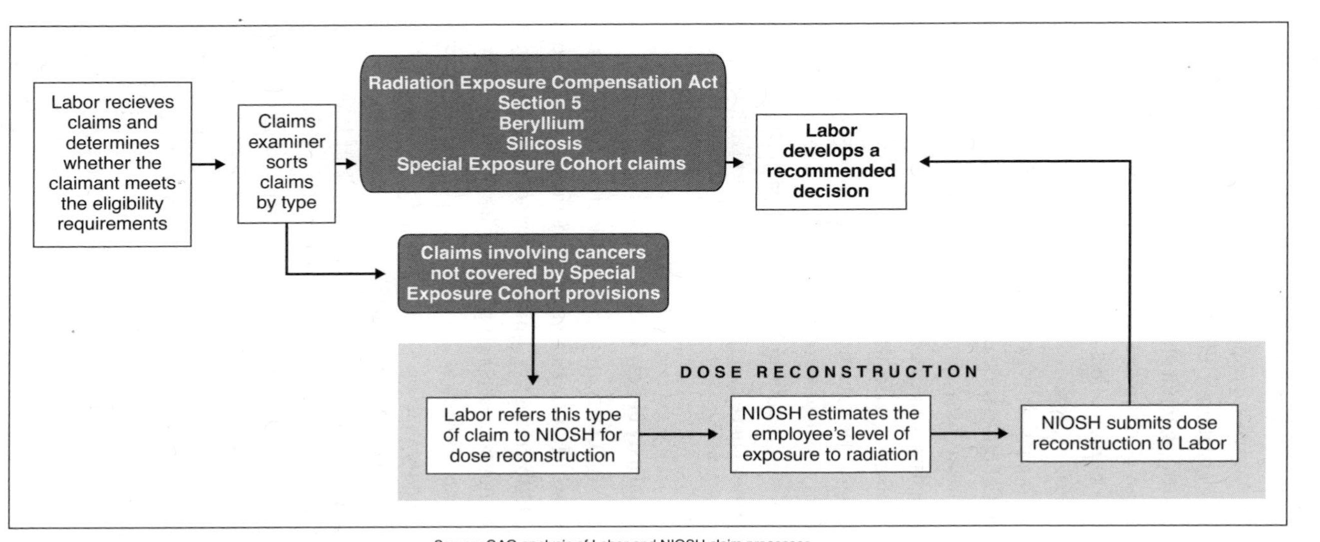

U.S. Government Accountability Office, "Energy Employees Compensation: Additional Independent Oversight and Transparency Would Improve Program's Credibility," GAO-10-302 (Washington, D.C., 2010), 8. Note: misspelling in original.

has been so complex, even convoluted, that it has been referred to as a "strange beast" with "weird appendages" by those who first designed and advocated the legislation.[16] EEOICPA claimants are assigned, channeled, and graded, without the federal government assuming their custody. Individual claimants bear the burden of proof, and, where there is no documentation of work history or exposure, the knowledge gap is filled by dose reconstruction, scientific evaluation, interviews, and mathematical expressions of statistical uncertainty. The EEOICPA legislation essentially arranges for data production about exposure and dose—resulting, namely, in a *spectacle of uncertainty,* which dehumanizes nuclear natures, splitting workers from sites and from each other, and cleanses the land of evidence by obscuring the links and geographical relations between bodies, sites, and materials. Workers, then, are *doubly remaindered*—as the remains of the Cold War and as the remains of compensation; unpaid, allowed to die under the sign of benevolence and government accountability.

Summary of How EEOICPA Works

The relevant parts of EEOICPA are Part B and Part E. Part E replaced the original Part D, which was deemed to be ineffective.[17] The 2000 law originally sought to create a technical assistance program run by the DOE to help former contractor employees file state workers' compensation claims for occupational illnesses. Part D covered illnesses that were caused, aggravated, or contributed to by exposure to any toxic substance while working at a DOE facility. Radiation-induced cancers were to be separated out and funneled into Part B. By 2004, DOE had granted very few claims filed under EEOICPA Part D, doing nothing for the vast majority of these workers, even though the DOE had spent $92 million in total program funds through September 2005 to run the compensation program.[18] A series of Government Accountability Office reports over the years since EEOICPA's founding revealed the extent of DOE's ineffective management and deficient controls, and confirmed constituents' complaints of delays.[19] In hearings on the issue, serious design flaws in the DOE compensation development process of Part D were evident, including DOE's failure to provide clear and consistent guidance on establishing the "probability of causation" for occupational illnesses, overloading and underpaying the Physician Panels that had been coordinated to determine causation, stonewalling efforts to find a willing payer for compensation claims, refusing to streamline certain cases, and subcontracting out the entire Subtitle D claims process to an underqualified contractor.[20]

In response, Congress revamped this section of EEOICPA in the 2005 National Defense Authorization Act: Part D was abolished, and Section E erected in its place.[21] Control was completely removed from DOE and placed under the Department of Labor (DOL). DOE's only role now is to provide support in the retrieval of records. Part E continues to address and determine whether a worker's toxic exposure was "at least likely as not" a significant factor in causing, aggravating, or contributing to certain illnesses. Part E covers DOE contractors and subcontractors, and anyone potentially eligible for Radiation Exposure Compensation Act (RECA) section 5 compensation (such as uranium miners, millers, and ore transporters). Medical care is provided, and claimants can receive up to $250,000 in compensation for impairment and lost wages.

Part B covers current and former DOE workers who have been diagnosed with cancers, beryllium disease, or silicosis, and whose illnesses were caused by exposure to radiation, beryllium, or silica while working directly for the DOE, the department's contractors or subcontractors, a designated Atomic Weapons Employer, or a beryllium vendor.[22] Individuals or their survivors found eligible may receive a lump sum compensation payment of $150,000 and medical expenses for covered conditions. Part B also includes an added benefit for those who have already qualified for RECA section 5 compensation. Compensation is supposed to be fast-tracked for beryllium disease or silicosis, or for particular site workers who have any of twenty-two specified cancers (e.g., bone cancer, renal cancer, or leukemia at least two years after exposure; lung cancer; various radiogenic-induced diseases, provided onslaught was at least five years after exposure).

Employees with "Special Exposure Cohort" (SEC) status do not have to undergo dose reconstruction or probability of causation. To be considered, claimants need to prove that they worked 250 aggregated days of employment in a class or classes of workers at a designated SEC site, and have any of the twenty-two designated types of cancer. An

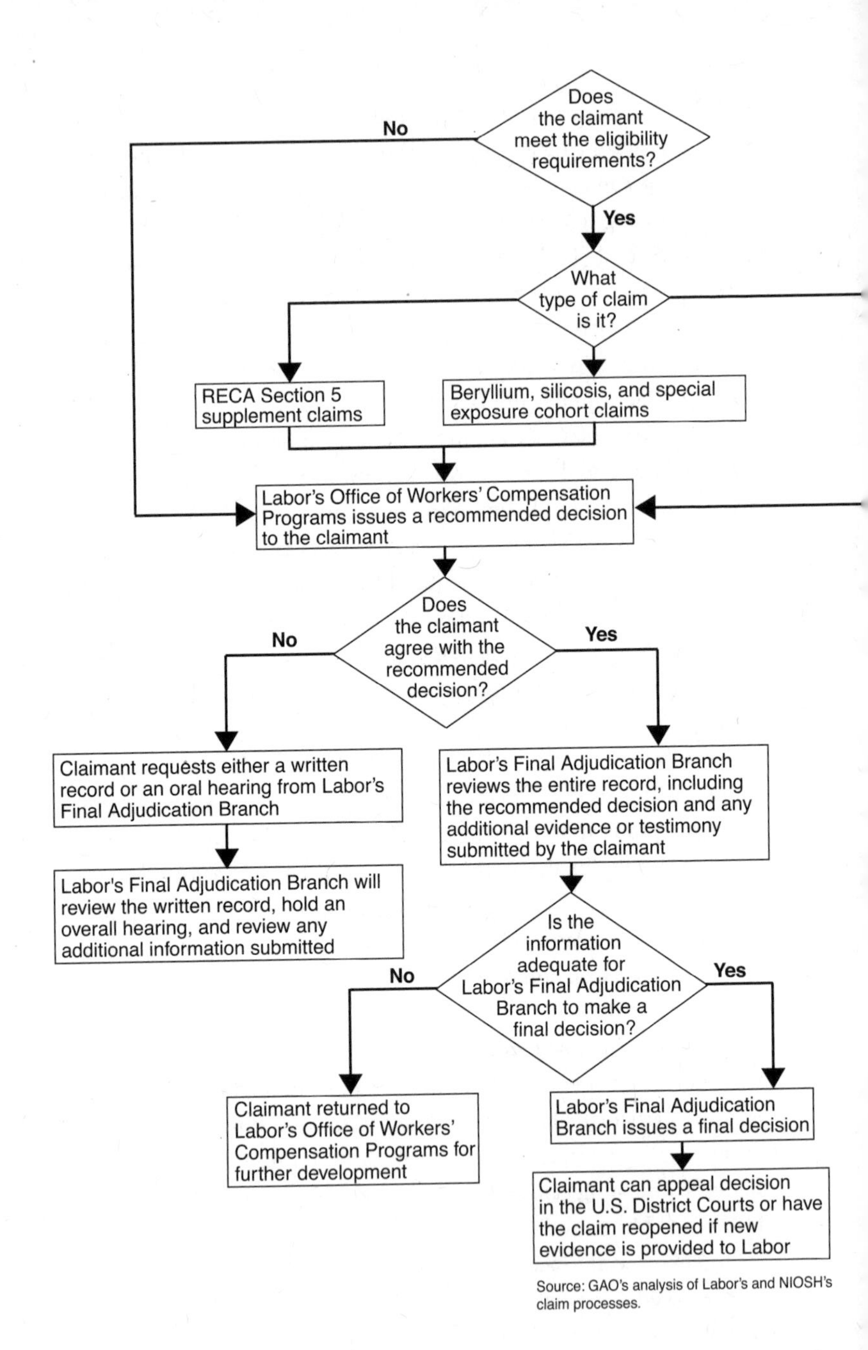

Source: GAO's analysis of Labor's and NIOSH's claim processes.

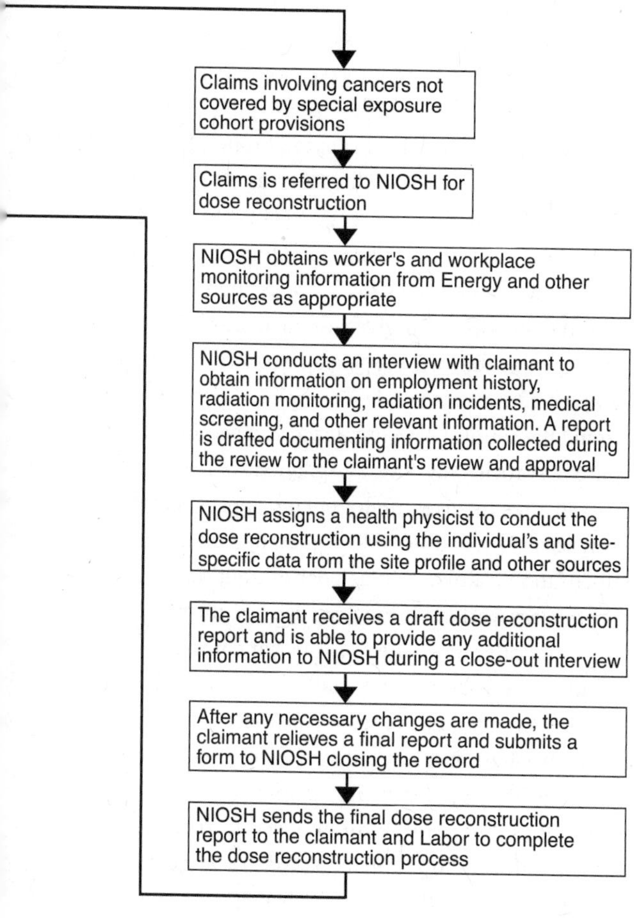

Claims involving cancers not covered by special exposure cohort provisions

Claims is referred to NIOSH for dose reconstruction

NIOSH obtains worker's and workplace monitoring information from Energy and other sources as appropriate

NIOSH conducts an interview with claimant to obtain information on employment history, radiation monitoring, radiation incidents, medical screening, and other relevant information. A report is drafted documenting information collected during the review for the claimant's review and approval

NIOSH assigns a health physicist to conduct the dose reconstruction using the individual's and site-specific data from the site profile and other sources

The claimant receives a draft dose reconstruction report and is able to provide any additional information to NIOSH during a close-out interview

After any necessary changes are made, the claimant relieves a final report and submits a form to NIOSH closing the record

NIOSH sends the final dose reconstruction report to the claimant and Labor to complete the dose reconstruction process

U.S. Government Accountability Office, "Energy Employees Compensation: Many Claims Have Been Processed, but Action Is Needed to Expedite Processing of Claims Requiring Radiation Exposure Estimates," GAO-04-958 (Washington, D.C., 2004), 34. Note: misspelling in original.

SEC class is a group of workers who were employed at a particular facility, sometimes a particular building, during a specific time period.[23] A given SEC work site may have classes of employees for only certain dates of operation. For example, at Rocky Flats, the former plutonium production facility near Denver, SEC status is only open to those employees of the DOE, its predecessor agencies, or DOE contractors or subcontractors who were monitored or should have been monitored for neutron exposures (which has been interpreted to include anyone working in a building where plutonium was used) at the Rocky Flats Plant for at least 250 workdays from April 1, 1952, through December 31, 1966. The ruling covers buildings 86 (the same as 886), 91 (991), 701, 771, 774, 776, 777, 778, and 779.[24] Basically, SEC status is a hodgepodge definition that can incorporate disease, building or site, and/or time period. Because SEC status allows for a streamlined compensation claims process (i.e., workers are given the benefit of the doubt and do not have to show that they were exposed), workers employed at Rocky Flats in more recent decades and DOE employees from numerous other sites around the country have organized and petitioned to be included under the SEC provision.[25] SEC designation was designed to address radiation exposures incurred under circumstances that make dose reconstruction scientifically infeasible or morally unnecessary as the basis for compensation claims, such as poor or nonexistent exposure records. Many nuclear workers argue that such circumstances are not the exception but the rule within the U.S. nuclear complex.

Cancer claims not covered by SEC are statutorily required to undergo an exposure assessment. This mandate was interpreted to entail radiation dose reconstruction and scientifically informed causation determinations per individual cancer claim. Controversial within the scientific community, these procedures are intended to yield reliable results using methods familiar to occupational health professionals and, by extension, toxic tort cases that draw on strategies of epidemiology. Methods involve reconstructing past exposures from interviews and existing documentation, and using confidence intervals to express statistical uncertainty.[26] The National Institute for Occupational Safety and Health (NIOSH) is the key player here, charged with reconstructing radiation doses and site profiles, and gathering information such as recorded radiation dosimetry practices at each DOE site.

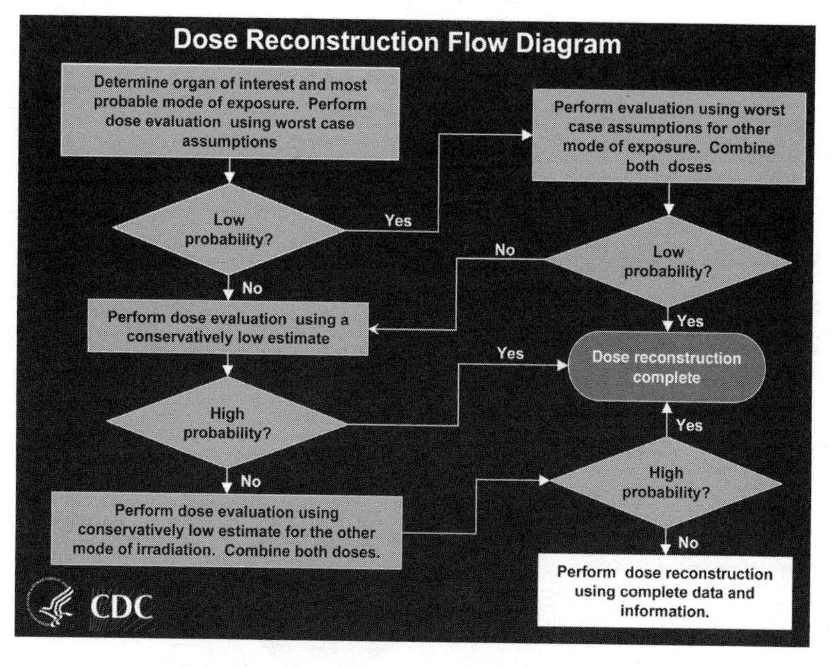

James W. Neton, Technical Program Manager, Office of Compensation Analysis and Support, National Institute for Occupational Safety and Health, "Implementation of the Dose Reconstruction Rule 42 CFR Part 82" (PowerPoint presentation, no date), http://www.cdc.gov/niosh/ocas/pdfs/pubmtgs/hpsp032.pdf (accessed September 10, 2011).

NIOSH produced what were considered to be scientific guidelines for determining whether a worker's cancer was related to occupational exposure to radiation: chiefly, the probability of causation. Using various strategies of estimation and a mathematical model of quantitative risk assessment that determines the probability of causation and that allows for an increasingly complex analysis of uncertainty, NIOSH has since developed a user-friendly software system that does all of the necessary calculations and assigns each claimant a probability that his/her disease is job-related. After data have been collected from various site profiles, workers' files, radiation monitoring records, and qualitative information from worker interviews, a worker's "dose" is reconstructed—an estimated type and level of radiation exposure received by the worker and the associated radiation dose to each organ affected by cancer. This information is entered into a set number of

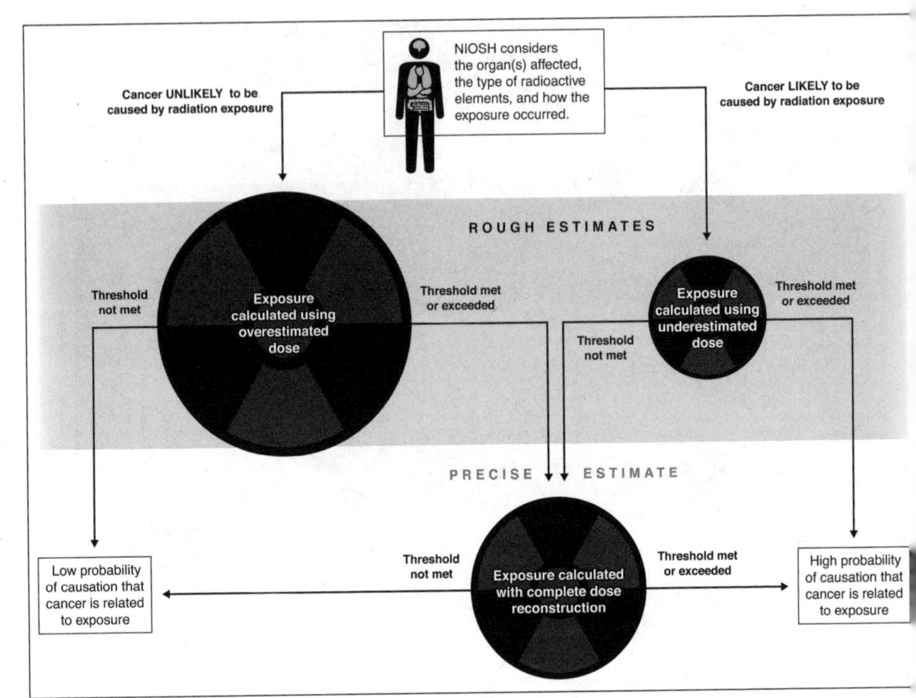

U.S. Government Accountability Office, "Energy Employees Compensation: Additional Independent Oversight and Transparency Would Improve Program's Credibility," GAO-10-302 (Washington, D.C., 2010), 20. Note misspellings in the original.

data fields of the software known as Interactive RadioEpidemiological Program (IREP).[27]

IREP is based on the radioepidemiological tables originally developed in the 1980s by the National Institutes of Health to help adjudicate compensation claims for those living downwind of atmospheric nuclear testing and for those veterans exposed to radiation during nuclear blasts in World War II or bomb tests during the Cold War.[28] Although use of the tables was contentious and ultimately struck from the downwinders legislation, it provided a foundation for the IREP mathematical model. IREP requires the input of data about a given radiation dose—the type and energy range, as well as information about the work process and the worker, from age at exposure, number of years of exposure, and age at time of diagnosis, to the claimant's sex, race, and history of smoking. After entering these numbers and other statistical evidence, IREP then produces a final figure, a percentage—what's considered the probability of causation—of the likelihood that the cancer was radiogenic. The results are sent back to the DOL, which handles the final ruling and communication with claimants. The basis of whether one's claim for compensation is accepted or not is expressed in the terms of whether the cancer was (in the official phrasing) "at least likely as not" to have been caused by exposure to hazards at work—meaning the claimant must prove 50 percent probability or greater that the illness or death was caused by on-the-job exposure to radiation. Otherwise, no compensation.

The final rulings, methods, and attempt to employ science to guide policy decisions and administer EEOICPA are deeply contested owing to profound institutional mistrust and resentment.[29] EEOICPA requires that a balance be struck between scientific accuracy and expediency in resolving claims, yet, for many workers and their advocates, the methods of dose reconstruction and probability of causation make the entire process arduous, unfair, and ultimately more costly than simply awarding compensation. As one analyst astutely warned at the outset of EEOICPA's implementation:

> While this approach seems reasonable on paper, it is impractical and will likely result in widespread injustices. It will take a long time to set up procedures for calculating individual doses, [to] validate models, and extract and validate data. The expense will be great and the results controversial and highly uncertain. At the end of this lengthy process, the government—despite a good faith effort and huge expenditures—may find itself less trusted and embroiled in more litigation than before.[30]

In many cases, questioning the scientific accuracy of risk assessment is moot: systemic problems in the data, such as insufficient records, fabricated zeros, or measurements of infinity, cannot be resolved through scientific methods, only assumptions. How the DOL is to be held accountable for its decisions also continues to be unclear; arbitrariness characterizes the entire administration process.[31] Basically, "holes" in data serve as epistemic objects of rule that can render workers immaterial, as the raw data of a biopolitical venture that makes "uncertainty" about nuclear natures the perceptible norm. Calculations of "uncertainty" about exposure and dose obscure and depoliticize the materially linked but differentially inhabited risks that remain from the domestic production of nuclear weapons and geographies of the Cold War.

Biopolitics and Uncertainty

How do radiation and toxic exposures materialize? How are they granted existence or not? How is the journey of errant chemicals or radiation from the environment to body discernible? What counts as evidence—clinical accounts, statistical calculations, or personal testimony of injury? How can one quantify exposure or historically determine environmental pathology in the absence of monitoring records or in the face of an archive laden with errors?[32] The nuclear age has been as much about the regulation of uncertainty as documentation of biological effects. "State power is as concerned with making bodies and behaviors ever more predictable and knowable as it is with creating—both intentionally and inadvertently—spaces of nonknowledge and unpredictability."[33] In the case of EEOICPA's probability of causation requirement and the science of dose reconstruction, what appears to increase the welfare of former nuclear workers often forces claimants to individually face random instantiations of scientific measures, black boxes of obscure procedures (often through complex networks of privatized subcontracting), utter voids of data, and constantly changing compensation criteria, biomedical categories, and databases. The physical reality of contamination is refracted through a series of informational omissions, errors, approximations, proxies, and models. Essentially, technical unknowns become part of a new biopolitical regime, wherein holes of information—whether accidentally produced or purposely generated—are instrumentalized by knowledge practices, state agencies, and private industry.[34]

EEOICPA's administration mobilizes *holes* to assess exposure.[35] Historically accumulated unknowns about radiation exposure serve as the basis of the production of knowledge. The lack of recorded data necessitates dose reconstruction and, consequently, probability of causation determinations.[36] An exposure that is *not* easily determined to be "at least likely as not" can be extremely productive from the perspective of dose reconstruction as a biopolitical venture. Lack of evidence and uncertainties are employed to project order, efficiency, intelligibility. Elisions of information and lacunae in the data from state-scientific-bureaucratic histories of mismanagement are used to expand clinical and bureaucratic regimes, and to intensify market-economic opportunities in the field of biomedical risk assessment.[37]

EEOICPA's administration uses workers—their data—as the raw material for a kind of knowledge production based on the remediation of irrevocable misinformation and historically accumulated voids. On the one hand, nuclear workers are viewed as an abnormal burden on society; they are seen as a drain on resources, as obsolete laborers from the United States' shifting nuclear complex and economy, and a superfluous population that must make way for new institutions, contracts, and labor practices. On the other hand, EEOICPA's implementation, particularly dose reconstruction, generates information and a cascade of subcontracting to private industry and portions of the nuclear weapons complex that are transitioning to biomedical services. The industry-promoting operation of EEOICPA suspends nuclear workers in an ambiguous position between life and death; nuclear workers filing EEOICPA claims occupy a kind of death-in-life. *The compensation procedures simultaneously seek to keep workers alive yet not grant them much.* EEOICPA administration enables an accumulation strategy based on the reprocessing ad infinitum of worker records, site documents, and holes of data, without reference to worker livelihood. As many workers have remarked, the compensation process drains their energies, adds stress to their relationships, devalues their life's work, and could even be said to bring on the premature death of some claimants.[38]

Reconstructing dose and determining the probability of causation that one's cancer was caused by exposure to radioactive materials are procedures intended to render exposures perceptible and measurable. Yet EEOICPA arguably accomplishes something else: the system to redress exposures actually refines uncertainty in order to deny claims, and "solves" the ethical problem of Cold War misinformation and gaps in knowledge through data substitutions and arbitrary decisions. EEOICPA's methods hone the quantitative expression of uncertainty in the probability of causation; "uncertainty" is built into every part of the process of estimating causation (i.e., dose reconstruction, cancer risk models, dose-response assumptions, etc.). Uncertainty is a term typically used to describe the lack of precision and accuracy of a given estimate on account of the amount and quality of the evidence or data. Here, however, uncertainty does not refer to that which is rendered imperceptible and/or unknowable by the calculation; EEOICPA's methods fold that which is ontologically indeterminate *with* historically produced voids of information *into supposedly knowable, calculable epistemic uncertainties*—into "certain uncertainties."[39] Although every method of exposure assessment inevitably involves a selection or depiction of reality and is therefore an exclusionary act, one that creates the imperceptible as much as the perceptible and/or predictable, in this case uncertainty is purposefully pursued and refined to create a supposedly "complete representation of uncertainty."[40] NOISH's methods of dose reconstruction and probability of causation seem to suggest that everything can be incorporated under a dynamic and expanding "uncertainty umbrella," which covers statistical uncertainties (e.g., risk estimates, measurement errors, amounts of exposure) and various forms of subjective uncertainties (e.g., default assumptions, efficiency measures, transfers of excess risk from using other models of exposed populations) used to adjudicate EEOICPA claims.[41] This quantification of uncertainty effectively scrambles what is unknown; it is difficult to determine whether "we don't know" because there are data but they are being hidden, or because there truly are no data. If there are no data, this might be a purposeful strategy: EEOICPA shows that investments have been made into maintaining lacunae of information about exposures; some key references used to adjudicate claims document the lack of knowledge, confirming inconclusive evidence of disease-illness links as proof of no evidence. Data can also be substituted: whatever is determined to be missing can be filled in, using template data, models, and proxies. The relevance and transferability of substitute information are then accounted for by a mathematical distribution of uncertainty.

The implementation of EEOICPA has meant that any sense of responsibility about the absence or poor quality of the evidence available for dose reconstruction is disposed of by the mathematical operations of an increasingly masterful picture of uncertainty. Simultaneously, voids of data serve as the foundation for new contracts and exposure assessment opportunities to plug holes. Monitoring records about nuclear site operations and workers' dosages are often inadequate, because of secrecy, purposeful deception, and the limitations of existing scientific tools and measurements. This situation has justified the development of the science of dose reconstruction, which draws on numerous assumptions, disease models, and instrumental substitutions to assemble evidence where there is little to none. The use of substitutes and the shifting around of information makes it difficult to contest the outcomes of these methods. The origins of the data inputs are often obscure, and dose reconstruction, as an industry, involves several layers of contracts. Essentially, profits are generated through obscurantist subcontracting, blurring state functions with private enterprise and demonstrating the extent to which holes of data can function as "cash machines."[42]

Holes perforate the population and individuals, simultaneously foreclosing and opening up different knowledges and political possibilities. By assigning content and/or filling in the void with models and methods, holes allow for the production and extraction of value. Strategic investments in maintaining holes as blanks of data can lead to the acceptance of "lack of evidence" as a form of evidence, in order to shut down inquiry; in such cases, the absence of information is produced as all there is to know.[43] Holes can also legitimate and/or obscure the arbitrariness of decisions and the constructed nature of both truth and uncertainty, such as assigned medical risk or selective evidence of safety and background levels (as defined by risk assessment). Furthermore, holes mark sites of "no access" and secrecy, a situation often encountered by nuclear workers seeking information about their own exposures, the materials they handled, and the sites where they worked.

Multiple forms of power are at work with regard to EEOICPA: that of regularizing, optimizing, and ordering at the level of benefits and resources; the effects of this on individual bodies, namely, the disciplining of claimants' bodies and their relations; and sovereign power, which, under EEOICPA administration and nuclear regimes more generally, is strategically deployed to shut down organizing efforts and "say no" to claims for information or benefits. In the case of former nuclear workers, sovereign power cuts in at various moments to undercut participation in the biopolitical sphere because these workers are not considered ideal subjects; rather, they are "allowed to die" by the very legislation designed to assist them. EEOICPA shows that "law cannot help but be armed, and its arm, *par excellence,* is death."[44] The link between legal institutions and the use of epidemiological models that assign individual risks and exposures based on statistical evidence can be lethal. Law functions by formulating norms, and by quantifying, appraising, hierarchizing, and distributing.[45] Under EEOICPA, this takes the form of arbitrary rules and assignments, such as the "at least likely as not" probability of causation requirement for determining exposure. Other compensation schemes could have been realized, such as a sliding scale of compensation, calculations that include years of life lost or lost chances, or blanket category–determined benefits. EEOICPA's compensation measures for occupational illnesses and exposure are biopolitical instruments that provide benefits and recognition, regulating and extending life, and that simultaneously distribute, assign, and justify allowances of death, letting numerous workers suffer illness and death without state responsibility or care. Put another way, while EEOICPA's public face suggests that compensation helps make workers live, it also *lets them die.* Although nuclear workers are not blamed for becoming ill, EEOICPA's system puts the burden of proof on them and, therefore, allows for—and exacerbates—their illnesses and deaths in its very procedures, leading to what one worker described as a feeling of "depleting citizenship."[46]

Here, the notion of "perforation" helps to assert the penetration of the body by radiation and toxic exposures, the compromising of bodily integrity, and the experience of vulnerability to illness and death.[47] In contrast to the rational-technical process of EEOICPA's compensation process at the level of the state, where perforation designates "no access" and "no info" or lubricates obscurantist policies and subcontracting, many nuclear workers refute the way holes of information and misinformation are used to make exposure something that no one, including EEOICPA legislation, can fix.[48] As legal and bureaucratic structures interact with scientific practices to calculate and assign medical risk and liability, assigning value ultimately to people's lives, claims to benefits often seek to make perforation visible, as an enabling condition against a background of denial. If the state strategically uses voids of information and totalizing calculations of uncertainty to discipline the actions of individuals and to produce knowledge that invisibilizes exposure and denies responsibility, then workers struggle to make their illnesses visible, as primary sources of evidence and irrefutable proof of exposure–disease linkages. Sometimes this takes the form of circulating their personal testimonies and experiences as living documents; many workers and their advocates also assiduously track documents and amendments, collecting an alternative archive of documents and data. Still other efforts debate the uncertainties inherent to exposure assessment methods. The desire to produce certain truths and legitimate science at the level of EEOICPA bureaucracy is matched by counter-truth claims at the level of the social body. Workers are very aware that politics shapes what they know and don't know.[49] Therefore, many reclaim various knowledges of a site, environment, or technology by calling on their bodies as witness and evidence. Such counterknowledges seek to perforate the black box of dose reconstruction and the probability of causation, insisting on the inseparability of their health from the legacy of their own Cold War labors.

Nuclear workers have pointed out that the stakes are a matter of life and death. The Department of Labor's 2010 Annual Report to Congress lists the statistics of EEOICPA's Part B:

- 113,609 claims filed, which, after eligibility requirements were taken into consideration, were reduced to 95,353 (that is, already more than 18,000 claims rejected)
- Out of that total, 47,907 were approved; more than 20,000 were denied for cancers not work-related (probability of causation was not greater than 50 percent); well over 8,000 were rebuffed owing to insufficient medical information to support the claim; still other cases remain in the reprocessing loop of dose reconstruction

As of November 2012, the numbers reported on the DOL Office of Workers' Compensation Programs (OWCP) Web site for Part B had increased to 134,799 claims filed, with 60,028 approved and 37,947 denied.[50] The OWCP also breaks down the statistics for the number of cases that have received dose reconstruction results and a final decision regarding the probability of causation: 8,995 cases accepted with a probability of causation (POC) 50 percent or greater, and 25,296 cases denied with a POC of less than 50 percent.

The following sections further investigate the epistemological politics of exposure assessment under EEOICPA, using a performative approach that scrutinizes the legislation and its material effects. The collective figure of the "Hole in the Head Gang" perforates the text to poke holes in EEOICPA's production of knowledge about exposure. It is named after a comment made by Charlie Wolf, a former Rocky Flats nuclear worker and activist who battled brain cancer, and inspired by the many support groups of former nuclear workers across the country who regularly meet to share information in their struggle to receive compensation. Portrayed by radiation masks that are regularly worn to position patients' heads during radiation treatment, the Hole in the Head Gang graphically gestures toward the experiences and ongoing physical reality of former nuclear workers who were exposed to radiation and hazardous chemicals. For those with brain cancer, it indicates the stitches, wounds, and scars on their heads memorializing the multiple surgical removals of continually reappearing brain tumors. It simultaneously marks the denial of linkages between brain cancer and exposure to toxic substances, such as plutonium and/or mixtures of

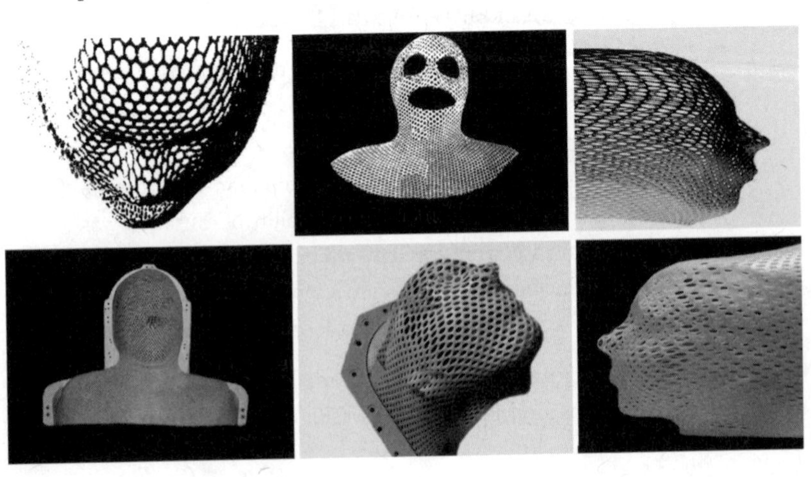

"Now, I've got a big hole in my head."
Charlie Wolf, ex-Rocky Flats worker and activist who suffered from brain cancer, which was denied any association with plutonium exposure. Legislation to reform EEOICPA, which bears his name, is currently under consideration.

HOLES

"Claimants will in no case be harmed by any level of uncertainty involved in their claims, since assumptions applied by NIOSH will consistently give the benefit of the doubt to claimants."
"Methods for Radiation Dose Reconstruction under EEOICPA," *Federal Register* 67, no. 85 (2002)

chemicals and radiation, under EEOICPA; in spite of the presence of such disease clusters linked to particular DOE sites, the EEOICPA system was set up *not* to study these groups but to assess individual exposures according to the tradition of the radioepidemiological tables.[51] "Hole in the head" also refers to the corrosive legacy of government secrecy, misinformation, classified materials, technical errors, and omissions of biological effects. As a nonindividualized group or Greek chorus that simultaneously comments on the dramatic action and contests the individualizing procedures of the claims process and the sanctity of the "whole" body, the Hole in the Head Gang calls out some of the ways EEOICPA produces denial and obscurity and allows for the depletion and death of claimants, as inherently an outcome of the legislation to compensate nuclear workers. The Hole in the Head Gang performs a kind of forensics that leaks information, locates administrative loopholes and catch-22's, detonates criticisms, and collages charts, cartoons, and contradictions. The overall "workbook" quality of the text serves up a hodgepodge of collected quotations and logic exercises that point to the absurdities of the legal, statistical-epidemiological methods and overall administration of EEOICPA.

EEOICPA legislation grew out of and in response to a culture of Cold War "privileged knowledge" that, under the cover of national security interests, stymied critical and independent evaluations, delegitimated data that suggested a positive association between nuclear exposures and subsequent development of cancer, and usually cited only those studies that did not find an association or link.[52] This culture purposely produced research that substantiated holes of knowledge, and refined methods for determining evidence to be inconclusive. Those scientists who deviated from the status quo on radiation and health, reporting positive results, often suffered professionally, and their work and concerns were labeled groundless or inconclusive.[53] This subjugation of knowledge was advocated in order to protect the population and avoid undue alarm. With this context in mind, the indeterminacy of causal mechanisms with regard to exposure is inextricably enmeshed with the legacy of denial and the official position of a lack of association.[54] Rather than "make workers whole again" by supporting possible disease and occupational exposure linkages and taking responsibility for the historic production of insufficient data, EEOICPA's performance has continued to suppress exposure links and rationalized the indifference of the state and government contractors to historic injustices through a cost-benefit approach to compensating harm.

Under EEOICPA, the burden of proof for all cases, in establishing that one's illness was caused by workplace exposure, is on the injured; the compensation process grades and assigns individuals without having to assume their custody. "Care" is contained, and the burden of liability is systematically diminished by eliminating the presumption of entitlement to compensation from the start. The logic at work is to *reduce complexity and cut costs through the distribution of benefits*. The process also seeks to minimize errors in determining causation as a proxy for minimizing legal costs and lowering the number of cases. This imperative was exposed in the infamous "Passback memo" between the DOL and the Office of Management and Budget. The brief document stipulated that Labor was to "contain the growth in the cost and benefits provided by the program," outlining a plan to base SEC status approvals on budget concerns rather than the scientific basis mandated by law.[55]

Risk assessment for each claim that must undergo dose reconstruction and probability of causation encourages a social imaginary of individual burden and privatized loss that buffers systemic harm. While exposure risk is predicted and/or determined collectively through statistics at the level of the population, EEOICPA assigns shares of risk to individuals and, subsequently, transfers responsibility for social causes and the impact of the nuclear weapons complex to the individual, without effective mechanisms to share the hardships or address suffering other than through this atomizing technical process. Workers who are claimants, then, not only deal with their illnesses and treatment as the embodied remains of industrial weapons/war production, but they must also labor to figure out the claims process, even though filing a claim by no means guarantees recognition of compensable harm, or support for the maintenance and extension of life.

EEOICPA's implementation has *more often than not* accepted a lack of evidence as proof of no evidence and encouraged the fine-tuning of inconclusive findings to reject claims. Responding to the complexity and number of claims under Part E of EEOICPA, several means were developed to expedite the claims adjudication process, among them criteria for the presumption of causation in certain situations *and* guidelines for handling claims where there was allegedly "no known relationship between certain illnesses and occupational exposure to toxic substances." The DOL's Department of Energy Employees Occupational Illness Compensation (DEEOIC) developed a list formally titled "Medical Conditions with No Readily Known Associations to Occupational Chemical Exposures."[56] Popularly referred to by workers as the "no pay list," the entries of diseases and cancers with "no readily known link" (conflated with "there is no link") served as a reference for the screening of Part E claims—that is, for denying *possible* relationships between some illnesses and occupational exposures from the outset. The DOL contends that the list was justified as scientifically peer-reviewed because it had been culled from authoritative scientific publications, medical literature, and occupational exposure records. It rescinded the resource in 2008, stating that it had never intended to

THE "NO PAY LIST"

Bulletin 06-10 Attachment 1

Medical Conditions with No Readily Known Associations to Occupational Chemical Exposures

Medical Condition	ICD-9 Code
Acquired Cyst of Kidney	593.2
Acute Reaction to Stress	308, 308.0-308.9
Amyotrophic Lateral Sclerosis (ALS)	335.2
Arthropathy	716, 716.0-716.9
Aspergellosis	117.3
Asphyxiation, Strangulation, Lightning	994, 994.0, 994.7
Atherosclerosis of Aorta	440
Bell's Palsy	351
Breast cancer (female)	174, 174.0-174.9
Breast cancer (male)	175, 175.0-175.9
Carpal Tunnel Syndrome	354
Central retinal vein occlusion	362.35
Conductive Hearing Loss	389
Carotid Vein Bruit or Weak Pulse	785.9
Cryptococcoisis	117.5
Cysts	528.4
Degeneration of Macula and posterior pole	362.5
Dermatomycosis	111.9
Diabetes Mellitus	250, 250.0
Diabetes Mellitus Type 2 without complication	250.2
Diabetes Mellitus without complication	250.1-250.9, 648.0
Diabetic Retinopathy	362
Digestive Peritoneum cancer	159.8, 159.9
Diverticulosis of Colon	562.1
Duodenal Ulcer	532
Esophageal Reflux	530.81
Eye cancer	190, 190.0-190.9
Female Vulva cancer	184.4
Gallbladder/bile duct cancer	156, 156.0-156.9
Glaucoma	365, 365.0-365.9
Gum cancer	143, 143.0-143.9
Herpes Zoster Infection	053.9
Histoplasmosis	115
Hyperplasia of Prostate	600
Influenza	487
Kaposi's Sarcoma (Lung) cancer	176.4
Late Effects of TB	137
Lip cancer	140,140.0-140.9
Major Depression	296.2
Migraine	346, 346.0-346.9
Mononeuritis of Lower Limb	355.8
Mononeuritis of Upper Limb	354.9

Note: Page 1 of 2 only

disqualify claims automatically based on medical conditions with no known causal link to toxic exposure.[57]

"Insufficient evidence" would now replace the language of "no known causal link." The list was rendered operationally obsolete by the implementation of the Site Exposure Matrices (SEM) Web site, which is now hailed for representing the most current, accurate, and comprehensive information regarding toxic substances and their known health effects directly linked to sites within the nuclear complex. Whichever message of denial is deployed—"no known link" or "not sufficient evidence"—these references demonstrate that "no liability" is assumed from the start. They betray a logic that absurdly asserts that there are no links between certain conditions and exposures because we don't know whether such links exist; because we don't know whether a particular link exists, we assume that it doesn't; because we don't know whether it's true, it therefore isn't true.

"No" takes a more diffuse form under the SEM. The SEM Web site identifies toxic substances used at each site and their chemical profiles, along with site-specific data such as labor categories, work processes, and locations.[58] It also contains information about causal relationships between exposure and disease. The SEM recognizes three distinct categories of illnesses: medical conditions for which a causal link to toxic substances has been identified, conditions for which no known causal link exists based on the National Library of Medicine's Occupational Exposure to Hazardous Agents database (Haz-Map), and specific medical conditions for which there is no established causal link based on Haz-Map but which a claims examiner will still investigate. Although not the sole resource of the DOL, the SEM is important to the initial determination of whether a claim qualifies for consideration. This is a form of screening based on what links are produced by the database; the database serves as a compendium of "known link" evidence.[59] Yet, *the absence of SEM-published research linking certain toxins to diseases can be used as a form of evidence that such links do not exist.* Claimants, then, must submit evidence to support the links necessary for a favorable decision.

Understandably, the sources, comprehensiveness, scientific legitimacy, and use of the SEM are of concern to many claimants, lawyers, doctors, journalists, and claimant advocates. The SEM incorporates

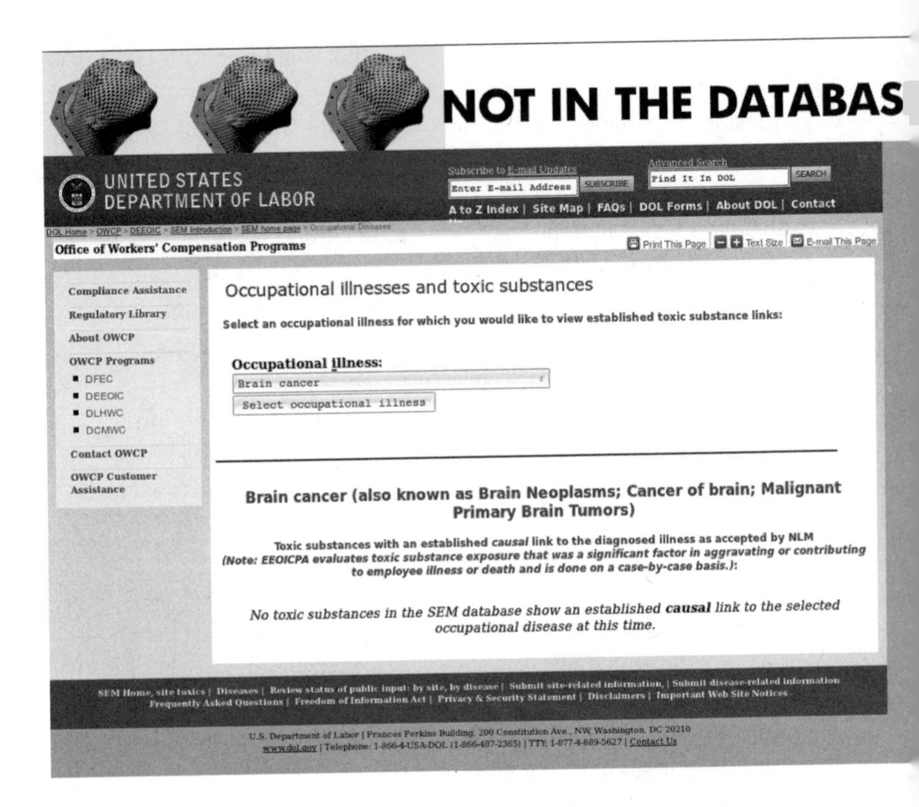

data from thousands of DOE documents regarding toxic substances at covered facilities, important unclassified information from DOE workers, and various publicly available resources.[60] Yet, as Labor warns users of the Web site, the complete documentation of toxic substances used at each site is not available to the DOL and may not even exist, owing to classified materials and historical secrecy measures. Although the DOL alleges that it consulted with workers from DOE facilities all over the country to gather input on hazards at these sites, workers complain that the database is missing too much information to serve as a reliable reference; it should not be exercised as the basis for the determination of a "known link" or, conversely, "no evidence."[61] Claimants may submit evidence that might lead to changes in the database, but it is unclear how such evidence is vetted by the DOL.

How data are incorporated into the SEM raises questions about the oversight and scientific validity of the database. The Site Exposure

Matrices have not been admitted to formal peer review. The DOL maintains that because the SEM inventory of illnesses and associated toxins is drawn from the National Library of Medicine's Haz-Map database, which draws on peer-reviewed scientific articles, a formal review of SEM is not necessary and would create delays for Part E claims.[62] That the DOL relies on Haz-Map to secure scientific validity falters on the same issues raised about the SEM. Haz-Map has not undergone peer review, and the ways it was assembled, and the ways it is updated, are not readily transparent. Haz-Map is a surveillance system of medical literature on occupational health that, without external oversight, is at risk of a conflict of interest: the National Library of Medicine researcher who developed the database is working under contract with the DOL to continually update the SEM to reflect current research into the relationship between toxins and disease.[63] What constitutes current research—or how it is selected—is not clearly spelled out. Because the SEM database relies on Haz-Map and lacks any external review of its own, there is no way to ensure that it is comprehensive or scientifically sound. The DOL was reportedly poised to remove from the database more than one hundred toxic exposures once considered to be linked to specific diseases.[64] Such allegations of disappearing links where evidence once existed are disturbing and difficult to corroborate; the DOL states that it tracks changes in the database, yet there has been no oversight of the way the data are revised, updated, expanded, or removed from the SEM. A 2010 Government Accountability Office (GAO) report pointed out that no external review of the whole Part E of EEOICPA had yet taken place.

Instead of "no known link" as the basis of evidence against a claim, "no" in this case involves a database of "known links" that can be used to produce documentation of "not in the database"—against claims of causal linkages between disease and exposure.[65] Furthermore, a database that has *not* itself undergone peer review can delegitimate claims of exposure–disease linkages—that which is not in the database—as that which has "no known peer review."

EEOICPA Part B enlists further epistemological practices of denial. Claims filed under Part B that involve cancer and radiation exposure must undergo dose reconstruction. The resultant data are necessary in order to determine the probability that the cancer's cause was radiogenic. The DOL relies on NIOSH's estimates of the type and level of radiation exposure received by workers and the associated radiation dose to each organ affected by cancer. Such exposure information is extrapolated from documentation of internal or external radiation doses, differentiated by type of radiation and energy range, dose pathway (i.e., inhalation of particles, contact with skin), and the way radiation interacts with each organ or body system. Workers were potentially exposed to external sources of ionizing radiation, as well as internal emitters of radiation from inhalation or ingestion. Reconstruction of doses from internal emitters is especially difficult: radionuclides, such as plutonium, distribute nonhomogeneously within the body and are understood to deliver time-dependent organ-specific dosages. Recorded internal exposure had to be detected from within the body and assessed using bioassay measurements, such as urinalysis, a labor-intensive exercise requiring knowledge of the solubility of the chemical form of the radionuclide taken into the body and a metabolic model to relate urine concentrations to tissue burdens.[66]

In order to reconstruct both internal and external doses, NIOSH must develop a basic understanding of the work process and aggregate the ways that a worker's presence and activities varied in time and space with respect to the radiation source. A structured telephone interview is a standard feature of the reconstruction process; if the worker is deceased, NIOSH personnel attempt to elicit names of former coworkers who might still be available for qualitative data collection. Priority is given to individual dosage monitoring records and bioassay data, placed in the context of research on the historic monitoring procedures of a given work setting. Dose reconstruction methods also include identifying events or processes that went unmonitored, evaluating production processes and safety procedures, and applying certain assumptions that supposedly overestimate exposures while achieving efficiency.

Despite efforts to claim the scientific efficacy of dose reconstruction methods, Part B faces numerous difficulties. Epistemologically and methodologically, "doses/dosages," in general, are very contested in the scientific community; there are ongoing debates about the effects of different types of exposure (gamma, neutrons, etc.), threshold limits and the shape of "dose response curves," the treatment of

"The essence of dose reconstruction is to fill the voids in the monitoring records by using a combination of science and professional judgment to generate a value of the appropriate dose parameter that is adequate for an unambiguous compensation decision."
Richard Toohey, "Scientific Issues in Radiation Dose Reconstruction" *Health Physics* 95, no. 1 (2008): 26

"It appears dose can be reconstructed."
James Lockey, member of Presidential Advisory Board on Radiation and Worker Health, quoted in the *Rocky Mountain News* (June 13, 2007)

argumentum ad ignorantiam
a specific belief is true because we don't know that it isn't true

"The only issue is whether adequate information exists for the reconstruction of dose to be done in a reasonable manner. We've heard no information whatsoever that it doesn't."
Wanda Munn, member of Presidential Advisory Board on Radiation and Worker Health, quoted in the *Rocky Mountain News* (June 13, 2007)

contaminants as single agents versus their combined multiplying effects, and the idea of a linear, trackable "exposure pathway" (which assumes that the body is well defined and separate from the environment, that exposure was finite and discrete rather than ongoing, and that a direct line can be established between an environmental residue and worker dose).[67] Typically, the environment is abstracted into a set of harmful chemicals or radioactive materials that produce different types of radiation (e.g., alpha radiation, such as uranium 235 and plutonium 239, which cannot penetrate the skin but can enter the body through a cut or by ingestion or inhalation and can do great damage to DNA). Dose establishment focuses on human contact points and targeted organs, and employs an internal/external dichotomy of the body. Yet the likelihood that a specific dose can be established, at one time or another, is slim.[68] Carcinogens might have acted independently, additively, synergistically, and/or indeterminately in producing a disease. The relationship between disease rate and dose level is by no means

"Trying to reconstruct a dose from hard data is difficult enough. Reconstructing a dose with data that's absent of hard records is somewhat of an art. There were unexpected events that were not set up for monitoring."
Michael Gibson, member of Presidential Advisory Board on Radiation and Worker Health, quoted in *5280* (November 2007)

"It's a black box."
Wayne Knox, nuclear engineer and health physicist, commenting on NIOSH's IREP software, quoted in the *Rocky Mountain News* (July 22, 2008)

"It's still garbage in, garbage out."
Laura Schultz, former Rocky Flats engineer, quoted in the *Rocky Mountain News* (November 8, 2005)

"How can you do a dose reconstruction from nothing?"
Cliff DelForge, former Rocky Flats worker on radiation safety, quoted in the *Rocky Mountain News* (June 12, 2007)

"[Dose reconstruction is] basically a giant slot machine. . . . The bottom line is the program is too ambitious for the science."
Joseph Batuello, physician/lawyer/aerospace engineer, assisting claim for ex-nuclear worker Don Gabel who died from a brain tumor, quoted in the *Rocky Mountain News* (June 4, 2007)

certain, as evidenced by the debate over low levels of exposure to ionizing radiation; much is still unknown about how cells react to very low doses of radiation.[69]

A further challenge to dose reconstruction efforts is presented by the absence of quantitative estimates of radiation dose and monitoring records that could be used by NIOSH to assess claimant dose-response relations. Radiation dosimetry records of the past were not organized for epidemiological study. DOE records have large gaps; site profiles are incomplete and uneven. Many workers claim that radiation exposure went unmonitored, accidents were not always reported, and erroneous information was sometimes put on file in the name of national security. Even where records exist, it is impossible to know the precision of the monitoring equipment; past dosage monitoring instruments have proved extremely inadequate. Although there is nothing about ionizing radiation that inherently disallows it from being evaluated in an epidemiologic way, there are so many voids of information and, as many workers contend, grossly inaccurate and altered data that any radiation monitoring records from the DOE, its contractors, and vendor facilities are of dubious value—if not patently unsound—for making compensation decisions.[70] NIOSH contends that for cases where dose monitoring records are not available, it can sufficiently reconstruct that information by using reasonable scientific assumptions that give the claimant the benefit of the doubt.[71] NIOSH contends that it can scientifically reconstruct doses by assembling and substituting evidence.

Because of the elisions of data about certain sites, as well as for moral reasons, the EEOICPA legislation allows for the establishment of SEC cohorts that qualify for compensation without dose reconstruction. However, NIOSH claims that it is feasible to estimate a radiation dose—which means that the claimant cannot be allowed into an SEC—if NIOSH has access to what it considers to be sufficient information to estimate the maximum radiation dose that could have been incurred in plausible circumstances. NIOSH refers to this as having enough data to "cap the dose." NIOSH asserts that data from personal dosimetry or area monitoring are not essential to capping the dose if there is information on the types and quantities of radioisotopes to which workers were potentially exposed. *Dose reconstruction, then, has become a compilation of default assumptions and substitutions* that, in the absence of recorded dosimetry badge readings and other records, come up with a potential maximum dose. For example, NIOSH augments missing badge information with stand-in data: the records of coworkers with comparable exposure risks, or an area's air monitoring data provided by the site contractor, are used to perform dose reconstruction if a claimant's records are not available or do not exist. Values are often interpolated between data points for nearby time periods. In some cases, NIOSH uses data from other facilities to fill in the gaps.

The assumption of "missed dose" also informs dose reconstruction. When analyzing missing bioassay data for internal emitters of radiation, NIOSH assumes a "missed dose" that represents the upper limit of internal dose that a worker is suspected to have received by a particular building or site area. This default was conceived by NIOSH as a way to give the claimant the benefit of the doubt and to account

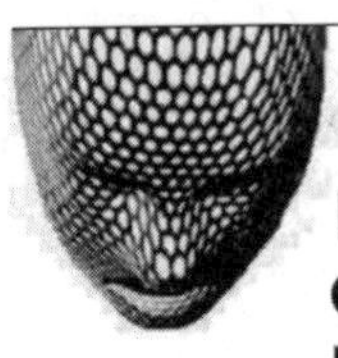

BENEFIT OF THE DOUBT, OVER-ESTIMATION & OTHER EFFICIENCY MEASURES OF DOSE RECONSTRUCTION

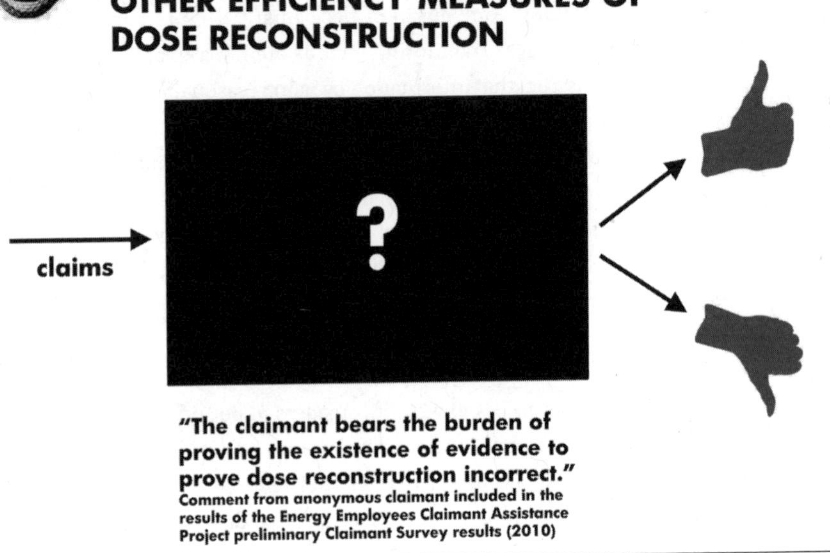

"The claimant bears the burden of proving the existence of evidence to prove dose reconstruction incorrect."
Comment from anonymous claimant included in the results of the Energy Employees Claimant Assistance Project preliminary Claimant Survey results (2010)

for the fact that bioassay sample programs of the past had higher detection limits than current technology. Other default assumptions are supposed to provide worst-case dose estimates; when faced with equally plausible radiation dose and cancer risk scenarios during dose reconstruction, NIOSH states that it will assume the scenario that results in the highest exposure to the worker. This can entail yet another kind of substitution: the swapping of the cancerous organ under consideration. When a worker's cancer is not in an organ for which NIOSH, in collaboration with the National Cancer Institute, has developed risk models, NIOSH contends that it will select, among comparable organs, the one that is most susceptible to cancer risk from the type of radiation exposure the worker was likely to have had. NIOSH says that *a dose estimate is sufficiently accurate if it is "reasonably certain" to be at least as high as the highest dose that the comparable organ could plausibly have received.*

According to EEOICPA rules, at any point during the steps of dose reconstruction, NIOSH may determine that sufficient research and analysis has been conducted to complete the dose reconstruction; that is, *a dose reconstruction is considered complete when additional analysis does not alter the finding*. NIOSH ends the reconstruction process when it knows that further research would not change the end result. To hasten that end, NIOSH applies "efficiency processes" upon receipt of claims to determine whether a worker's radiation dose was clearly high or clearly low, which allows NIOSH to issue a timely dose reconstruction where attempts to refine the exposure estimate would not result in a compensable claim.[72] For example, the practice of rough estimation, described in NIOSH's regulations, allows NIOSH to more quickly complete dose reconstructions for certain kinds of claims based on the likelihood of compensation; it allows for dose reconstruction only insofar as to provide an unambiguous compensation decision to DOL. NIOSH also applies overestimated exposure values that are supposed to favor claimants and speed up the process while maintaining the veracity of the outcome. When compensation seems unlikely, NIOSH overestimates the potential dose received by an employee based on the type of cancer and length of employment; if the probability of causation does not reach or exceed 50 percent with an overestimated exposure dose, then NIOSH justifies determining the case uncompensated and closed. Whatever the values, educated guesses, assumptions, and substitutions, and however they are implemented, NIOSH can conclude dose reconstruction at any time for the purposes of efficiency.[73]

In spite of NIOSH's efficiency measures, cases that require dose reconstruction have taken up to three or more years.[74] A GAO investigation discovered that, in 2008, the cost to run the dose reconstruction program meant that Part B of EEOICPA—at $106 million—was nearly twice as much as the cost of Part E. The same investigation, upon examining NIOSH's administration costs, found that NIOSH did not keep tabs on the money spent on claims denied versus claims approved. Against the backdrop of distrust in the government resulting from the history of secrecy, errors, and deceit tracing back to the Manhattan Project—precisely what prompted the legislation of EEOICPA in the first place—dose reconstruction remains opaque in

Developed under contract with the
National Institute for Occupational Safety and Health (NIOSH)

SENES Oak Ridge Inc.
Center for Risk Analysis

User's Guide / More Information / Contact NIOSH

Interactive RadioEpidemiological Program
NIOSH-IREP v.5.6

For Estimating Probability of Cancer Causation for Exposures to Radiation

To begin by manually entering required inputs	click here
To begin by using a NIOSH-provided input file	click here
To calculate PC from multiple primary cancers	click here

NIOSH-IREP was created for use by the Department of Labor for adjudication of claims in accordance with the Energy Employees' Occupational Illness Compensation Program Act of 2000 (EEOICPA). NIOSH-IREP was adapted from the National Institutes of Health's (NIH) Interactive RadioEpidemiological Program (IREP) developed by the National Cancer Institute (NCI) to update the NIH Radioepidemiological Tables of 1985. (The version of IREP developed by NCI is known as NIH-IREP.)

NIOSH-IREP v.5.6, introduced on 9/21/09, includes a corrected algorithm that has the potential to slightly increase the PC results for special cases of acute lymphocytic leukemia. Also, an algorithm has been corrected that models the uncertainty related to the effect of time since exposure and attained age for selected solid cancer sites and for the NIH-lung model. Click here for more details about the modifications made to version 5.6 and to other recent versions. Comments and suggestions should be communicated directly to NIOSH.

"It should be emphasized the intake dates, scenarios, and intake levels were based upon mathematical models and do not necessarily prove that such intakes occurred on the given dates."
NIOSH, quoted in packet of information on POC sent to a former Rocky Flats worker pursuing a claim under Part B (2007)

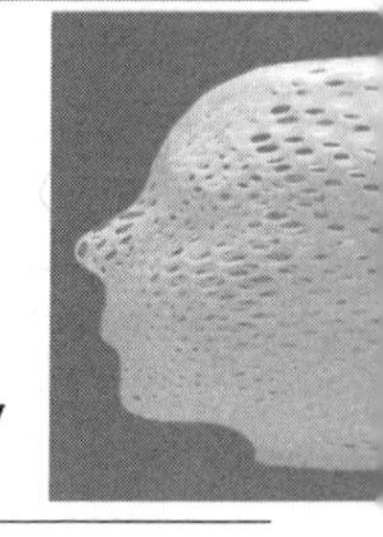

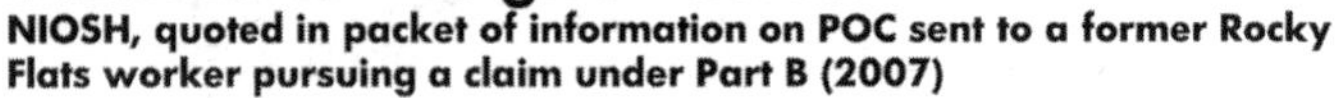

procedure and barely credible, *if not downright unbelievable,* to many former workers.

The connection between exposure and injury is difficult to perceive; causation mechanisms are complex, with long latency periods between the original exposure and signs of disease or illness.[75] EEOICPA's compensation structure of "at least likely as not"—one has to prove 50 percent or higher probability that one's cancer was caused by exposure to radioactive materials—emerges out of a history of legislation and court cases involving veterans and the Department of Veterans Affairs, as well as downwinders. It also draws on the "more likely than not"/"but for" sine qua non convention of toxic tort law.[76] The 50 percent probability of causation (POC) uses a threshold criterion that assumes "no liability" from the start and an "all or nothing" result: one side is not acknowledged (no compensation), while the other side is acknowledged, after being assessed using statistical calculations of individual claimants as discrete units. It is worth noting again that there are other ways to assign causation, such as different probability-based models and applications of group-based data to individuals.[77] Among the alternatives, the UK's voluntary Compensation Scheme for Radiation-Linked Diseases allows for proportional compensation for probabilities as low as 20 percent.[78] This means that some amount of compensation is granted, without the burden of proof on the injured; proportional compensation also eliminates the threshold requirement that essentially requires the exposure to a toxic substance to at least double the frequency of injury occurrence in order to award the claim. Some epidemiologists advocate compensation schemes based on number of years of life lost; other compensation structures have incorporated lost chances as a form of legal injury.[79]

While the choice of a threshold POC scheme for EEOICPA's compensation program was arbitrary, in that other models could have been used, it is based on a technically sound mathematical interpretation of the burden of proof. What *is* debatable is the appropriateness and justice of using epidemiological estimates of the POC for compensation rules, such as in the case of nuclear workers.[80] EEOICPA's Part B applies a formula to calculate the probability that injury was caused by exposure, using population data generated by epidemiological methods. If the disease rate is significantly higher in an exposed population than in an unexposed group, then one can infer that the exposure has done some harm. However, it is impossible to tell who exactly were those who had been affected; epidemiological probability works at the level of the population but not at that of the individual. Yet, under Part B, the POC does the work of parsing out who is to be compensated and who should not. This is achieved by ignoring the empirical-material grounds of epidemiology. Because of the differing traditions and concerns of law versus sciences that are concerned with stochastic events, the POC confuses "chances" with epistemic certainties/uncertainties, correlations with causes, strength of association with strength of evidence, population data with individual "assigned share."[81] Traditionally, the law requires proof that an exposure was a condition sine qua non—a condition without which the plaintiff's injury would not have occurred; it is assumed that either the exposure was necessary for the cancer or it was not, and if it was not, then it could not have been the cause. Necessity, not chance, is the explanation. Yet that is precisely what epidemiologists are interested in: *ex ante* probabilities or chances for predicting future events. *Ex ante* probabilities allow for the formulation of preventive strategies—how injuries might be avoided in the future—not what caused a particular person's injury or disease. "A chancey cause is one that is neither necessary nor sufficient to produce an event, but makes that event more likely to occur."[82] Accordingly, radiation exposure does not merely cause cancer as a onetime event, it initiates changes that make one more prone to develop cancer; exposure raises the chances of developing cancer. These "chances" are physical properties—correlations—that tell us something about material transformation.

Epidemiological evidence provides information about correlations, not singular causes, which often puts statisticians, who are interested in the mathematical relationships among variables, at odds with tort lawyers or the DOL, who must determine the causal connections among variables and come up with a summary statistic or coefficient that reveals whether and to what extent a particular actant or exposure caused the harm.[83] "Chance" cannot be reconciled with the POC formula because of the mandate to establish *ex post,* not *ex ante,* causation. Additionally, probabilities are viewed as epistemic, rather than physical correlations with predictive value, and are mistakenly used to designate the strength of evidence available. In more mathematical terms, two different quantities get conflated: there is the assumption that the *ex post* probability of the effect (cancer) is 1 (i.e., there is no doubt that the claimant was injured), and, therefore, it is then assumed that its *ex ante* chance of occurring was 1 (i.e., given all circumstances, it had to happen).[84] The POC is essentially a statistical probability mapped directly onto a standard of proof continuum with the accompanying idea of a quantifiable threshold; a continuum of probability is posited with stronger evidence imparting supposedly greater epistemic probability. This is an arbitrary, even erroneous, way of *presenting evidence that does not inherently exist,* both in that probability as chance and probability defined as epistemic certainty/uncertainty are conflated, and in that the POC does not—cannot—actually measure the strength or probative value of evidence. The strength of evidence in EEOICPA claims cannot be equated with the strength of causal association in many cases because of the vast holes of data, inaccurate records, and gaps of knowledge.

Furthermore, "when degrees of causal efficacy are treated as degrees of proof, the 50% threshold becomes an arbitrary boundary, passing or excluding plaintiffs according to the power of the causes to which they are victims rather than by the strength of their evidence."[85] As a result, the enforcement of the threshold rule in EEOICPA legislation is an arbitrary epistemic move, not one based on physical correlation; it is one that *treats causes as instances of uncertainty of "true causation," or, uncertain cases of sufficient causation,* rather than as true instances of multiple or joint probabilistic causation.[86] If scientific descriptions of cancer progression and other toxic tort phenomena are persistently stochastic, suggesting that these chances are a fundamental feature of the world, the EEOICPA compensation POC threshold criteria of Part B denies the actual physical relations, chances, and "might be's" represented by epidemiological data and confuses the issue of partial causation with that of uncertainty. This is a form of biopolitical governing whereby *uncertainty is universally assumed (as insufficient causation) in order to reject outright responsibility,* and probabilities are confused with proof as the reason for refusing compensation. *Such data production serves as tactical spectacle; it converts the material geographies and effects of work and war into abstractions—into abstract relations of quantifiable uncertainty.*

Enough complaints were registered about the use of "probability of causation" that official documents began to change terminology, from the POC to "Assigned Share." Yet some epidemiologists argue that Assigned Share is still a logically flawed concept *and* is subject to substantial bias and therefore unsuitable as a guide to adjudicate compensation claims in cases of potential radiation-related cancers. For example, the actual fraction of the Assigned Share—the fraction of total disease incidence in the exposed population that can be linked to the cause in question—only includes *excess cases,* where it is clear

POC = R/R+B

- Where R is the risk of given cancer due to radiation
- Where B is the baseline risk of given cancer due to all other causes

POC is calculated from estimates of excess relative risk due to given radiation doses to organ in which cancer was induced

POC = ERR/(1+ERR)

- Where ERR=R/B – the rate ratio comparing exposed and unexposed populations

Equations cited from "Report of the NCI-CDC Working Group to Revise the 1985 NIH Radioepidemiological Tables"

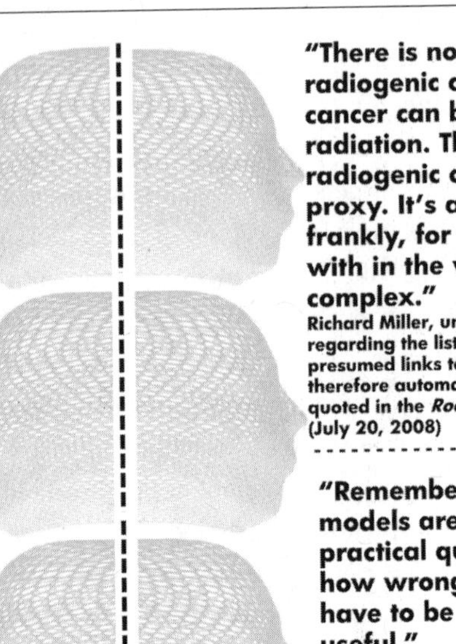

"There is no such thing as radiogenic cancer. Every cancer can be caused by radiation. The 22 radiogenic cancers is a proxy. It's a lousy proxy, frankly, for what we deal with in the weapons complex."
Richard Miller, union policy analyst, regarding the list of diseases with presumed links to worker jobs and therefore automatic compensation, quoted in the *Rocky Mountain News* (July 20, 2008)

"Remember that all models are wrong; the practical question is how wrong do they have to be to not be useful."
George E. P. Box, in George E. P. Box and Norman R. Draper, *Empirical Model-Building and Response Surfaces* (New York: Wiley, 1987), 74

that exposure caused the cancer; there is no consideration of cases in which work processes hastened the onset of disease. The POC or Assigned Share, even if correctly estimated, is insensitive to the impact of exposure on disease timing; it cannot distinguish accelerated from unaffected occurrences. If it is possible that a radiation-caused cancer could have occurred without the occupational exposure, then it is excluded from the calculation. The calculation of attributable risk simply assumes that there are no cases where a cause brought about an earlier effect than otherwise would have occurred. It's an "all or nothing affair."[87] The "at least likely as not" clause requires that a cancer be attributed to one cause or another but not both, and it assumes that an operating cause must be a necessary component of sufficient cause. Either the cause is "fully determinate" or it doesn't contribute at all. According to this universal causal determinism, an individual's disease has a single mutually exclusive cause. Multiple disease mechanisms and synergistic effects are not considered; causation boils down to a single event and a single agent. And, if more than one cancer is present, then they are determined separately and added together. According to NIOSH, when new cancers are added to the equation, then dose reconstruction must be refined using "more probable and precise" exposure estimates rather than the original overestimates. Claimants, in some cases, have been shocked to find that their POC for one initial cancer was significantly higher than a returned POC wherein several additional cancers were tabulated and included. Here again we can see that all cases are treated as instances of uncertainty of true causation from the start.

Furthermore, many exposures do not double the risk of disease to pass the threshold at which exposure causation is deemed to be "at least likely as not"; exposure doses at which the POC exceeds 50 percent may fall well below the dose at which the incidence of disease doubled.[88] Basically, the POC leads to the *systematic underestimation of the potential range for probability.* The threshold criterion serves as a disposal mechanism that works against many claimants by *dumping liability for associations that may really exist.* The POC is suspect in the assumptions made about causation, particularly in that "no single radiation-induced mutation makes the difference between health and disease."[89] Even though there is currently no way to know the essential difference at the molecular level between those whose cancer was caused by radiation exposure and those whose cancer was caused by some other source, the determination of POC under Part B assumes that normal background exposures and vulnerabilities can be separated from those exposures that took place at work. This bounds the work site, and fixes workers and work environment in place and time. Conversely, the blanket application of the POC formula treats all members of the population as if they were governed by identical cancer risks, when in fact individual vulnerabilities widely vary because of differing social, environmental histories. To complicate this, the IREP software that is used to determine the POC includes factors of age, sex, race, and so on that are associated with different diseases, as part of the calculation of "normal" background vulnerabilities; that is, certain backgrounds and disease susceptibilities are attributed to certain populations when assessing worker statistics to produce the POC. Some epidemiologists have pointed out the potentially unjust result of this:

"They've spent their careers creating models for each site, out of the ambiguities. This program is providing full employment for them. It's a fat trough." Richard Miller, union policy analyst, quoted in the *Rocky Mountain News* (July 20, 2008)

"Our workers died while NIOSH desperately dabbled in science under the threat that its program would be eliminated spelling an end to the multi-million dollar business that is dose reconstruction—a NIOSH and HHS jobs program at the expense of the workers and the taxpayers." From the Formal Letter of Appeal to the ruling against the petition to grant SEC status to Rocky Flats workers who were exposed to radiation dose from 1967 to 2005, and to radiation dose other than neutron from 1952 to 1966, filed by the Rocky Flats United Steelworkers of America, Local 8031 (September 25, 2007)

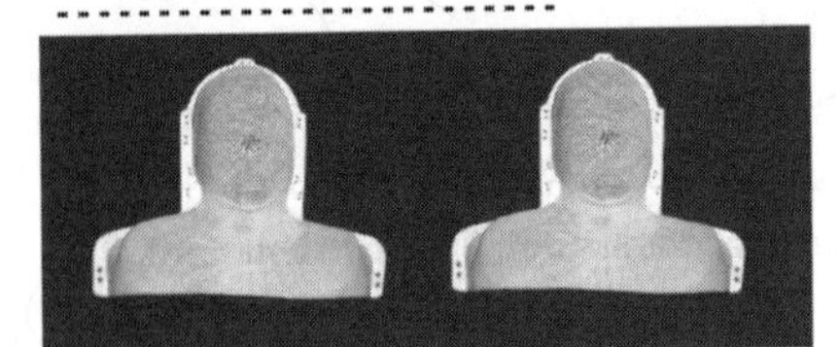

privatization as value-generating obscurantism

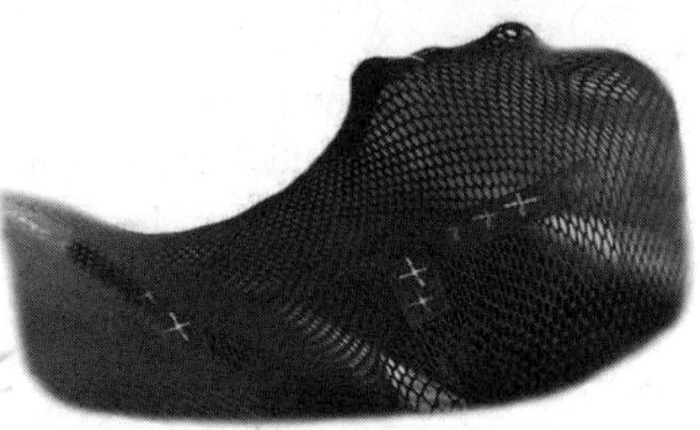

"If you poll most of the claimants, you will learn we believe you are waiting for us to die off, then you can say your program at least tried, yet more importantly, you all were given our money as paychecks while proving we were all wrong." Comment from ex-Rocky Flats worker, included in DOL survey of EEOICPA (2008)

differential vulnerabilities stemming from social-environmental relations that, for example, might involve gender or racial inequities and that impact health are quantified, inputted into the software matrix, and potentially used to argue that the cancer is as likely to be caused by outside factors as by work exposures. When faced with the unknown degree of heterogeneity in the background disease risk of an exposed population, the POC cannot be estimated without making unverifiable assumptions.[90]

Estimations of POC call on mathematical models of disease biology and models of the way radiation and other risk factors affect the course of disease development. Biologic models of diseases are necessary for the estimation of the probability that an individual's injury was caused by radiation or a toxic chemical; POC calculations cannot be made on epidemiologic data alone. The probability of causation, therefore, is a function of the underlying biologic models.[91] There is much discussion about the biological bases of chronic disease and background factors, for example, whether the development of cancer is a linear process or cumulative and collaborative. Even though "different biologic models can lead to different probabilities of causation even when they lead to the same epidemiologic data and population dose-response curve," such models are not readily transparent in the EEOICPA program.[92] Yet disease mechanisms underlie the IREP software—used by the DOL to determine the POC—and serve as the basis for calculating the POC of claimants.

Moreover, embedded in the IREP program are data based solely on radioactivity exposure studies from atomic bomb survivor studies. The dose response data from the exposed human populations of the Japanese Life Span Study (LSS), who were impacted by the very bombs the U.S. nuclear complex manufactured, have served as the template for determining the biological effectiveness of different radiation types and, in this case, for determining the compensation claims of U.S. nuclear workers.[93] The devastation of the atomic bombs dropped on Japan and their documented effects have provided the primary basis—the basic raw evidence—for making judgments about radioactivity exposure and health effects. The survivor studies, funded in part by the DOE Office of Health, Safety and Security, are widely used for risk estimation and modeling and, as the "gold standard" for judging other epidemiologic evidence, have shaped the scientific culture of radioactivity exposure studies, including peer review and the acceptance of evidence from other studies and subjects.[94] The LSS was considered useful in that it studied a large number of people, many of whom received relatively large doses of radiation, and were tracked through follow-up study that continued over a long time period. However, the validity of extrapolating cancer risks from studies of A-bomb survivors who experienced a radiation blast, to nuclear workers who experienced chronic exposures, is highly contentious to some scientists and ethically suspect. The most vehement criticisms target the fact that Japanese bomb survivor data involved high doses, not low ones received over longer time intervals (as addressed in studies of previously exposed workers or of persons exposed for medical reasons), as well as

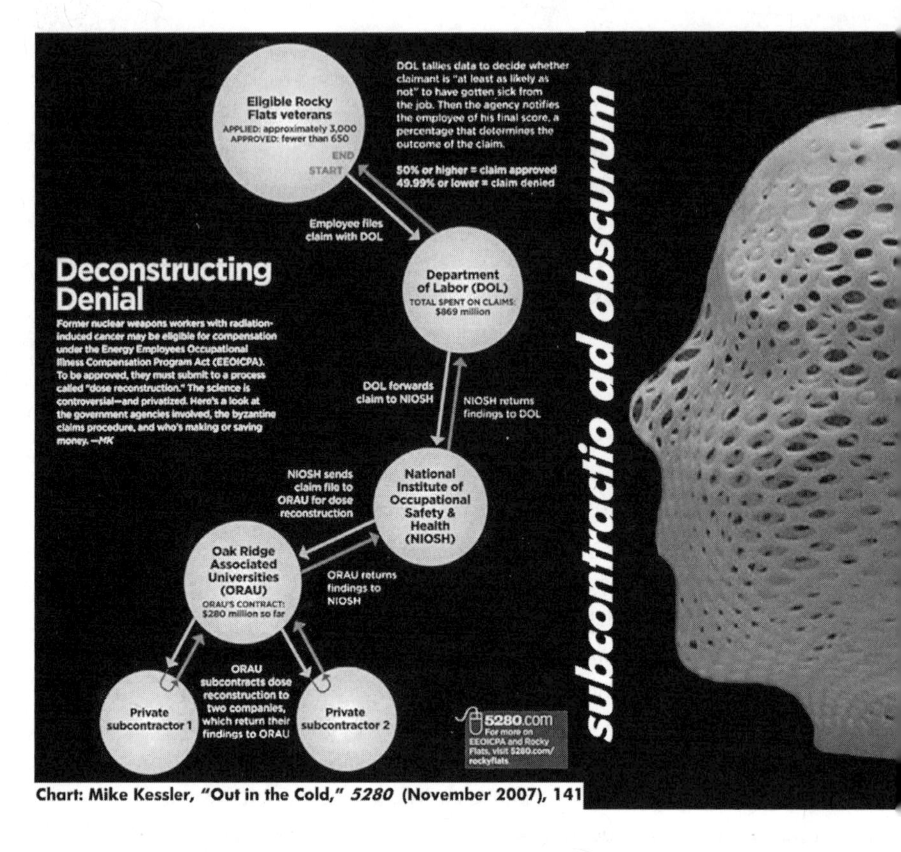

Chart: Mike Kessler, "Out in the Cold," *5280* (November 2007), 141

selection bias, in that the mortality that followed the bombings led to a remaining group of select survivors who were then entered into the study several years after the initial impact. Contrary to the treatment of LSS data as a kind of "pure" template, the study involved extensive estimations of radiation releases—essentially dose reconstructions—based on interviews *conducted in an occupied nation by a scientific team funded and directed by the U.S. government.*[95] Supporters of the use of radiation risk models derived from the study of atomic bomb survivors point out that the data have been modified to account for the numerous and substantial uncertainties involved in the transfer of risk estimates between Japanese bomb survivors and nuclear workers. As a result, the NIOSH-IREP model provides a range of values to reflect the overall uncertainty, with the requirement that the value most favorable to the claimant is used.[96]

All inconsistencies, errors from default assumptions, transfers of excess risk from Japanese populations to exposed workers, and issues owing to disease models or cancer model assumptions are reportedly capable of being incorporated into an uncertainty distribution, as *mere measurement errors*. Such data production *profoundly depoliticizes and obscures the geography of exposure, mortality, and war that links Japanese bomb survivors with U.S. nuclear workers: one set of data is extracted from place and history to be applied as a template of data to another.* This cleanses the land of evidence on both sides through the conversion of local/environmental, occupational, and geopolitical relationships into atomized numbers that are vulnerable to control, substitution, and template conscription. The *refinement of calculations of uncertainty to produce the truth effect of disembedded/disembodied data transfer diffuses any form of responsibility for the life-and-death consequences of nuclear weapons* and *compensation*. The effort to produce a total picture of "certain uncertainty" demonstrates that specific policy interests are in operation with low tolerance for risk and social responsibility; they seek to manage epistemic risk by supporting a knowledge industry that amasses a black box of inferences and assumptions, substitutions and deferrals, models and templates that repudiate the fraught geopolitical and biomedical contexts of the conditions of their own emergence.[97]

The state essentially fails to discharge its obligations to claimants by following the law. This has enabled a shift in former sections of the bomb complex; it could be considered an industry that generates something out of (historically accumulated) "nothing"—holes in data, ignorance, erroneous records, and so on—by remediating such voids and unknowns, refining uncertainty, and relentlessly reprocessing its own methods. Blanks and black holes of information essentially serve as value generators in this biopolitical venture; *they support a knowledge economy of denial, ranging from crude assertions of "no links" to the more technically accurate statistical exegesis of uncertain causations.* Disavowal characterizes the entire legislative system of EEOICPA, rationalizing and legitimizing indifference by the state to historic injustices, and permitting ongoing damage to health, environment, and lives. EEOICPA has systematically sanctioned unprovability, guaranteeing no compensation in many cases.

Whether one points the finger at DOL or NIOSH, the EEOICPA

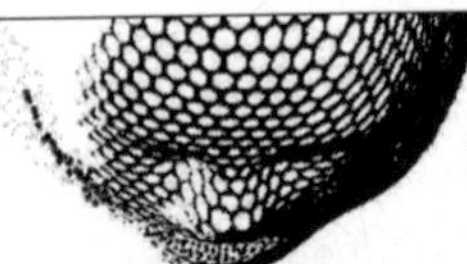

"LET₊DIE"

NO GOs | WHO KNOWs

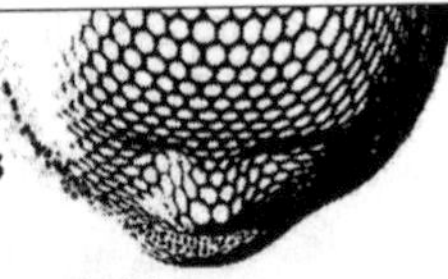

"We don't know what we were exposed to. It was on a need-to-know basis."
Laura Schultz, former Rocky Flats engineer, quoted in the *Rocky Mountain News* (March 2, 2005)

"What we have in our bodies is like a stick of dynamite that seems like it's going to explode at any time."
Bob Carlson, ex-Rocky Flats worker, quoted in the *Denver Post* (May 3, 2007)

"Yes, we are just statistics to
you smug, arrogant guys,
But from this experience,
at least we got wise,
To ALL nuke workers,
we say BEWARE!
When you need your
Government's protection,
Guess What? IT'S NOT THERE!
Judy Padilla, ex-Rocky Flats metallurgical operator, "The Rocky Flats Legacy" (Memorial Day, 2007)

"We're expendable."
Mike Logan, former Rocky Flats plutonium handler, quoted in the *Denver Post* (April 27, 2007)

"I don't want to have to die – don't make me have to die."
Former Rocky Flats worker during meeting about failed Rocky Flats petition for SEC status (June 2007)

"They're letting him die. They murdered him at Rocky Flats. His own government murdered him and they are still murdering him."
Terrie Barrie, advocate for sick nuclear workers, discussing her husband who was a machinist at Rocky Flats, quoted in *Rocky Mountain News* (July 20, 2008)

"Record-keeping was lousy, nonexistent and sloppy. They ignored their own scruples and ignored their ethics. In doing so, they sentenced a lot of people to sickness and death."
Jim Kelly, former United Steelworkers Union leader, quoted in *Westword* (September 28, 2000)

program has been described as an "industry of denial," one that is difficult to navigate because of the obscurantism of its subcontracting.[98] NIOSH contracted out parts of dose reconstruction to Oak Ridge Associated Universities (ORAU), a university consortium, to provide telephone interview support and health physicist services, and to evaluate SEC petitions. This federal contract is reported to have ballooned from seventy million to several hundred million.[99] In partnership with national laboratories, government agencies, and private industry, ORAU subcontracted the work to two private companies, MJW Corporation and Dade Moeller and Associates.[100] The complex layering of contracting and subcontracting to the private sector can diffuse responsibility, obscure incentives, and rationalize the reprocessing of claims while doing nothing for workers. There are calls for an investigation into the potential financial interest of NIOSH's contractors in stalling the completion of dose reconstructions, lowering estimations, and denying petitions for SEC status in order to extend the length of their lucrative contracts. By this logic, contractors postpone a final ruling to "improve" the exposure assessment and upgrade the methods used, in order, ultimately, to facilitate the growth of the biomedical data remediation industry. Concern over such financial conflicts of interests has led to a more general questioning of bias: the implementation of EEOICPA has in numerous instances employed those who once defended the government and its contractors against claims of occupational illness. The ORAU team in particular has been dominated by consultants entwined with the DOE and its former site contractors. "In some cases, these consultants have pre-EEOICPA records of working as expert witnesses against state workers' compensation claims at the very sites where they now perform dose reconstructions."[101] EEOICPA's administration has provided new appointments—particularly in the biomedical arena—for portions of the transitioning nuclear weapons industry, ironically handing over key practices to former DOE-affiliated players who were once a part of the programs that necessitated the passage of EEOICPA in the first place.

Workers complain that important decisions about their claims are made by questionable contractors with no possible review or input by claimants, and that the DOL routinely loses records, requiring people who are ill to repeatedly submit the same documentation, which delays adjudication. In some cases, workers die before a final ruling.[102] Workers must relentlessly "confess" their illness and furnish documentation—what Michel Foucault would refer to as "the examination"—a disciplinary form of power that operates through documentary techniques: "The examination combines the techniques of an observing hierarchy and those of a normalizing judgement. It is a normalizing gaze, a surveillance that makes it possible to qualify, to classify and to punish. It establishes over individuals a visibility through which one differentiates them and judges them."[103] The turning of real lives into a procedure of documenting and submitting evidence is a method of domination that makes the individual a "case"—an effect and object of power, an effect and object of knowledge. *The procedure also functions as a kind of value-generating obscurantism* that works via bureaucratic convolution, inertia, absurdity, and dead ends.

The DOL complains that NIOSH's dose reconstruction is time-consuming and stalls Part B. Delays are further betrayed by the DOE's notoriously slow response time in furnishing important documents necessary to process claims. A particularly convoluted "catch-22" process can happen if classified information is involved in the judgment of a claim or petition for SEC status. Such information cannot be accessed by the claimant—it is a site of no entry; it can be used to deny a claim or petition, even though claimants cannot know this information—even though it might contain information about materials that continue to permeate their lives and damage their bodies. Worse, in such a situation a claimant's or petitioner's due process rights to an appeal are limited to the absurd situation of the U.S. government both representing him/her in a closed hearing with a judge *and* defending the claim's denial. One worker has summed up the impact of EEOICPA on workers thusly: "Half these people are giving up because they can't survive, they can't do all of this information and . . . a lot of people die than do anything else."[104] Even claims that are accepted often face insurmountable obstacles: delays in receiving their compensation checks or medical insurance cards, many physicians do not participate under the EEOICPA program for various reasons including lack of incentives, and more.

Given these logics, contradictions, and consequences of EEOICPA, nuclear workers have fostered a range of responses. Briefly recounted here, these responses include discourses of heroic sacrifice, activities that work with holes of information to contest and reform the legislation, and news kinds of communities and citizenship practices centered on document sharing, occupational memory, or disease/illness identity. The following gestures toward this "work of life" under the conditions of the compensation system and the biopolitical position of nuclear workers. It gives a brief impression of the labor involved in living as the remains of U.S. bomb production, and considers the tactical possibility for nuclear workers to reappropriate their own conditions of vulnerability and illness, to practice an ethics of "living on."[105]

Nuclear workers have developed several strategies to improve EEOICPA legislation, and/or to provide documentary and watchdog services over the claims process, make the public aware of the harm perpetuated by the compensation legislation, and support each other. Nuclear workers have frequently drawn on the patriotic imagery of

Cartoon by Ed Stein. Copyright 2007 Ed Stein.
Reprinted by permission of Universal Uclick for UFS.

"Cold War Warriors" to symbolically assert their right to compensation. The "Cold War Warrior" is a celebratory narrative, oft promoted by the DOE, that portrays nuclear complex employees as having won the Cold War by building weapons that defended the United States against the USSR and other threats of the time period. Many nuclear workers utilize this discourse to position themselves as worthy of compensation and honorary treatment as Cold War veterans for having sacrificed their health for the nation. The logic and rhetoric of sacrifice asserts a form of accountability that citizens can use to ground claims against the government. If the state recognizes deaths as sacrifices, then those deaths come to represent something significant or sacred.[106] Along these lines, the United States observed the first Cold War Patriots National Day of Remembrance on October 30, 2009, dedicated to those who served their country working in jobs related to the nuclear weapons program.

While nuclear workers often rhetorically access the idea of national sacrifice, they are not able to fully call upon it in ways that are effective to their overall struggle. It is difficult for workers to establish and maintain heroism, in that overcoming exposures requires that one can claim innocence before exposure in order to emerge as victorious hero. This is not always possible for nuclear workers because they are frequently blamed for their exposures, even those derived from circumstances beyond their control. As civilians in the domestic sector of the military–industrial complex rather than formal military personnel, they do not qualify for veteran status on a practical administrative level, and their status as workers for the government (though often under private defense contractors) has, on the part of industry and government, resulted in the strategic dismissal of responsibility for their welfare. Anticompensation arguments are made on the grounds that workers themselves are at fault for their suffering because they made weapons of mass destruction, or on the neoliberal basis that they are solely responsible for becoming ill because they took on such dangerous jobs. Because workers willingly participated in a hazardous industry for better pay than other jobs, they are vulnerable to criticisms that accuse them of performing as contracted mercenaries with a profit motive. As a result of these rebuttals, nuclear workers and, subsequently, their illnesses and deaths are frequently disqualified from being understood and valued as sacrificial. They are seen to be complicit with their own deaths, their compensation ridiculed as another "handout."

If nuclear workers cannot fully access the logic and rhetoric of sacrifice due to their status as workers, then the sacrificial logic of the entire nuclear complex makes their calls for assistance appear dependent, even parasitic, because of the changing state–labor relationship and the fact that many of these workers cannot be made well. The nuclear complex married extensive accumulation to a mode of paternalist regulation.[107] Over four decades, "an internal legal code set down worker rights and responsibilities around the idea of 'industrial service' arising from lifelong employment continuities, with promotion, pension, sickness, leave, and sometimes housing rights accruing over time, and the internal legality of joint labor/capital grievance and disciplinary procedures."[108] The paternalism of the nuclear industrial regime made many workers come to rely on the protection of the state; in turn, workers ensured that the United States could maintain productive capabilities, if needed, for the Cold War. The dependency that was created also worked to overshadow the wall of separation that could be strategically erected between military and civilian to deny responsibility for the fallout of occupational exposures or the inability to assimilate vast numbers of these former workers into the restructuring economy or "greening" of the military. "The current restructuring of the defense industry is designed both to scrap jobs and to change those remaining. As defense companies attempt to increase the flexibility with which they can redeploy labor between tasks and projects, they have moved into the forefront of the confrontation with organized labor."[109] In this context, nuclear workers' calls for compensation reach for an outmoded paradigm of industrial citizenship that now leaves them exposed, in need of the protection of the state, yet abandoned.

Essentially, these workers are denied both military status *and civilian status;* they are suspended between supposedly separate military and civilian worlds, without access to the full privileges of either. The term "military–industrial complex" captures the political-economic blurring of industry, science/academia, and the military, but not the strategic military–civilian divisions that characterize the historic treatment of nuclear workers—that have repeatedly enabled the abandonment of their care and, simultaneously, antagonized the positioning of workers with respect to downwinders, downstreamers, various humans and nonhumans subjected to fallout and/or directly bombed by nuclear weapons. As a result, nuclear workers are positioned as compromised subjects—as "impure" civilians—not "pure" enough to be real victims of the Cold War. They are blamed for their overreliance on the state and the "inconvenience" posed by the reminder that civilian infrastructures and everyday life are imbricated with military developments.[110] Nuclear workers were not morally deficient; they operated under national security policy as the public's surrogates in an expanding industrial-economic base of the nation.[111] Such nuclear workers challenge the way the figure of the civilian disavows its investment in state power and economies of violence.[112] In a nation based on productivity, nuclear workers are also traumatic figures in that, as ill workers, they can no longer function in the economy, and they cannot count on the former moral economy to provide for them. In this way, they are seen as failed subjects and are often relegated to public displays of impending death or the abject materiality of disease, pitted against images of flourishing wilderness in the midst of contamination, as

Cartoon by Ed Stein. Copyright 2007 Ed Stein.

delivered by the rise of postnuclear/postmilitary nature refuges.[113] Such spectacle, while it might stir up and consolidate sympathy, focuses on the body, not the work site or work relation, and endangers recognition of the everyday drift of these workers toward death and of the everyday health risks, banal illnesses, such as cancer, and environmental toxicity accrued from the Cold War that we now experience as part of the general background conditions of ordinary life.[114] Making worker deaths into a spectacular event, with the ethical dictate of empathic identification, also works to dispose of larger social accountability for the lethal forms of secrecy, unjust uses of uncertainty, and the convoluted remedies that these workers continue to face. Despite the masculine demand of workers for respect and acknowledgment of sacrifice, the spectacle of pain effectively aligns sick nuclear workers with stereotypes of the feminine, the excessively emotional state of the hysteric, or the stubborn physical body that refuses to get well.[115] In doing so, their suffering is framed as nonredemptive—economically and socially.

As the American public has struggled over what to think or do about sick nuclear workers, workers with the help of legislators have collectively pushed for the revision of EEOICPA legislation. EEOICPA's inclusion of measures to establish SECs, which means those who qualify are exempted from the burden of dose reconstruction, has encouraged numerous workers to petition for automatic inclusion in the benefits program. Although petitions have been denied, SEC-designated sites expanded from the original four to encompass certain classes of workers from seventy-three facilities, as of November 2012.[116] The idea of exposure cohorts has inspired the "Charlie Wolf Act," named after the former Rocky Flats project manager and sick nuclear worker advocate who died of brain cancer in 2009.[117] The Charlie Wolf Act proposes amending EEOICPA to expand eligibility and further improve the procedures for providing compensation.[118] Enlarging the category of qualified people, to encompass those living near weapons plants and families who were exposed to secondary radiation brought home by car or in clothes, would attempt to better account for the sweeping scope of exposure and human cost of the Cold War, and the intimate relations and permeability among work sites, homes, cars, and bodies.[119] Granting special exposure status to all workers and possibly beyond, however, would likely necessitate a complete revamping of the current methods of claims processing. Regardless, SEC cases, although exempt from dose reconstruction, still take significant time to process; cases in 2008 reportedly took, on average, two and a half years to process, with individual case completions ranging from three days to just over seven years.[120] Deferral proceeds in various forms.

Motivated by institutional inability to aid sick nuclear workers, some workers and their supporters have formed advocacy groups, often drawing on the help and representational capacity of the unions. Two notable examples are the Cold War Patriot Advisory Committee, a nonprofit organization that aids former nuclear and uranium workers, and the Alliance of Nuclear Worker Advocacy Groups, which serves as an affiliation of many advocates from across the nation to monitor the implementation of EEOICPA and represent and assist claimants under the program.[121] Worker advocacy Web sites provide emotional outlets and tremendous resources to help workers and/or their survivors with their claims. Countering the gaps of information, secrecy, and elisions of evidence, these groups practice an intensely documentary ethics and politics. They find, track, and salvage key documents and exhibits, fostering community through information sharing. “Contests over citizenship rights are waged by documentary means and strategies, especially through the use of the Internet and web to mobilize, maintain and strengthen associations between victims.”[122] Contextualized within a large political history of citizenship projects that now assemble around illness and proliferating categories of vulnerability, suffering, risk, and susceptibility, these worker groups also share medical information and scientific expertise, and they act as *public bodies* in virtual space.[123] They acknowledge the experiences of workers, provide a way for such knowledge to be shared, legitimate the actual bodies, illnesses, and accounts of workers as primary sources of evidence, channel energies toward collective goals, and, in doing so, help to maintain hope in spite of the many ways the EEOICPA claims process enervates lives.

Once privileged as patriotic industrious workers but now remaindered by a benefits program intended to help them, nuclear workers are aware that EEOICPA, while seemingly the remedy, also “kills.”[124] Conversely, that which allows workers to die by only creating certain amounts of care also serves as a reason for living—to prove exposure and pathology and to achieve recognition. “Life” becomes highly politicized and acquires new potential value because it is negotiated through the counterknowledges and counterpolitics of illness. Such identity-based illness movements and interest groups form through contesting denial at the level of the state, and by engaging critically with the new forms of commerce and political economy encouraged by EEOICPA. EEOCIPA channels the voids of knowledge and the injustices of formerly denied claims (by government and industry alike) into new value-generating interventions and administrative “competencies,” supposedly for the benefit of those who are sick or dying, or their survivors. The procedure might be seen as a political strategy that withdraws state responsibility for the legislation's efficacy. It achieves this by totalizing the uncertainties involved in exposure assessment, and by systematizing tautology. EEOCIPA's administration enlists the same inadequate knowledge practices and historic record of mismanagement that EEOICPA is intended to remedy, intensifying bureaucratic excess and inertia. It also implements cost-benefit logics that defer to secret budgetary caps, and outsources key decision making to a computer program, epidemiologic probabilities, and disease models.

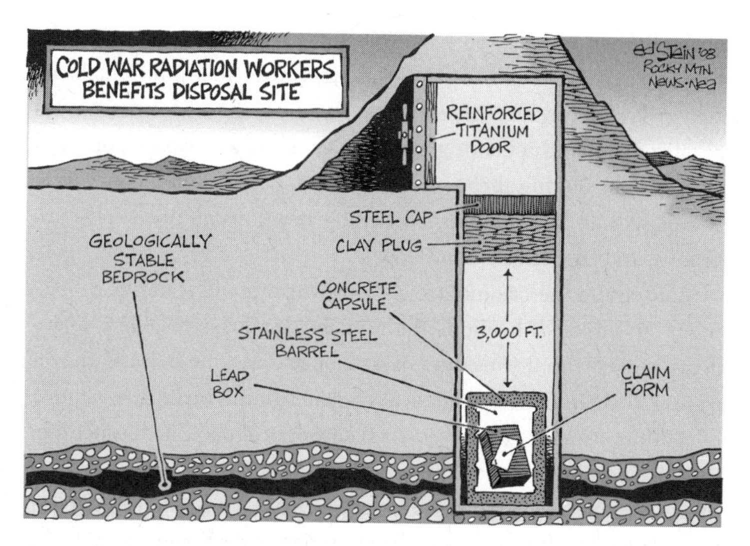

Cartoon by Ed Stein. Copyright 2008 Ed Stein.

The work of taming the juggernaut that is EEOICPA is then relegated to the private sphere of individuals, where moral claims of sacrifice made on the state largely fail.[125]

EEOICPA betrays "a biopolitics that has allocated to itself the right to give life and to give death, that is, to produce new remnants."[126] In significant contrast to the well-funded and advanced gerontology of the bomb and the care exhibited toward the United States' arsenal of aging nuclear materials, former nuclear workers must figure out how to live as remnants of the Cold War. As the anachronistic human remainders of U.S. bomb production, many know illness and death intimately. EEOICPA's burdensome paperwork, questionnaires, computer programs, and complicated administration are symptomatic of the subjugation of nuclear worker knowledge and experience in the post–Cold War period. Workers, with the help of their supporters, are left in the position of figuring out how to live with depleting bodies: bodies with holes from material exposures—exposures to radiation, hazardous chemicals, and waste—and from state secrecy, rejected links, and incentivized systemic denial of compensation. Even while cancer insists that we understand human bodies to be always vulnerable and continually perforated, on the physical cellular level as well as at the level of the human subject and social relations, other perforations divide: from classified information, to legal procedures that individually assign risk and privatize illness, to the abandoning of former modes of citizenship and rendering superfluous of certain expectations and ways of life. Putting these various perforations in relief, they might also be seen as opportunities to assemble new communities, collective actions, and cooperative knowledges.

Contrary to the original legislation's rhetoric of "restoring wholeness," an ethics of living as the remains—as inassimilable remainders—works with the residues of state operations, the holes of knowledge and uncertainties, to politically agitate around making new kinds of communities via disease clusters, networks of document watching, the sharing of occupational memories and bodies of evidence. This process, while involving discomfort and pain, attempts to cultivate new hope for living together by reappropriating death—or, more aptly, by expropriating individual bodies to *deprivatize illness and death*.[127] Death has become cast as private and apolitical; nuclear workers are left to die for the supposed greater good. Instead, what is needed are practices of care that can begin to penetrate the spaces of horrible loss left to workers and their survivors as they struggle to make life somehow livable—even as Cold War residues.

The expansion of the SEC cohorts, documentary ethics, and, on numerous occasions, the humor supported by nuclear worker groups exemplify this important labor of living as the remains of the U.S. nuclear complex. For some, the very physical properties of bodies are primary sources of evidence against state-sanctioned denial of compensation and the means for visualizing the unseen and intimate exposures that certain communities and environments have experienced.[128] From this perspective, the sick bodies of workers are an extension of the vast contaminated nuclear landscape of the United States; they are intimately connected to sites and work processes.[129] Nuclear workers are experimental subjects of those landscapes; they do not have the liberty to remove themselves to a privileged distance of critique. Instead, they work with their own remains—memories, anecdotes, body counts, personal records, medical information, physical pain—and collectively amass evidence to make arguments about the program, in the process often cultivating expertise in epidemiology, law, and policy.[130] Organizing around the radical expansion of special exposure cohort designation might be understood, then, as a strategic reworking of EEOICPA legislation and forming of "waste collectives" that, in refusing individualized burdens of proof, epistemologically bring together bodies, technologies, and environments against state efforts to dispel the human costs of the atomic project. This involves more than a politics of injury, recognition, and sacrifice defined in relation to the state; it does not merely recite the violence and limit political and ethical responses to that of individual rights but reestablishes connections between buildings, bodies of land, and workers.

Worried about the legacy of denial and abandonment of history, some have called for public stewardship of the remains of the nuclear complex, after decommissioning and cleanup have left many former facilities bereft of any remaining signs of the mission that once took place there.[131] Workers are involved in oral history and environmental memory projects about former nuclear weapons facilities; others, as the "hole in the head gang" suggests, take part in support groups that address the ways exposures shoot through spaces, bodies, and families. Such cases shift the disciplinary function of individual documentation and state-mandated confessions of illness (i.e., dose reconstruction, conferring truths about the self and work process, etc.). Just as EEOICPA places workers into the field of endless documentation to establish claim, burdened by the establishment of proof, then such projects foster multiple documentary publics and support networks and, in doing so, rearrange biopolitical governing relations to that of a minimum of domination by holes of information. While the truth is sought in many cases—evidence to substantiate claims, to make visible the absurdities and loopholes—it is the constant and collective recitations of truth and rejections of "certain uncertainty" or "proof of no proof" that minimize the harm of EEOICPA and of being relegated to the remains of war.

How can/will life continue among the remains of the U.S. nuclear complex and bomb production, among the downwinders and downstreamers, the atomic veterans, nuclear workers, uranium miners, and/or those who have received direct bomb blasts or fallout?[132] What does it mean to survive, when so many bodies are slowly and incrementally stolen by cancer, illnesses, and death—as part of the ongoing geographical and temporal outsourcing of the nuclear weapons complex's toxic effects?[133] How does one live with a depleting body? And what is to be done when regulation, recognition, and compensation, for all intents and purposes, augment the burden of exposures and long-term disparities in life chances? Many of those who once worked to make the bomb a reality are now left with bodies that "could go off at any moment." Nuclear weapons workers—supremely ambiguous figures—reveal the offloading of risk onto land and labor alike, and the accumulating holes of knowledge (often elusive, profitable, and purposeful voids) that bore through any taken-for-granted boundaries between bodies and environments, work and waste, life and death. "'Living on' aptly describes inheritances, traces, and half-lives whose quasi-existence overflows classical ontology. It names a realm where representation can no longer keep account of the difference between continuance and vanishing, positing an ambiguity within ordinary life."[134] For nuclear workers, to practice an ethics of "living on" is to take life by the holes: to find and contest lacunae and catch-22's, to work with data gaps—especially of one's own body—as the residues of the state, to insist that uncertainty should not negate possibility and legitimate denial, to recognize the suffering and vulnerability but also the collective opportunity of reconfiguring one's very continued existence on perforation, vulnerability, and ambiguity. This is an ethical labor to live through exposures temporally—there is no return or ending; spatially—as unanticipated relationships, traces, or clusters; and intensely—to resolutely keep the future open by refusing exclusionary (murderous) uses of uncertainty. The current struggle of nuclear workers insists that *"survival is not simply what is left, what remains; it is the most intense life possible."*[135]

04

TRANSNATURAL REVUE

IRREVERENT COUNTERSPECTACLES OF MUTANT DRAG AND NUCLEAR WASTE SCULPTURE

A vast and aging U.S. military–industrial complex gropes for funding and significance in the twenty-first century. In the purportedly post–Cold War era, some parts stagger on as "superannuated half-lives" while others command immense public funds for specialized and controversial sciences, defense contracting, stockpile maintenance, and environmental refurbishing.[1] Although nuclear weapons have largely drifted into the periphery of cultural awareness, judgments about which nation-states should or shouldn't have such weapons technologies continue to sprawl across news pages; they remain as threats, channeled through a vocabulary of statehood and national superiority. While atomic age memorabilia appears relegated to the past in antique malls or is critiqued as part of a remote cultural era via the reassuring distance of academic scholarship, the Cold War is by no means over, as evidenced by many areas and populations of the world; such places continue to grapple with local devastation wrought by large-scale nuclear catastrophes, among them Hiroshima, Chernobyl, the Nevada test site, Windscale/Sellafield, and other events of mass exposure. In the context of global warming and, most recently, the disaster at Fukushima, Japan, the discussion of nuclear energy today is an ethical minefield, because of concerns not only over waste storage, safety, regulation, and the international politics of the commercial nuclear industry but also the legacies of antidemocratic nuclear state secrecy and nuclear-energy colonialism throughout the world.[2] Controversy over contamination and its aftermath permeates the present. Many environmental justice groups and antitoxics campaigns in fact draw on the imagery and material imprints of "fallout" and the sciences of exposure, to demand recognition of the differential, ongoing impacts of the post–World War II chemical revolution and the nuclear age.[3]

This book has examined several spectacles of military remains. "Spectacle" here does not refer to representations as illusory or detached from "true reality" but rather to forms of population management and arrangements of collective experience that allow for disposal, displacement, endgames, abandonment, and exposure.[4] The preceding three chapters observed some of the nuanced ways that administration of land, animals, and laborers has seemingly given the United States "closure" on the Cold War: the return of nature and new creation stories, more government accountability for the legacy of waste and harmful effects of secrecy, recognition and compensation for the unknown risks taken by nuclear workers. Yet, as the chapters showed in detail, these spectacles obscure the ongoing ruination, death, suffering, and effects of domestic war making. Chapter 1, "Where Eagles Dare," suggested how the historical mass deaths and contemporary biomonitoring of different species of animals at the Rocky Mountain Arsenal, including the ambiguous death-in-life status of the bald eagle, secure nature preservation at the wildlife refuge. In the case of Rocky Flats, featured in chapter 2, "Alien Still Life," compliance measurements and various arrangements of distance allow for the abandoning of negative legacies of the nuclear complex and "doing nothing" to administer land and waste. Through the drama of the "Hole in the Head Gang," chapter 3 depicted the heavily bureaucratized denial—"let die"—of nuclear workers achieved by a program originally intended to recognize and remunerate their sacrifices. Such spectacles made of the remains and by-products of the nuclear complex legitimate restructuring and energize new industries and forms of value, including the science of dose reconstruction, the institutional invention of Legacy Management, and the genetic revival and preservation of the bison, to name but three. They also effectively inoculate responsibility for ongoing harm; hypervisibility can facilitate disposal of the negative legacies, by offering opportunities to consume the positive popular culture of environmental education, stewardship, and compensation.

Consumption of exposure is part of a burgeoning entertainment war culture.[5] Contradictorily a historical subject on the one hand, and an urgent contemporary reality on the other in the form of inhabited and embodied risks, the Cold War era is increasingly the subject of popular memorials, museums, and tourist sites. As past forms of pop culture of the atom helped to domesticate controversial knowledge by channeling it into popular culture modes, the current tourist trade of 1950s Americana kitsch and "heritage" opportunities scattered across the U.S. landscape—of Cold War research labs, decommissioned missile silos, the first plutonium plant, testing sites, and even backyard bomb shelters—defer the toxicity and negativity of nuclear weapons work into easily consumed commodity forms.[6] Various ground zeros serve as consumer opportunities for "authentic presence" and proximity to apocalypse; nuclear-themed events, Cold War museums, and atomic tourism transform nuclear weapons into objects of excitement and national pride; even nuclear waste in this context can be converted into patriotic excess.[7] Essentially, the recreational culture of capitalism draws out signs and sites of war in everyday life, and various publics collaborate by consuming exposure as somehow over, in the past. Nostalgia-inducing attractions rally a revivalist desire for the Cold War and World War II as a "simpler, better time," while allusions to a still-present material threat can produce an affective register that ranges from bad jokes and condescension to the sublime or "stuplime," the latter referring to the exhaustion and boredom that follows overwhelming dread.[8]

The first three chapters explored the celebrated postmilitary nature refuge, high-performing nuclear waste management office, and the struggle of nuclear workers who still labor to survive even under compensation legislation and visibility. Such consuming of exposure in these cases suggests the affinities between the phenomena of popular nuclear nostalgia and Cold War kitsch with disaster tourism and neoliberal "atmosfear,"[9] a neologism for the widespread geographies of insecurity, paranoia, and despair attendant to state retribution and violence, the vast assemblages of surveillance and control over populations and places that suffer from debt, poverty, and foreclosure, and the ever-present environmental threats of so-called "risk society."[10] Risk management and communications achieve a subtle extension of disaster consumption into everyday life under the auspices of "safety" and expert opinions: risk administers reality—risks manage the terms by which reality is performed—through norms of exposure that are popularized and consumed, such as the idea of "background radiation" and "acceptable background levels" of exposure, or the more humorous reference to the radioactivity of bananas. Toxicity, for example, is differentially experienced and subject to discriminatory visibility—much of it rendered forgettable and ordinary because the burden of risks falls unequally on the poor and those human ruins of war, atrophied industries, or residual institutions.[11] Spectacle organizes regimes of perception and sensation that are not only uneven but subjugate knowledge and disqualify certain communities. When consuming exposure reaffirms the safety of the consumer or normalizes exposure as part of everyday life, these arrangements of spectacle tend to obscure what has been abandoned, allowed to die, and made residual in order to recast forms of value and generate new industries. They do not foster ethical reflection on degraded environments and other forms of "slow violence" that produce subjects with limited possibilities: the afterlife of hollowed landscapes, gutted infrastructures, material debris, social ruination, resentment, and enduring damage.[12]

The chapters also considered how the residual can serve as a material source of rupture and potential in popular consumption.[13] The residual refers to that which remains, lasts on, and stubbornly resides in the present, albeit out of place or pace; the category includes anything from memories to waste. An important legacy of "postmodern" art and fiction has been the attention given to the residual, to produce potential critiques of consumer culture, technology, the Cold War, conventions of beauty or the sublime. Even where this engagement has merely been a figural gesture, it has in many cases drawn attention not only to the marginalization of people and places but to the radically changed ontology of waste since the chemical revolution and the Cold War.[14] Performance art has frequently involved direct work with garbage and waste as art media as well as explored the relations between maintenance services and the institution of art, toxicity and art making, war and landscape art, atomic memorials and earthworks.[15] There are numerous examples in contemporary fiction that dramatize the protracted quality of disaster of the U.S. military landscape. The gothic is explored in a wide range of novels that address the sacrificial logics that permeate the militarized U.S. West.[16] Often mixing irony and despair, myth and autobiography, and/or apocalyptic satire with nature and science writing, many of these novels allow for death, suffering, and exposure to rupture ideas about the pastoral landscape; unpredictable hauntings and vengeful mutations disrupt the uncanny interfacing of desert ecology with military blast zone or waste repository to reveal physical scars on the earth and genetic scars on humans and nonhumans alike.[17] The cultural performances of toxic tours often employ the residual—whether an abandoned site, ubiquitous waste drums, or unofficial accounts of toxicity, assault, and illness—to challenge the status quo. With full appreciation of the irony of inviting people to tour polluted sites, noncommercial expeditions organized and facilitated by people who reside in those toxic sites guide outsiders through the areas where they live, work, and play, providing stops and lessons along the way to highlight particular concerns, problem spots, physical ailments, and political struggles.[18] Adopting tourist conventions and maintaining some of the historical links between travel and social change, toxic tours attempt to politicize memories, raise awareness of the exploitative relationship between waste infrastructures and race—often referred to by activists as "environmental racism"—and, through educational exposure, alter the social and material patterns that have led to local devastation and widespread pollution.

Just as tourism is one of the most popular ways that people experience sites, numerous responses to military-to-wildlife conversions of DOD and DOE lands, and critiques of the overlapping yet effacing effect of military zones and wildlife refuges, have taken the form of countertours or countertopographies.[19] Among the many examples are documentary trails of downwinder trauma and ecocide,[20] artist-led fieldtrips that explore desert ecologies in relation to military blasting pads and dumping grounds, and grassroots forms of contamination tracking that attempt to reclaim space and scientific knowledge.[21] Such performances frequently combine environmental theater, performance art, and various genres of photography (landscape, portraiture, etc.) with cultural geography, anthropology, and in some cases an intense ecological sensibility. While they do not all incorporate the ironic playfulness that often characterizes toxic tours, they utilize conventions of environmental interpretation and experimental pedagogies as the means to mobilize further action for environmental justice.[22]

Another strategy has been to redeem and reinhabit toxic sites through subversive forms of dwelling and alternative economies of value—for instance, treating waste spaces as a kind of "commons." Implicating the role of the visual in permitting the abandonment of waste—particularly the ironic way that visual concealment *and* mass exposure have cloaked certain sites in secrecy—some artists and social activists have "burrowed into marginal waste landscapes, releasing their subversive potential, shedding light, focusing attention, and undermining social and aesthetic sensibilities that have caused these places to be seen as objects of cultural scorn."[23] Some projects have involved reflection on the relations between restricted public access and permanent insecurity, linking health and safety to environmental security and even military secrecy. Photography, in particular, has opened up a wide range of affective responses to the military–industrial complex and its impacts, including archival documentation and witness of nuclear science and industrial processes; the documentary moral realism of environmental fallout and social devastation; and magical-realist photomontages that arrange for an understanding of the mutable zones and interlinked materialities of region, race, culture, and toxicity.[24]

This chapter deems such efforts to be counterspectacles that use military remains and other residuals to contest the organization of life in/through war production, to refute the current administration of land or waste, and to refuse to be governed thus. As considered here, counterspectacles forward aesthetic practices that seek a more just politics by envisioning rearrangements of which ecologies, communities, or entities can enter politics. They draw on the remainders of permanent war and corporate toxic "open-door" provisions to arrange potent counterhistories and countertruths.[25] "Wastes" here encompass anything considered disposable, no longer of value; waste in my usage refers more directly to the hazardous chemical, radioactive, and material legacies—their lethality and uncertainty—of military and industrial productions. Under conditions of uncertainty, slow death, and toxic stealth, counterspectacles are necessary to nurture entanglements with waste and "a generative interruption called response"—as opposed to outright repudiation, disgust, or denigration.[26] Each of the preceding three chapters gestured toward possible ethical responses to the forms of "death in life" that result from military–industrial restructuring and war's changing domestic organization. These tactical ethics sought to engage the residual, challenge aesthetic orders, arrange other truths and relations with institutions, and regularize alternative forms of life. Chapter 1 suggested an ethics of "deadstock" to account for the dead and biomonitored animals that make possible the cultivation of life at the refuge. In chapter 2, an understanding of the land as an "alien still life" potentially challenged its abandonment by engaging the negative figure of waste as "the return of the repressed" to the past as inert remains *and* a more ambiguous consideration of "this is *still* life, but not as we know it." The third chapter advocated an ethics of collective "living on" under conditions of denial, compulsory documentation, extraction, and alienation experienced by nuclear workers seeking occupational illness compensation. Insisting that we are resolutely *not* postnuclear or postmilitary, these ethical responses attempted to create opportunities for the death-in-life ambiguities, mutations, and other "negatives" to emerge, which, as this chapter will continue to argue, can be occasions for exploring—demanding—ways of living otherwise.

Counterspectacles, therefore, are risky endeavors. They are tactical and ethical, and indispensable to politics; they insist on aesthetics as a form of ethics—meaning, ethics and aesthetics are not opposites—and explore what kinds of aesthetic practices encourage a more exuberant politics.[27] In this way, aesthetics can be understood to refer to not only what is perceived within the realm of art but also, more broadly, the practice of politics, including the ordering of knowledge and sense perception and the configuration of what is visible and/or possible within a given political order. Essentially, aesthetics gives rise to the very possibility of politics because it refers to the *biopolitical arrangements of relations and visibilities.*[28] Representations in general, then, are forms of practice that contribute to an ongoing politics of how we make and recognize the world.[29] Artistic practices are "ways of doing and making" that potentially question old structures and interpretations, develop novel kinds of political subjectivity, and make visible the biopolitical organization of life—what is protected, disvalued, brought back from the dead, or laid to waste as an unrepresented expenditure or figure of nonvalue.[30] Such practices can achieve this by intervening in the distribution of perception, affect, economies, and even materials that constitute social relations, the social field of action, and, therefore, political community.[31]

This chapter seeks to advance the ethical responses introduced in the three preceding chapters, and contribute to this larger effort to open up consideration of waste and to aesthetically redeploy waste. Such a contribution includes the way play and humor can lead to new kinds of social relations and politics, the possibilities that waste, as utopian residue, can be channeled into social critique with demilitarizing effects and environmental responses.[32] Just as the earlier chapters experimented with humor as a way of organizing and presenting analysis, this chapter considers the way play, in the form of camp or the slapstick of repeated failure, can serve as an important tactical means of drawing people in, interrupting routinized responses, and compelling action.[33]

Play and humor can potentially disrupt political and aesthetic orders, but they can also reconfirm existing consensus, neuter political challenge, and become another commodified form. No performance is necessarily pure efficacy or pure entertainment.[34] As Richard Misrach poignantly captured in his now iconic Bravo 20 project, which satirically presents a sacrificial landscape of desert apocalypse and deathly rapture at the former Bravo 20 bombing range in Nevada, there is no hard and true line between securing the memory of devastation and the recuperation of ecocide and disaster as evidence of righteous domination *or as consumer exposure.*[35] This ambiguity presents an opportunity to explore the ethical demands of the mounting remains of military–industrial restructuring. Through this chapter's development of a *transnatural* lens, the spotlight is placed on ways of *playing with the residual and negative aspects of military-to-wildlife conversions, nuclear waste containment, and mutation.* The following questions direct the discussion: How might new forms of political agency perform responsibly in the aftermath of military zones and nuclear facilities? How does rethinking ethics as an imaginative, embodied, relational, and aesthetic practice potentially lead to new forms of life not based solely on the promise of waste containment and supposed salvation of the wildlife preserve? What could this mean for those who live as the material remains of the nuclear complex—who live with nuclear sites in the form of cancer and other mutations? What counterspectacles of "living on" emerge from changed relations with waste? How can a radioactive drag queen or nuclear-waste sculptor potentially inspire new institutional and material relations with waste? How might such figures draw together experimental coalitions of queer, labor, environmental, and other activisms in response to the pervasive popular consumption of exposure?[36]

Entrance: Transnatural Ethics and Aesthetics

Following the preceding three chapters' detailed empirical analyses of the creation story at the Rocky Mountain Arsenal, the cleanup and legacy management of Rocky Flats, and the national nuclear worker compensation act, one way to understand the ethical framework that was employed in these cases—and that continues into the present—is through the spectacle of a waste/nature and human/waste divide. The cleanup and management of decommissioned "post"military and nuclear facilities effectively naturalize and normalize such sites through commonsense understandings of nature and ecology as preindustrial wilderness; as industrialization is seen to have impacted nearly every corner of the nation, there is intense desire for unblemished places, expansive views, and hidden nature. The spectacular projection of military-to-wildlife conversions as the return of nature to the public abets an understanding of such sites as clean or purified by technology rather than as always already "technonature," a complex and dynamic comingling of waste and nature, the social and ecology. Whereas nature was once considered to be waste for its unproductivity, here waste and industrial history are eclipsed by a circular nostalgia for nature as purity, as no longer occupied by military industry and therefore unmarked by war.[37] The value placed on the visible adaptability and presence of certain organisms to radioactive or contaminated environments at military-to-wildlife refuges affirms ideas about the survival of nature as the inert remains of war. It relegates questions about how waste was produced, is to be lived with, and will continue to manifest military occupation to irrelevance; it also obscures the hazards that remain and the material properties and ecologies of such sites, and divisively relegates former site workers and downwinders as anachronistic "living remains" of a military–industrial ecology that has supposedly ceased to exist. While such sites no longer have active missions, they are still affected by the chemical, radioactive, infrastructural, and other material properties of those missions, and, as such, they continue to matter; they are by no means over.

In short, the institutionalized ethics still guiding the treatment of military-to-wildlife conversions and postnuclear nature refuges has promoted a particular conception of nature as wild and external and, therefore, simultaneously, a celebrated residue of military presence (i.e., that because of military enclosure, nature was preserved) *and* cost-effective waste container (whether because cleanup can be curtailed in order to avoid disturbing habitats, or because outsourcing to natural attenuation reduces responsibility and legitimates disposal). As shown in the preceding three chapters, this binarism can be seen at work as the basis of new biovalue developments, such as real-estate incursions and flagship species, environmental education founded on settler fantasies, new sciences such as dose reconstruction, and, most ironically, the genetic preservation and "return to the West" of the decimated American bison. Furthermore, in spite of the appearance of more institutional responsibility, stewardship, and caretaking, this entrenched nature/waste binarism allows for limiting nuclear-worker compensation issues to a narrowly defined workplace separate from environmental consideration. Fence lines serve to circumscribe the material legacy of domestic war, annulling the linkages between the work-site and domestic sphere of workers and the embeddedness of the missions of these sites in everyday lives, health, and ecologies. As the "Hole in the Head Gang" in chapter 3 showed, occupational illness is essentially treated as separate from the ecology of the site. Epidemiological abstractions break the material links between bodies and buildings/sites, dematerialize the geographies of U.S. war making, and cleanse the land of contaminating human evidence.

The current decommissioning and conversion of U.S. nuclear facilities and military arsenals into nature refuges demands an ethical-environmental practice that does not collaborate with state efforts to produce "a return to nature." Joseph Masco asserts, "Life within a radioactive zone changes the terms by which we can evaluate the nuclear revolution; it cannot be narrated within a discourse of purity."[38] This "purity of nature" obscures toxicity and contamination and hides militarization of waste/nature and human/waste relations in plain sight. Former military sites and decommissioned nuclear facilities urgently call for an environmental ethics not grounded in nature as a pure category, but in its porosity and permutability with waste and the human—what I am naming *transnatural ethics.*[39] Transnatural ethics recognizes the polymorphous relations and interchanges of supposedly fixed categories and takes on Gay Hawkins's suggestion that "it might actually be *waste,* rather than 'nature' or 'the environment,' that triggers new actions, that inspires us."[40]

A transnatural ethical framework considers waste to be internal to that which is assumed to be natural. Waste is both other to and internal to the "normal" workings of society. The production of waste is in essence a production of boundaries.[41] The technoscientific instruments for designating and managing waste—technologies that inform the monitoring, controlling, and displacement of waste—are tools for political social management; *waste management is social management.*[42] Therefore, transnatural ethics considers how waste is materially, socially, and discursively produced. While institutions struggle to contain the semiotic and material turbulence of waste, transnatural ethical practices would seek to ignite public recognition of both the unstable world of discourse and ongoing material transformations, in order to encourage debate. Transnatural ethical practices seek material and semiotic thresholds that demand social change, often hidden by dominant narratives of environmental purity and/or militarized discrete categories maintained by the state.

Transnatural ethics is a *tactical, relational ethics.* "Tactical" means that there is no pure or safe position of critique; the response is generated in the action, with tools that are "contaminated." A tactical performance is a "relation of being implicated in that which one opposes, this turning of power against itself to produce alternative modalities of power, to establish a kind of political contestation that is not a 'pure' opposition . . . but a difficult labor of forging a future from resources inevitably impure."[43] As such, transnatural ethics does not endeavor to “tidy up” contamination against a naturalist backdrop, nor does it reject “nature” as being passé or ruined and therefore disposable, conceptually or materially. Instead, transnatural practices endeavor to utilize waste to remediate sites where nature is associated with fixity, permanence, stasis, and conservatism *and* to open up conversation about what the demilitarization of public lands in the United States actually means and how its current manifestation as waste containment and/or abandonment might be otherwise.

Acknowledging important theoretical work on “natureCultures”[44]—such as the dynamism of matter, the constitutive, transformative, and metabolic relations between bodies, environments, and technologies—transnatural ethics registers that boundaries between nature and waste, humans and waste, are not hard and fast but “continually in the making, and something for which we are obliged to take up responsibility.”[45] This responsibility entails experimenting with *the formation of subjectivity in waste*. Donna Haraway, who has done much to further a nonfoundationalist environmental ethics by developing various figures that question what counts as nature and what is at stake in the maintenance of nominative nature or culture, captures the ethical demand of radioactive materials to rethink the nature of subjectivity, kinship, and habitat:

> Located in the belly of the monster, I find the discourses of natural harmony, the nonalien, and purity unsalvageable . . . Like it or not, I was born kin to PU^{239} and to transgenic, transspecific, and transported creatures of all kinds; that is the family for which and to whom my people are accountable. It will not help—emotionally, intellectually, morally, or politically—to appeal to the natural and the pure.[46]

To delineate transnatural ethical practice is not merely to swap one neologism for another—the “transnatural” in place of “natureCulture”—or simply to extend natureCulture in new directions. Rather, transnatural ethics carries an implicit semantic critique and forwards an inversion of the metaphorization of mutations to consider the actual lived embodiments of nuclear mutation, such as bodies with cancer, and forms of tactical play under such conditions.[47] “NatureCultures,” as a term put to practical use, calls the binarism it critiques back into being. By contrast, “transnatural” seeks to simultaneously mark the natural and the indeterminate operation of "trans," as that which is always questioning and undoing the natural as a thing or category, *and* that which is emerging beyond the natural but still in relation to it. The transnatural draws on queer theory and the emerging field of queer ecology; "queer" operates both analytically as a verb that estranges, contests, and deconstructs *and* as a rich variety of historical generative practices of LGBT communities, supporters, and publics.[48] "Transnatural ethics" rejects the idea that the theoretical exercise of blurring "nature" and "culture" is emancipatory and instead acknowledges the extant blurring, such as transuranic waste or cancer from exposure to radioactive materials, in order to explore the intercorporeal effects of and critical embodied responses to such existing natureCultures. In the context of war and military or nuclear technologies, there is nothing inherently liberating about an ontology of mutation—the "monstrosity" of the event or the indeterminacy of mutations[49]—in response to humanist divisions between nature and culture, human and nonhuman. Actual mutations have been relentlessly produced in/through modernity and binaristic Enlightenment thinking; as the repudiated "outside," such mutations were called on to purify categories, such as the human.[50] There is a danger in metaphorizing the mutations experienced by bodies through advocacy of figures such as the cyborg or the "promise of monsters," a tendency sometimes found within work broadly categorized under natureCultures and the "posthuman."[51] The larger theoretical aim of implementing the transnatural, then, is not to invoke figures of transgression metaphorically or a temporal utopia of mutinous indeterminacy as theoretical intervention to surpass humanist ontology but to reveal *the material work that produces the separation of nature as pure and to attend to the remainders of this separation,* such as subjugated knowledges, "impure" cancerous bodies, perforated land, and abject materials, such as nuclear waste. The purpose is to counter truth claims by tactically exploring the mutations, porosity, and "crossings" in existence.

"Trans" here also draws on the concept of "trans-biopolitics"[52] to foreground power relations that constitute and determine "life" and that involve humans and nonhuman populations, including "cross-species" relations and animal-to-animal relations. Although not covered in this volume, the discourse of life increasingly features cells, tissues, organs, and so forth that challenge widely understood notions of natural limits and natural kind and that inform categorical biopolitical decisions such as "fit/unfit." From a related perspective equally invested in the notion of "trans," the human can never be anything but "transcorporeal," inseparable from the environment and part of the "messy, contingent, emergent mix of the material world."[53] Emphasizing movements across bodies and the interchanges between various bodily natures, transcorporeality acknowledges the often unpredictable material agencies of environmental systems, toxic substances, and biological bodies.[54] Transnatural ethics, too, seeks to open up space for materiality within environmental considerations; it does this, however, not so much as an ontological investigation but in terms of critical embodied practices that inspire inventive subjectivities and coalitions. Essentially, I am interested in what transnatural ethics can do or show. Such practices do not offer the decisive action and robust theorizing sought after by many academics and activists. As shown below through the chapter's sketches of two key figures, transnatural ethics is inclusive of ambivalent, ridiculous, and often failed attempts to reorganize life and relations with waste. We should expect no less if the aim is to challenge claims of epistemic mastery over the external world, political mastery over unruly and aberrant Others, or ethical self-mastery, which all "derive from and underwrite a reductive imaginary in which epistemic and moral agents are represented as isolated units on an indifferent landscape."[55]

Michel Foucault's work on ethics is instructive; he argued that the subject is both crafted and crafting, and the line between how it is formed, and how it becomes a kind of forming, is not easily, if ever, drawn. This involves a critical, active, and experimental attitude toward the self.[56] Contrary to an environmental ethics based on naturalistic and moralistic justifications, Foucault rejected the assumption that the alleged natural world should be the source of norms or directions that humans should obey.[57] Éric Darier has extended Foucault's framework of ethics to the environment, developing what he calls an "aesthetics of green existence." Darier proposes a queer and green ethics based on the permanent radical questioning of the broader conditions that shape how humans see the world, so as to think differently from the way we now think.[58] Practitioners reflect on and rework forms of embodiment, continually transgressing the limits and conditions that have constructed their current and past subjectivities, including understandings of and practices surrounding nature; this involves artful practices that take a constant critical attitude toward foundationalism and assumed categories, such as nature equated with purity, or a limited kind of vitalism set against the artificial, inorganic, commercial, and industrial.

In contradistinction to Enlightenment-based empiricism and ethical frameworks grounded in external moralism, a transnatural ethics pursues porosity, irony, and aesthetics in relation to the body, environment, and institutions. The stance is a playful one, marking a world full of creative and resourceful boundary breaching, at the same time as it is a political one. Experimentation with different modes of constituting subjectivity in waste and different ways of inhabiting our corporeality is a political-ethical activity. This does not simply amount to promoting individual management of bodily resources, health potential, and life capital, which is a main feature of contemporary neoliberalism. The formation of subjectivity in waste requires a more ontologically insecure position that asks: Who or what will be a subject here? What will count as a life?[59] Who/what is made possible and sustained, and who/what is sacrificed, contained, or rendered insignificant in order to preserve the vitality of other bodies?[60] What counts as environmental damage? This line of inquiry suspends judgment in favor of a riskier practice that seeks to yield artistry in new forms that address sovereignty and bodily integrity as a continual negotiation.

"Human" and "social" are no less subjects of ongoing fabrication and critical reworking than other material-relational entities.[61] As Jane Bennett states, "humans are always in composition with nonhumanity, never outside of a sticky web of connections or an *ecology* [of matter]."[62] The notion of composition is important, in terms of the attention directed toward both materiality *and* arrangement. Aesthetic practices such as bricolage are central to transnatural ethics; bricolage potentially employs waste for new critical ends, exposing the seams of construction instead of producing the effect of purity. The labor of rearranging relations with waste and attending to the form through which material relations become visible might stir affective responses that engage with the residual through humor and irreverence. Such affective responses must divest themselves of the frequently rehearsed binarism of theory that, on the one hand, affirms life—looks to utopian futurity and hope—and, on the other, meditates on pain, the abject, crudeness, suffering, and death.[63] Playing with the residual can make power relations and current institutional organizations of knowledge and bodies visible, and potentially, through tactical counterperformances, rehearse other ways to live. A "transnatural aesthetics of existence" could develop artful forms of endurance that utilize uncanny materials like plutonium for public-forming experiments. This aesthetic orientation might spur engagement with uncertainty and exposure, and intervene in the remaindering of humans and stockpiling of nonhumans alike.[64]

While great care must be taken when discussing the hazardous consequences of toxicity, this stance seeks alternatives to discourses of nature as purity and other norms of the natural that often operate within antitoxic and environmental stances; "while environmental justice groups have long argued that the claim of the natural is a particular normalizing discourse, it is still often used as a rallying cry around which mainstream environmental problems are mobilized."[65] A transnatural approach, by contrast, is a *queer ecological/ecocritical* one that seeks to address the movement and accumulation of toxins in the world by dramatizing the materiality of various bodies and their permeability, capacities, limits, possibilities, and failures. It does not endeavor to recuperate ideas of bodily wholeness or environmental integrity/purity, which so often allow and encourage various forms of *transphobia* to guide responses to toxicity. Dominant antitoxics discourses deployed in mainstream environmentalism frequently adopt the rhetoric that pollution is responsible for undermining or perverting the natural order, natural ecologies, natural bodies, and natural reproductive processes. As disability theorists and ecofeminist scholars within the field of queer ecology have shown, these discourses often mobilize cultural fears of mutations, perversions, monstrous productions, disabled and/or defective bodies, and assume—even advocate outright—a *compulsory* social-environmental order based on what is considered to be normal and natural.[66] Postcolonial critics, too, have been cautious about "preservationist discourses of purity, given the role such discourses have historically played in the racially unequal distribution of post-Enlightenment human rights."[67]

Transnatural ethics does not endorse disease, pollution, or damage; it seeks to engage with the impurifications in existence, grounding theoretical and aesthetic play in critical reflections on/of embodied experiences and responses, in an effort to amplify the coalitional possibilities of a wide variety of subjects and biological positions.[68] Transnatural experimental and diagnostic arts redraw conceptual boundaries intimately linked to material practices involving both human and more-than-human natures; diagnostic arts of connection seek to break from entrenched cartographies of conventional bounded landscapes and bodies in order to reverse the lack of choices we often face when dealing with toxicity and waste. Rather than insist on particular ways of living at the outset, such arts confront the ways the natural and pure—and by extension the unfit and abnormal—continue to be used to condemn, control, and stigmatize. They also endeavor to move critique outside the narrow realm of reason, pure knowledge, seriousness, and "rights"—which often leads to the militarizing of boundaries and bodies—for creative, often irreverent, embodied ecological projects.

An inventory of transnatural aesthetics takes stock of a variety of approaches—for example, critiquing representations of nature as an inert, determined realm apart from human culture, and revealing the phantasmal status of the natural through hyperbolic exhibitionism. Some methods focus on creating resources from unexpected devices—using waste, organic materials, or even institutions as prostheses to foreground bodily interfaces or to politicize disease. Other courses of action embrace and rework risk for subversive aesthetic and collective ends, such as testing critical "wearables" and carnivalesque designs for body security to satirize imminent threat. The repertoire also includes exaggerating normative environmental subjects—"good green subjects"—and parodying the impact of particular subjectivities on arrangements of matter/energy; implicating pedagogy and forms of memorializing in the landscape and implementing ways of remembering outside the regime of the gaze; playing between caretaker and "sick" and counteracting the isolating effects of disease through "waste collectives"; and inculcating irreverence toward the morality plays that typically accompany representations of toxicity and that underpin institutional control over relations with waste. This last technique opens up several additional possibilities, such as reenchanting waste as art material, restructuring disciplinary organization of waste matters, and *reconsidering the figures we use to think through contemporary environments.*

The following sketches of two figures unsettle dominant representations and organizations of nuclear waste. They also, in very different ways, draw on residual nuclear technologies and the ruins of war to challenge the ongoing effects and the secrecy, control, and abandonment of decommissioned sites of the U.S. nuclear complex. The polarity manifested by these two figures shows the range of tactics and possible "storytellers" that could challenge "the a priori forms that govern what is visible in experience and politics . . . [and] reconstitute our political aesthetics . . . with figures that make visible both the play of the world and the evacuation of that play from the world."[69]

Transnatural Act 1: Nuclia Waste

The triple-nippled drag queen comedienne Nuclia Waste radiates optimism. Permeating Colorado with her wit, costumes, and charitable causes, Nuclia Waste narrates her origins as the Princess of Plutonium in relation to the U.S. nuclear complex. She was born of two former workers exposed regularly to high doses of radiation at the Rocky Flats plutonium production facility near Denver, which produced the plutonium trigger of nearly every nuclear weapon in the U.S. arsenal for more than forty years. With a full head of green glowing hair, the ability to mutate animal-vegetable-mineral, and a love of the Day-Glo bioluminescent color of the Rocky Flats mesa, Nuclia Waste now employs her special plutonium powers to attract tourists to her transuranic world.[70] *Foiling schemes to convert the radioactive pad into a sterilized, pristine nature prairie, Nuclia Waste makes Rocky Flats over as her own private Plutonium Palace, often incorporating the nuclear power plant into the core design of her costumes. Her Web site offers an online tour of her appearances across the U.S. nuclear landscape, as well as photos of her own garden, which boasts produce with a half-life instead of a shelf life.*[71]

Photograph and persona copyright Nuclia Waste.

Nuclia Waste is a potent figure that offers a way of querying and *queering* the environmental imagination in response to Rocky Flats (featured in chapter 2), a site where active publics are needed to put transnatural ethics into action. Nuclia Waste's digital performances and her live-performance practice of mutant drag are provocative examples of a transnatural aesthetics in response to Rocky Flats, one that offers great potential for critical embodied coalitions of former workers, residents, and activists. Although putting this figure to work analytically does not do anything literally to remediate the site, she facilitates an experimental extension of queer theory into the ecopolitical sphere of Rocky Flats that models the relational thinking and ontological positioning of transnatural ethics on the drag queen, as the point of inquiry. Nuclia Waste offers a helpful epistemological starting place from which to interrogate the sexual fields of power that occupy Rocky Flats, explore the relations between contamination and green regulation, and speculate on some of the ways that a queer environmental aesthetics reveals relations of power, questions boundaries, and plays with transnatural relations.

Former nuclear workers may view such transnatural play with horror and lack of control rather than with the irony that can be produced rhetorically and representationally by Nuclia Waste. In addition, focusing on such a figure—a radioactive drag queen as trope of the transnatural—appears to run the risk of metaphorizing the very real bodies of workers or downwinders that have been compromised by unknown exposures to carcinogens and radioactive elements. However, employing Nuclia Waste might provide potential resources for bodies; her performance practices offer some tangible responses to the purifying discourse surrounding Rocky Flats, such as visibilizing the porosity of body and environment and the ways humans and nonhumans have been irrevocably altered by nuclear projects (which could strengthen claims for compensation to workers and wilderness alike). Instead of more militant tactics that invoke sacrifice and purity, rights and injury, she opens the possibility for different social practices in relation to Rocky Flats that acknowledge mutation rather than return the site or bodies to normal/natural in order to then abandon them. Further, she shows how human bodies might be used as sites of interface and play. This is useful for reflective embodied responses to mutation from exposures. If the nuclear era is supposedly now bygone and kitsch on the cultural level, then it is necessary to repoliticize the remainders, especially the human remains, of the nuclear plant and its cleanup. Although Nuclia Waste's performances are not explicitly political in relation to the cleanup of Rocky Flats or worker compensation, she reintegrates waste and popular culture and publicly plays with porosity and counter-truth claims about the site, irreverently suggesting that we might all be queer in this "postnuclear" age—our bodies, families, homes, domesticity, and fitness in general. Such camp humor can help sustain vital attitudes and prosthetic personas under conditions of diagnosis and crisis.[72]

So, who is Nuclia Waste, and what are her performances? A giant yellow waste drum, marked with Nuclia Waste's name in glowing letters, greets visitors to her Web site.[73] By clicking on the ominous waste can, the viewer enters the main menu of Nuclia's site. Set against a purple backdrop, a giant atom symbol lays out the Web site's various options, with Nuclia's own face and giant neon-green wig set in the middle, dominating the page. Various kitsch elements, such as lava lamps, are integrated into the scene, along with the banner "Welcome to the radioactive pad of Nuclia Waste." The color selection and design already prompt a questioning of the natural; something is already "off." Placed along the atom symbol, the link "Who is she?" takes the visitor to Nuclia's autobiography page, where she describes herself as a product of the Cold War era, born of Rocky Flats workers who were exposed to radiation. This easily accessible primary performance describes Nuclia Waste's mutations in everyday terms, queering the nuclear family and drawing on numerous kitsch artifacts to tell the story. The performance also personifies the cleanup of the Rocky Flats site and introduces a rivalry between Nuclia Waste and Miss Dee Contaminate, who "swore by her EPA approved prom gown that she would someday rid Rocky Flats of Nuclia Waste forever." The autobiography ends with a description of her dedication to Rocky Flats, which, we are informed, she has rebuilt as part of her Plutonium Palace, in spite of Dee Contaminate's relentless efforts to banish Nuclia. "Always searching to rid Rocky Flats of Nuclia Waste, Dee is quite alone in her sinister endeavors, for Nuclia constantly draws more and more admirers (of many sexes, races, and galactic origins) into her orbit. Her personal physicist, Dr. Miles Island III, attributes this to a previously unknown attractive force he has dubbed 'Groovity.'" The story narrates Nuclia as a beneficial inhabitant of the Rocky Flats area with special attractive powers, in contrast to the jealous and petty Miss Dee Contaminate with her relentless pro-cleanup and emphatically "anti-fun" schemes.

Nuclia Waste's digital performance also includes a brief tour of Rocky Flats as part of a suite of activities found in the "Fun Stuff" link from the main menu. Visitors are introduced to Nuclia's main living quarters within the Rocky Flats complex. The image of a Mars-like landscape links to a view of her inner chambers: "Inside the main structure of Rocky Flats is a city within a city. The green glow of

Photograph and persona copyright Nuclia Waste.

plutonium radiation permeates the air, keeping Nuclia looking fresh as a daisy." By clicking again on the image, one further penetrates the complex for a peek at the nuclear reactor core: Nuclia's inner sanctum and regeneration chamber. Another option is "Nuclia's Strange Roadside Attractions" tour. This activity icon invites the viewer to travel with Nuclia Waste through the continental United States in her customized Weiner Mobile—a giant automobile shaped like a hot dog on a bun (essentially a photo of the promotional Oscar Mayer hot dog car but branded with "Nuclia Waste"). Upon clicking the hot dog car, the viewer is transported to a large glowing green map of the United States and, by clicking on individual states, can view documentation of roadside oddities in the form of tourist photos, which are photomontages of Nuclia humorously embedded in tourist scenes. Some of the strange attractions are made/sculptural; others are found/grown. Blurring the line between organic and inorganic, the real and surreal, the attractions include giant mutated vegetables, building-sized donuts, corn palaces, strange creatures, the world's largest peanut and rubber stamp. In the state of Colorado, where the Rocky Flats Plant is located, the viewer can see Nuclia caught in the gaping pinchers of a huge beetle sculpture; the caption reads: "Apparently, the effects of my Rocky Flats radioactivity has *[sic]* had some impact on the insect population as far south as Colorado Springs. I can only imagine what the mosquitos look like." By accessing the state of Florida, one confronts a postcard-type image of Nuclia next to the world's largest lobster. The caption includes another reference to radiation: "I had no idea how far ranging the effects of my Plutonium Powers are." A third example, the state of Arkansas displays a photo of Nuclia with a farmer and "One Really BIG Watermelon." Here, Nuclia's annotation states: "Ronnie and I had a good time when we visited his Produce Market. He grows his melons with radioactive dirt shipped down from Rocky Flats. We spiked this one later that night with some vodka!" The interactive tour in effect encourages visitors to view the entire United States as a nuclear landscape, and to consider the presence of nuclear waste in everyday life (although present in some places more than others). This is "an ecology that embraces deviation and strangeness as a necessary part of biophilia."[74] The delight and amusement generated by visiting mutated and giant organics also stirs excitement for the products of

Nuclia's own garden for sale. From the main page, visitors can go to her "Plutonium Store" and buy delectable-looking honey in jars with a label that jokingly states the product is from Rocky Flats and contains "Honey fit for a Queen!"

In terms of Nuclia Waste's live performances, the aspect most central to my framing of transnatural aesthetics is her practice of "mutant drag," namely, her performance in a visual register, and how this might be expanded. Nuclia Waste's mutant drag typically includes giant neon green hair, three breasts, a sparkly beard, sometimes greenish skin, green and pink eye makeup, various kitschy accessories, and neon party clothes. While there are multiple forms of drag, from celebrity impersonations and "gender illusionists" that often naturalize gender through successful mimicry to more-than-human theatrical personas with amplified dimensions of elegance/fashion, mutant drag falls into the category of comical high-camp drag that does not attempt to "pass." With clownlike values of satire, ribaldry, exaggeration, and kitsch, Nuclia flouts the glamour of "good drag." Instead, she mixes male and female signifiers, and plays with artifice and bad taste, from Cold War–era nuclear-family kitsch (bowling pins, frying pans, Tupperware, ice cream cones) to popular icons of nuclear waste, radioactivity, and mutation (neon green, hazardous waste cans, atom symbols, proliferating body parts). Nuclia takes mutation one step further by integrating stilt walking in parades as evidence of her mutant ability to grow extended appendages.

This form of drag shows potential in its playful insistence on impurification and irreverent critiques of nature/culture and waste/human binarisms.[75] Such drag does not so much denaturalize as it embraces the natural as the "always already transnatural." Relishing the impossibility of purity, mutant drag queers nature *because* nuclear waste and radioactivity are queer. In place of an environmental politics grounded in a purified version of nature, and rather than the heterosexual family and home as the place for environmental practice, mutant drag directs attention toward the impurifications already in existence as that which requires responsibility, creative ethical responses, and broader kinship. The camp aspects of Nuclia Waste's visual register flout conventions understood to be human, suggesting that "the human" has been irrevocably changed by nuclear technologies. In spite of the danger of mutant drag potentially normalizing mutant bodies as outrageous and consumable spectacle, this performance practice opens up possibilities for unruly biological and social exchanges to serve an environmental politics that does not respond to toxicity with purifying logics, hidden eugenics, or transphobia. Mutant drag begs the question: *Why respond to toxicity with austerity?* This is particularly interesting in the present context of neoliberal cost-benefit discourses that limit environmental performances to individually mindful consumption and rational recycling. By contrast, mutant drag plays with an overload of kitsch and celebrates life through boundary-breaking yet lighthearted engagement with the "grotesque body" of the irradiated mutant and nuclear waste. Such queer attachments work to "celebrate the excess of life and to politicize the sites at which this excess is eradicated."[76]

More specifically, Nuclia Waste's mutant drag potentially cultivates public memory of the Rocky Flats National Wildlife Refuge as the residue of an earlier era of nuclear technologies, and "pure nature" as retroactive construction. As nostalgia for the Cold War intensifies, and as the bomb production period appears outdated and available as retro, the drag queen plucks Rocky Flats from the dustbin of history to serve as both domicile and sartorial inspiration. The drag queen's use of camp rediscovers history's waste in the former plutonium plant, amplifying the excesses of the nuclear regime of power as riotously "bad taste." Her radioactive retroactivity calls attention to the weird temporality of waste—the continuing presence of the leftover, the elusive, the remnant—potentially anything that has escaped progress and the technological reproduction of the nature refuge.[77] Rocky Flats, in the margins of a spent historical mode of Cold War industrialization, manufacturing labor, and the nuclear family, becomes irradiated with glamour, a recurrent theme of camp humor.[78] Rather than paralyzing fear of fallout or apocalypse, Nuclia suggests a risibility politics that starts with the ridiculous or oppositional meanings of exposure.[79] Such an approach might serve to cultivate plural modes of constituting public culture rather than resort to containment and conservative preservation efforts. Camp in this sense offers a modality of "counterpublicity," wherein "getting a laugh" provides an opening for critique that grapples with the contradictory and absurd ways waste is produced and lived with.[80]

Nuclia Waste's spectacular habitation of a national sacrifice zone and the suggestion of the porosity of her body and the environment also offer a performance-based "transuranic queering" of the post-nuclear nature refuge, fantastically exaggerating dominant environmental norms and challenging discourses of fitness and progress that often underlie the "return to nature." "Queers"—as theoretical and academic-political framing of LGBT communities—have often been positioned as boundary creatures, neither fully natural nor fully civilized. Queers have responded to their positioning as unnatural/artificial/pathologized in different ways: from sexual authenticity and the living of one's nature, to a transgressive embrace of their supposed perversion and the radically inauthentic.[81] *Queer* as a verb has been used by environmental ethics and environmental education to question the adequacy of dualisms and to reevaluate nature.[82] Catriona Sandilands argues, "to queer nature is to question its normative use, to interrogate relations of knowledge and power by which certain 'truths' about ourselves have been allowed to pass, unnoticed, without question . . . to queer nature is to 'put out of order' our understandings, so our 'eccentricities' can be produced more forcefully."[83] Queer theory, as it informs the emergent field of queer ecology, strives to disrupt the "naturalness" of nature and sexuality, transforming, destabilizing, and subverting sexual identity and sexual norms as they inform environmental positions and practices—particularly the enmeshment of heteronormativity and econormativity.[84] This operation provides a way to reframe the DOE's environmental cleanup actions as practices that contribute to the reproduction of a particular sexual order.

In the context of the changes occurring in the U.S. warfare economy, including a way of organizing life that historically eroded the distinction between civilians and soldiers and founded the nation-state on military–industrial capitalism, the nuclear complex has had "a presence in our lives far beyond the explicit debates about energy generation; it has a strong hand in defining who we are and what we should become."[85] The cleanup and management of Rocky Flats as a wildlife refuge is a highly visible project within this context; it is praised as the first nuclear weapons complex to be completely cleaned up. The figure of the drag queen illuminates a crucial aspect often overlooked in the discussion of militarization and nature: any production and management of nature is intimately part of a genealogy of regulation of sexuality. Discourses of nature have been employed to formulate and regulate specific sexual identities; likewise, understandings of sexuality have informed ideas about nature. The Rocky Flats National Wildlife Refuge—its naturalization as a wild space—is largely successful because of the traditions of landscape aesthetics and pastoralism, and the United States' history of environmental conservation and recreation, which have historically essentialized and racialized nature as a space apart, as a space of intensive moral and sexual regulation.[86] Further, the nature refuge rehearses a fantasy of self-enclosure that permeated Cold War America. As Peter Coviello has argued, power in the nuclear age was thoroughly sexual and "queer"; it could penetrate anywhere, knowing no bounds and creating anxiety over radiation and sexuality alike. Sex and sexuality, therefore, were an intimate part of militarization during the Cold War, constructed to manage the population. Militarization entailed more than the material-economic organization of society around the production of weapons; it also molded social institutions seemingly not connected to weaponry or directly to war, from fields of knowledge to (re)definitions of proper masculinity, femininity, and sexuality.[87] As many scholars of the Cold War have noted, this period involved intense efforts to ensure biological and ideological reproduction, through the containment of women in homes, suspicion of the homosexual as the weak link in the domestic order, and the defense of white heterosexual masculinity.[88]

The DOE's recent turn to the traditionally gendered role of cleanup demonstrates a reinvestment in Cold War sexual politics. During the cleanup of Rocky Flats, the DOE and Kaiser-Hill occasionally invoked the feminine, supposedly taking on the qualities of a "good housekeeper" in frugal spending on the cleanup and in the spectacular removal of the site from public burden.[89] This "gender bending" may appear to be preferable to the many practices previously noted at Rocky Flats. However, the DOE's cleaning actions can also be understood as a continuance of Cold War masculinity, which militarized the boundaries of the domestic and the natures of sexuality, and obscured women's labor with the wonders of technologically advanced kitchens and homes.[90] In the contemporary instance, the rhetoric of domesticity attempts to alleviate public scrutiny of the cleanup effort and reclaim safety; it also mobilizes long-standing ideas about the resilience of nature and the U.S. nation to naturalize the bounding and policing of the site and thereby maintain the institutional ethical framework—*ultimately effacing the cleanup labor.* Nuclia Waste's mutant drag runs contrary to heterosexualized containment and reproduction of nature. If "home is radiance," then Nuclia Waste devotes her energies to subverting the DOE's rhetoric of domestic responsibility and security by playfully committing to the reirradiated restoration of the Rocky Flats facility as her hip Plutonium Palace.

Nuclia Waste's mutant drag is creative in ways that provide an alternative to procreation premised on the visual spectacle of the site's return to wilderness and, essentially, the ongoing military occupation of nature. Her love for radioactive habitation provokes a productive tension between the radical negativity of the queer, as a figure, and the queer as regeneration, plenitude, and excess. In place of a reactionary ecological fundamentalism, or a turn to mutation as liberatory, which potentially makes Rocky Flats nuclear workers who are ill and seeking compensation into failed "mutant subjects," Nuclia Waste shows how to utilize bricolage and kitsch—aesthetic practices that are central to transnatural ethics—in tactical ways that might reject claims to sovereignty, natural or political, in favor of exploring the openness of the world without romanticizing or depoliticizing it. Camp, in particular, serves as a way of creatively assembling/expanding/adding/amplifying as a means of sustaining selves and communities.[91] Camp suggests that the superfluous is needed to sustain us.

Nuclia's exhibition of radioactive contamination and her metaphoric play with plutonium can also be read as parodies of the practice of "knowledge as exposure." In the case of former Rocky Flats workers, pedagogy of exposure, no matter how well-intentioned, can have the effect of reinforcing the visibility of imminent death, or a promised invisibility. While Nuclia's use of camp produces obvious tensions in relation to workers (e.g., how funny is "mutation" when cancer is considered?), such a strategy could also prove useful. Instead of promoting themselves as Cold War warriors, who are symbolically valued for their sacrifice to the nation but not materially compensated for occupational illness, workers might cultivate a taste for ironic glamour in

Photograph and persona copyright Nuclia Waste.

place of nostalgia for martial values, to assert a critical attitude toward diagnosis/cancer and dramatize the body—its abjection—as a site of possibility and parodic inhabitance capable of converting the fearful into the comically livable.

Waste continually crosses the boundaries of bodies, spilling into the cellular matter of humans and nonhumans, creeping into the future.[92] Nuclia has played with documentation of the removal of waste from Rocky Flats, annotating newspaper clippings and narrating herself as the waste shipments in first person. Her online tour of the U.S. nuclear landscape also suggests that while nuclear waste is kitschy in this era, mutation is in fact the norm, differentially experienced through the social inequalities and spatial divisions of the nuclear landscape. Her postcards rhetorically perform radioactivity's promiscuity, in the process parodying the usual NIMBYisms (she is disappointed when she is escorted out of your backyard barbecue).[93] There is no pure private or civic space in Nuclia's transuranic world, which calls attention to the differential ways waste is materially and semiotically contained and the body and backyard protected. Her relishing of the permeability of waste and nature, human and waste, indicates the extent to which our daily lives deny that permeability through intrusive codes of environmental managerialism, and the ways knowledge of pollution and the vulnerability of the body have led to the increased fortification of bodily boundaries, identities, and communities. This work implicitly pokes fun at the bodily security of liberal environmentalism, inciting questions about who specifically has the privilege to claim more security than others. Given this, Nuclia Waste might inspire a body politics that resists such "environmentality" to actively explore bodies in the throes of materialization, shaped by corporeal histories, everyday practices, norms, and biophysical processes of Rocky Flats. This too is potentially useful for workers: the link she marks between nuclear apocalypse and immunology discourse in constructing "a body that must be defended" challenges the too-often-assumed binarism of environment and epidemiology that limits how we experience the world and are affected by it.[94] The idea of certain bodies as abject toxins underlies modernity. Nuclia Waste is hopeful because she "takes on this contamination" in a fuller recognition of the ecologically abject, the uninhabitable—what Sandilands calls a transgressive and committed desire to rehabituate, refamiliarize, and rematerialize the body in relation to the world.[95]

Bodies and nuclear technologies collide at a powerful intersection where it is possible to explore questions of citizenship, environmental justice, and everyday life, to invent new kinds of institutions, and to reconceptualize an environmental ethics that attends to waste, workers, and downwinders.[96] What I'm calling "transnatural ethics" allows for an alternative response to radioactive natures, opening discussion of the subjectivities, ethical practices, and aesthetics that emerge in response to the "postnuclear" nature refuge. Challenging and taking responsibility for places such as Rocky Flats requires continual experimentation with creative cultural-political forms and alliances in the face of the United States' ongoing toxic open door. Strategies associated with a queered transnatural ethics and aesthetics can be of utility to people studying landscapes as well as people engaged with the politics and processes of changing them. Social processes, political institutions, and cultural norms affect both how landscapes are perceived and how they are materially reproduced and managed. Further, bodies are arranged and ordered in relation to landscapes; as places are naturalized, so too are certain bodies, social relations, and identities. Exploring the centrality of waste to the constitution and ordering of nature and bodies encourages a less dichotomy-obsessed ethics toward landscapes and subjects, contaminated or otherwise, and an ontological interest in active matter and intercorporeal porosity. The aim of this shift is to queer a whole range of environmental philosophies and institutions, simultaneously fostering critiques of heteronormativity and notions of normalcy operating within environmental concerns and toxic discourses *and* working toward new coalitions that take better care of the environment by welcoming "unknown readings, new claims, provocative analyses—to make things happen, to move fixed positions, to transform our everyday expectations and our habitual conceptual schemas."[97] Critiquing the normalized body and naturalized environment demands a commitment to "seeing beauty in the wounds of the world and taking responsibility to care for the world as it is" while maintaining critical perspective on the changing materialities of various bodies and environments as part of the embodied reality of the nuclear project.[98] The promise of this *embodied ecological politics* is a political project of justice that tactically and aesthetically practices interdependence, vulnerability, and creativity.

Intermission

There is much opportunity and work to be done to address the ways that cancer and other mutations from exposure are the material supplements of the military–industrial complex. Furthermore, the material underhistory of World War II throughout the Cold War to the present—especially nuclear waste—remains, enduring as a brute force of old industry and conflicts, as a multiplying by-product of our inability to do anything about it, and as repositories of vulnerabilities and impaired states that produce subjects with limited possibilities and who are burdened unevenly by what is left.[99] Produced in its most massive quantities during the buildup of the American nuclear arsenal, nuclear waste persists beyond management, affecting physical landscapes and bodies and, as "the nation's garbage," sullying domestic associations of the nation with untrammeled wilderness, the western frontier, and manifest destiny.[100] The ecologies and people of the U.S.

West have especially borne the brunt of decades-long geopolitical struggles and the radioactive remainders of the nuclear complex, while the nation, more widely, is also subject to continued postwar cultural excess, hosting dumping grounds and sanitary landfills—typically out of sight—that manage the load of discarded materials with their radically changed material compositions.[101] As the chemical revolution of the 1930s made its way into industrial production for domestic use in the form of plastics, toxins such as pesticides, and synthetic materials, the results were and continue to be nonbiodegradable, toxic, and hazardous wastes; combined with new remainders from the burgeoning electronics industry, household garbage was taken over by materials with extended future trajectories.[102] Waste's new staying power, as a result, generated new toxic ecologies of the nation domestically and internationally. The "becoming waste" of domestic life was administered by industry and society through burial, physically and culturally, and rejection of waste's stubbornly resistant materiality.[103] This has subsequently led to terminally impermanent domestic storage, de facto wilderness wastelands, international dumping, and/or denial.

A future archive of Cold War toxicity, waste is evidence of the deep contradictions of military–industrial production and the supposed "environmental turn" of contemporary U.S. war. As chapter 1 investigated through the figure of the bald eagle, World War II and the Cold War have produced wastes that cannot be consumed or accorded value except through the management of waste dumping outside the social and through burial; the materiality of such waste is significant and provides an important source of criticism of the material stakes of production and consumption that sustain war. In the case of the Rocky Mountain Arsenal, this is crucial to understanding the relations of humans and nonhumans. Waste is materially linked to places "where not everything solid melts into air" and "where only eagles dare." As the chapter revealed, the bald eagle is tragicomically positioned as symbol of the nation and war, and as an animal that is materially sustained on lands disposed of by the military; yet, instead of being allowed to inspire meditations on death or waste–nature relations, the bald eagle has been enlisted to sponsor a national genesis story of the military's preservation and re-creation of nature. The Arsenal has essentially become a kind of consumable disaster. The discovery of an abandoned vial of flesh-eating gas or military munitions debris only seems to amplify the aura of the place, lending tourist-desired authenticity rather than reflection on the domestic material legacies of war. Through a similar process, the nuclear arsenal has seemingly receded from many American imaginations, exiled as a relic of a supposedly bygone nuclear age. Concerns over nuclear proliferation, political and cultural containment, and mutually assured destruction have largely moved into the realm of nostalgia or return as political-rhetorical pastiche in the Cold War's aftermath. Moreover, the conversions of nuclear facilities to nature refuges reaffirm old habits and attitudes toward waste; not only is the materiality of waste eclipsed by the supposed return or resilience of nature, but material relations with waste and the material stakes of the wastes of war remain unaddressed, displaced, or denied. This is not simply ideological greenwashing; it is about perpetuating particular biopolitical arrangements of bodies, ecologies, and special materials—specifically, relations with waste as the ultimate figure of negation.

Residues of the Cold War are as persistent in the attitudes toward waste—and, by extension, organizations and economies of waste materials—as in the waste materials themselves.[104] Pragmatic efforts to generate knowledge about toxicity and to make exposures visible have made important interventions in the relations between waste and society, particularly the uneven material geographies of waste in low-income communities. Moreover, toxic discourses have challenged traditional understandings of what counts as an environmentalist movement and foregrounded the interdependence of humans and ecology and the importance of social justice to health. Antitoxics campaigns and other forms of ecopopulism, for example, have rebuked the errors of a whole epoch of industrialism and militarism, diagnosing toxicity and threat, and tactically drawing on images of a ruined national pastoral ideal and a world without refuge from toxic penetration.[105] These approaches, important as they are, continue to invest in diagnostics based on Cold War–era nuclear fear. Although ideas about a ruined world have a longer history, ecocatastrophical fear can be considered a significant by-product of nuclear fear and the Cold War. The frustrated indeterminacy of exposures energizes approaches that use the rhetoric of blame, threat of infringement, and outrage to produce moral melodramas and counterspectacles that often, in a Virgilian mode, reveal the existing toxic infernos here on earth.[106] Fear of fallout from nuclear detonation undergirds contemporary toxic discourse. Radioactivity was historically framed within toxic discourse; this has played a significant part in shifting "apocalypse" from the vivid and horrifying spectacle of the mushroom cloud to the administration of life (and death) through responses to cancers and diseases indeterminately resulting from unseen and often everyday exposures.[107]

Toxic discourse has also been shaped by the domestic awareness campaigns about toxic chemicals and products in the home. While this effort highlights the material relations and harmful impacts of certain everyday items and encourages people to make other arrangements, the individual bears the burden of this reorganization. This neoliberal approach to domestic health effectively entrenches the Cold War–era military production of the private sphere as a space of self-reliant survivalism against contamination and as an accumulation site of military–industrial products. If fallout shelters are campy remainders of an earlier age of anxiety and paranoia, the domestic antitoxics trend often reveals how Cold War–era relations with waste live on, through the legacy of militarized private spheres and actions that continue to separate private from the public spaces where waste is to be disposed of, and, therefore, perpetuate uneven geographies of vulnerability and inhabited risks. In addition, these toxins, at the heart of the home and everyday consumption, are themselves the ongoing remainders of a military–industrial–*toxins* complex; this fact, however, is left unquestioned through the insistence on individual rational practices.

Reviewing some of the connections between toxic discourse and U.S. nuclear military production does not blunt the importance of these approaches. Rather, it opens up discussion of *how to intervene in the legacy of war that informs material relations with waste, and therefore ongoing toxicity and the consumption of exposure.* Certainly, waste should be critiqued as the excess and bad by-products of military–industrial capitalism and consumerism, but what are the possibilities for tactical counterspectacles that reappropriate waste and, more important, work with the residual material properties of waste? Must we always be at war with waste? How might we inspire a tactical material ethics that rejects war's ongoing organization of material relations, institutions, infrastructures, and understandings of waste? In the case of nuclear waste, the focus is typically placed on fear of radioactive materials and somehow making exposures visible. Although this work is essential, how else can we imagine relating to the stubborn materiality of nuclear waste? Rather than repudiating the actual material wholesale (e.g., plutonium is bad or evil) in a morality tale that cannot do anything about the material existence of these materials, how might we live with these leftovers of the military–industrial complex under terms different than banished matter or abject fetish of the nuclear age? Other than the "safety" of distance, denial, or containment, how else can we respond to waste, and how does its ambiguity—as inanimate materials that appear to be alive—encourage a relational-material waste ethics? For example, could the long-lived duration of nuclear wastes be converted into artful forms of endurance? Is such a reappropriation of nuclear waste even possible?

Imagining a new materialism that would transform our relations with waste is a central aim of the transnatural aesthetic agenda, and an important way of imagining the world differently. The pursuit would demand no less than a reordering of the processes and institutions that bind us to waste and enable exposure—including spectacular distance (as this always entails that somewhere and someone are experiencing exposure) or by simply not seeing waste (such as the complicit inattention to military productions in the home). It would also demand significant divestment in waste as the central character of disenchantment stories, which are perpetuated in the extreme by military-to-wildlife conversions and their attendant creation narratives and settler/frontier nostalgia. Hawkins explains that disenchantment stories presume a fundamental dualism between human culture and nonhuman nature, and they assert that waste can only function as that which destroys the purity of both sides of the opposition rather than as that which undoes the opposition, insists on relationality, and foregrounds value regimes, material practices, and properties.[108] The treatment of waste through relations of distance, denial, and disposability persists; moreover, the ethics and affective responses to waste continue to have recourse to guilt or moral righteousness, blaming waste for contaminating both culture and nature, and often neglecting the complexity and materiality of waste. The central question for transnatural aesthetics, then, is how do we learn to live otherwise with waste and foster change in the organization of our daily lives and our consumption, desires, aesthetics, knowledge practices, education, and institutions, which are all, in part, the residues of war?

Discussion of the following figure—James Acord, a sculptor of nuclear waste and, by association, the nuclear industry—is intended to show a second example of transnatural aesthetic possibility. In this case, the artist seeks to reenchant transuranic waste directly through art making and, in doing so, foster a material ethics that suspends moralism and judgment to reorganize the nuclear complex and knowledge of nuclear processes and materials. In contrast to Nuclia Waste's highly visible and entertainment-oriented performances of mutant drag and playful toxic tourism, this figure—focused and serious, yet also humorous—shows how counterspectacles do not need to perform predominantly in a visual register to have an impact. In fact, this sculptor has rarely exhibited. Publicity has mainly taken place through presentations in the art world and nuclear engineering circles, and through several biographical treatments, photographic portraiture and documentation, and even a fictional representation.[109] It is at times difficult to assess fact from fiction with respect to this figure; his life could be interpreted as a kind of performance art that puts into practice ideas that this chapter has associated with transnatural ethics and aesthetics.[110] My narrative largely draws on and retells the experiences he shared in an art lecture as well as several secondary accounts. The discussion lends additional materialist support for transnatural aesthetics but also maintains a "tall tale" quality. I will focus on his sculptural mission of transmutation and intersections with the Department of Energy (DOE). Through what I refer to as his "transmutation follies," I read Acord's aesthetic political project as a kind of "queering" of the nuclear weapons complex and the administration of nuclear waste.[111] Following this narrative analysis of his performance art, the chapter explores Acord's transmutation sculpture project as the basis for a *materials-oriented* transnatural aesthetics. The chapter ends with a triptych of three manifestations of his sculptural work. The triptych serves as a reenchantment story of nuclear waste, one that figuratively reveals and rearranges the moralism, economy, and containing of waste that administer fundamental aspects of social life.

The sculptural projects discover—in their very making—the ambiguities of proliferation and containment, and the contradictions and contingencies of nuclear regulations and institutions. While the figure Nuclia Waste opens up political possibilities via a counterspectacle of embodied porosity, mutation, and nuclear retroactivity, this sculptor's efforts seek to generate access to nuclear materials and processes and, in so doing, demonstrate an alternative administration and "use" of nuclear waste as "artful endurance." Although his work is controversial, brief exploration and discussion here allows me to expand on the potential material ethics of the transnatural and the political possibility, however ridiculous, of converting the nuclear complex and nuclear technologies into something else—not by repudiating or banishing them from memory, but through "sculpting" relations with nuclear matters differently.

Transnatural Act 2: James Acord

James Acord—"nuclear sculptor"—was an artist with few sculptures but a life story that leaves the impression of performance art.[112] *He created events that probed the history of nuclear engineering and the material aesthetic affinities between sculpture and nuclear processes. He sought to incorporate radioactive materials directly into his work, and endeavored to deploy nuclear technologies in the service of his art making. The latter was a feat that required no less than the attempted annexing of a nuclear reactor and nuclear engineering to his vision of converting highly radioactive waste into inert metal for use in granite sculpture. This herculean task led him to give numerous presentations to the nuclear industry, seek an artist-in-residency at a DOE nuclear compound, and acquire, astonishingly for a private individual, a license to handle radioactive materials. Acord, described as an artist with a radioactive imagination, asked questions about the long-term storage of waste and, in a cat-and-mouse game with nuclear regulatory authorities, shed light on the bureaucracy, secrecy, and security still cloaking the nuclear industry and the huge social disparities and knowledge gaps between art and nuclear science.*[113] *With few victories and many unrealized plans, Acord lived out the principle that the population should be able to access nuclear technology and nuclear waste. This made the nuclear industry as well as many within the arts and environmentalist communities uncomfortable with his efforts, revealing social investments in particular forms of struggle and unresolved questions about the nuclear age that might not have been visible otherwise.*[114] *Drawing together canonized art in uncanny collaboration with reactor designs and nuclear waste storage, Acord proliferated the conviction that art making could wrest power from the closed club of nuclear science by inspiring engineers and validating material knowledge, increase openness and circulate facts about nuclear processes, and, finally, through inhabiting inhospitable institutions, change them for the better.*

Acord, as he explained it, took a traditional approach to sculpture: metal and stone. Although he worked with a variety of materials throughout his life, among them salt, gold, silver, and steel, he was predominantly interested in working with granite, because of its durability, a trait necessary for art that endeavored to last within contaminated environments. He learned stone carving in Barre, Vermont, home to some of the finest granite mines in the United States and also to a community of stone carvers. Barre granite, like all granite, contained a significant amount of uranium. Acord became interested in nuclear issues, both in response to nuclear events, such as the Three Mile Island nuclear accident,[115] and in terms of the way that uranium formed an integral part of the stone he was learning to carve and made granite a potential source of uranium to be quarried for U.S. Cold War purposes.[116] Acord aimed to successfully blend metal and stone in a sculptural work and, to do so, studied uranium (and its radioactivity) as the best metal to use with granite. As Acord experimented with ways to contain uranium inside a granite sculpture, he discovered nuclear waste and containment strategies. At the time, internment of nuclear waste was performed by solid-granite batholiths, the favored material containment technology for the storage of unwanted nuclear materials. Acord began to channel his skills into the artistic pursuit of nuclear waste containers as fine-art sculptures; the core idea was that waste storage should not be done by science alone, and that the fine art of sculpture would make a valuable contribution to society's comprehension and handling of nuclear matters.

Acord's interest in the materiality of both container and waste, through his craft of granite and uranium, meant that he suspended judgment and the logic of accusation in favor of sculpture and the

James Acord—nuclear sculptor. Photograph copyright Arthur S. Aubry.

skilled trades, such as metallurgy. He stated publicly that, as a sculptor, he was not for or against technology in initial orientation: "It's [nuclear technology] an advanced technology that exists, and sculpture is an art that uses advanced technology. So I felt that it was logical and inevitable that sculpture should address nuclear issues."[117] Also seriously interested in alchemy, Acord considered nuclear science not inherently evil but expressive of a long-standing human desire to transmute one element into another. Accordingly, the real secret of the nuclear age was that nuclear physics had made the transmutation of elements possible. Thus, Acord embarked on a quixotic quest to transmute radioactive transuranic by-products from nuclear weapons production into stable, harmless, and eternal works of art.[118] In particular, he wanted to transmute the long-lived radioactive waste element technetium (atomic number 43)—a silvery-gray radioactive metal that, as nuclear isomer technetium 99, has a half-life of 211,000 years—into ruthenium (atomic number 44), a nonradioactive member of the platinum family. Technetium was an element discovered by nuclear processes: only minute amounts naturally occur as a spontaneous fission product in uranium ore by neutron capture in molybdenum ores; technetium exists in quantity in spent reactor fuel from uranium 235 fission.[119] Nuclear scientists corroborated Acord's principles and methods for his transmutation project, intensifying his resolve to incorporate live and transmuted nuclear materials—later weapons-grade plutonium—into his sculptures.

A passage from an extensive and well-known biographical piece on Acord written by Philip Schuyler for the *New Yorker* captures the material ethics underlying such a project:

> Acord recognized that the idea of transmutation, central to alchemy, was equally applicable to radioactivity and also to sculpture, both of which involve the mysterious transformation of elemental materials. In addition, alchemy provided a moral underpinning for all of his work. "The base man who desired only for wealth would always fail," he [Acord] said. "The higher adepts, the true alchemists, knew that what they were really transforming was themselves."[120]

When asked in interviews how he would reconcile sculpture, a primarily visual art, with the materials he wanted to use—namely, that radioactivity could not be seen with the naked eye—Acord explained that transmutation of one elemental substance to another using the nuclear process of neutron capture *was sculpture*.[121] According to Acord, nuclear science and sculpture both, in the most basic sense, had a material sensibility and were practices interested in the ordered arrangements of materials.[122] And while many nuclear scientists responded to Acord's work with doubt about its impact on a general public that would not be able to understand or see Acord's use of 316 low-swell steel, let alone the process of transmutation, Acord reasoned that his very doing of the work would make a difference. He was particularly keen to use an actual nuclear reactor and the knowledge gained from nuclear science to convert radioactive wastes into something benign through his transmutation project.

Acord moved to Richland, Washington, to obtain access to his ideal sculpting materials and tools, which were in ample supply at the Hanford Site.[123] Hanford offered an extensive complex of nuclear reactors as well as an unfathomable amount of nuclear waste. Established in the early 1940s as part of the Manhattan Project in the south-central section of the state near the Columbia River, Hanford manufactured the plutonium used in the first nuclear bomb test at the Trinity Site in New Mexico and in the bomb—referred to as "Fat Man"—detonated over Nagasaki, Japan, in 1945.[124] During the Cold War, site operations were expanded to include approximately nine nuclear reactors and several large plutonium-processing complexes, which produced the plutonium used in most of the U.S. nuclear arsenal.[125] Although Hanford weapons production reactors have been decommissioned, decades-long manufacturing has left tens of millions in gallons of high-level radioactive tank waste, as well as solid radioactive waste, contaminated groundwater, and a significant amount of low-level waste.[126] Currently, Hanford is considered the most contaminated nuclear site in the United States; essentially, the whole area is a de facto waste site. The vastness and remoteness of the area, originally constructed through federal land withdrawal by executive order, now covers approximately 575 square miles.[127] This remote landscape is one of the most federally managed of the United States, and is mostly closed to the public.[128]

When Acord landed there in the early 1990s, security clearances were still the norm. Acord attempted to breach this secrecy and the closed club of nuclear scientists and industry, embarking on what in retrospect might be conceptualized as a kind of research-based process art or performance art that at times bordered on slapstick. Although this doesn't sound like an especially effective approach, particularly in that Acord was ultimately unable to perform transmutation at Hanford, he did manage to lend visibility to the compound-like nuclear facilities and the area's contamination, even frequently serving as unofficial news media liaison to Hanford. He also successfully reached out to both scientists and artists on nuclear issues, organized conferences, gave classes on the importance of art and the ways art and science were connected, and became involved in several local projects, including a time capsule that would communicate the volatility of nuclear issues and his vision of constructing a modern Stonehenge at Hanford that would endure to mark the contaminated land. His efforts also demonstrated that one human lifetime was too short to outlast the regulations and bureaucracy that grant access to nuclear technologies and nuclear waste.

In his quest to transmute technetium into ruthenium for sculpture, Acord sought to enlist Hanford's Fast Flux Test Facility (FFTF)—an experimental reactor regarded as one of the most advanced reactors/machines on the planet and also highly controversial. FFTF was capable of creating its own fuel, manufacturing medical isotopes, and transmuting isotopes from one element into isotopes of another. Acord considered FFTF to be the ultimate tool of alchemy. Furthermore, Acord had been highly impressed by the craftsmanship that had gone into the making of FFTF and reportedly compared the attention to detail exhibited in the very display structures of the facility's visitor center to be on par with that of medieval cathedrals.[129] Although his attempts to formally establish an artist-in-residency with the DOE failed, he put a proposal together that requested access to FFTF and DOE collaboration, because the DOE was attempting to diversify the way FFTF was used. Acord's proposal is said to have been accepted for an impossible $74 million, but later he was told that it would take an act of Congress for him to move forward with his plans.[130]

Persistent in his pursuit, Acord prompted a series of encounters and intimate pitfalls with nuclear regulatory authorities—what I refer to as "transmutation follies." A German company offered him a ton of depleted uranium in the form of breeder reactor assemblies, which he envisioned using in a colossal monument to the nuclear age, one that would simultaneously make a statement about complex technology and its impacts, and serve as both educational resource and warning marker. However, in order to receive the waste offering, he had to further master nuclear bureaucracy and technology: working with radioactive/nuclear materials required several layers of protection, complex training, and a lengthy administrative process to establish a license to work with radioactive materials. After passing classes in nuclear materials and engineering at the local community college, as well as figuring out all of the necessary paperwork, Acord was eventually granted a radioactive materials handling license—no small triumph for one lone individual—which he celebrated by tattooing the number on the back of his neck. Although the license gave Acord the right to own and inspect the uranium, it did not allow him to possess it; it had to be stored at Hanford with one of the site's licensees.[131] Furthermore, he was now responsible for the material, which meant keeping up payments on the license, acting as his own radiation safety officer, and other time-consuming activities. Folded into these events was a residency as a community artist with the physics department at Imperial College in London and lecturer work under the aegis of Arts Catalyst, an organization dedicated to melding art and science. Acord expected to gain access to a small-scale nuclear reactor, but a series of highly publicized nuclear accidents in Europe and Japan made the collaboration too much of a media risk for those managing the reactor and Acord returned to the United States. Eventually, he was unable to keep up payments on his license, and his uranium stored at Hanford was confiscated.

After spending approximately a decade legitimately and legally working with nuclear materials with a license and in communication with the DOE and the nuclear industry, Acord had little work to show. Although he was unable to produce his transmutation sculptures, his stalwart efforts resonate as both an unbelievable confrontation with nuclear authorities to gain access to a nuclear reactor and an equally ridiculous grassroots effort to make nuclear science a more ordinary form of knowledge and nuclear waste more accessible to the public. In contrast to the overt camp humor of Nuclia Waste and the possibilities of ludicrous parodic inhabitances of nuclear waste to dramatize the body and ecology, Acord's transmutation "follies"—in the sense of repeated failures and mishaps—were not purposely organized as humorous, visual, queer-oriented, or formal performances, and yet they operated in a slapstick mode with the nuclear industry, with queer effects: "every attempt . . . to comprehend the 'natural order' or 'logic' of the encompassing system . . . culminates in a kind of collapse."[132] Through his failures, Acord irreverently shook up what was understood to make sense; by moving to Richland and seeking so many venues to convert nuclear scientists to an appreciation of the arts, rally support for his transmutation dream, and transform—queer—set ways of thinking by showing the connections between nuclear process and sculpture, Acord endeavored to disrupt the "business as usual" of the nuclear complex and community. If nuclear waste is matter considered to be out of place, then Acord was even more so, and the humor of this impossible quest and its many failures can be read as an effective political intervention, albeit not the kind one might expect.

Acord was deeply serious about transmutation, and, rather than overtly transgressing regulations and flouting authorities, this part of his story suggests that he followed the rules in a literal way to abide by the law. Even "playing by the rules" had the effect of compelling negative reactions to his very presence, requests, and earnestness. The catchphrase "transmutation follies," then, also refers to the institutional responses to Acord, which were often more ridiculous than his initial provocations, thus pointing out the power relations and lunacy at the heart of institutions and laws even as he failed at his ostensible purpose.[133] In this serious form of play with nuclear rules and agencies, he reckoned with various investments in nuclear waste: the regulatory and geopolitical schemes that control waste materials and objectify waste, the legacy of secrecy and misinformation of the Cold War that lives on in the control of nuclear waste, and the social reliance on affects such as horror, abjection, and blame in response to waste. Recounting Acord's story also throws into question the boundaries between functional and gratuitous use of nuclear materials. In an interview, Acord remarked, "There's a sense in the scientific and engineering community that artistic use of the nuclear process is frivolous. But this makes me more determined."[134] This judgment, that nuclear waste should not be put to frivolous use, leads to the stunning realization that nuclear waste evidently has a range of potential uses, some of which can evidently be considered "good" and others simply not allowed. The fact that this decision can even be administrated shows how productive nuclear waste is with respect to sovereign state power; its use and placement are strictly controlled, and this has many social effects. From a perspective attuned to the connections between waste control and social control, nuclear waste exists somewhere ambiguously between governing the population through "sovereign immunity" (through secrecy, arbitrary decisions, etc.) and more biopolitical forms of administration concerned with safety, welfare, productivity, and the enhancement of life.

A second, no less important, understanding of transmutation in Acord's story is as a resolutely *materials-directed* transnatural aesthetics. Such a tactical material ethics insists that waste should not be considered mere spectral haunting or absent present; it looks for the literal materials categorized as waste and confronts the ways that the composition and decomposition of these materials and their physical organization are part of the larger governing relations of society—their administration informs institutions and even arranges everyday life. Following Acord's lead, this material ethics would also experiment with a stronger material orientation and response to nuclear waste, adding "nuclear waste sculpture" to the queer ecological program and its repertoire of critical possibilities. Just as nature or the body are not passive, neither is matter—particularly nuclear wastes. However, Acord's project of transmutation does not enlist any simple ethics of affirmation of nuclear waste, nor does it stop at acknowledging the ontological "liveliness" of such materials; these two responses are not on par with the irreverence of Acord's relations with the nuclear industry or popular culture. Instead, I read his material aesthetic labor as an attempt to rearrange dominant nuclear affect, such as the spectacular threat of waste or the spectacle of technology/the machine, through a certain *sensibility to materials and through craftsmanship;* his sculptural work inherently called on the nonhuman agencies of stone and metal and a human way of "knowing materials" through the sensual/empirical arts and trades of stone carving and metallurgy.[135] What this offers transnatural aesthetics is twofold: a rejection of the division of form and matter (i.e., neither spirit nor language cuts out forms from a passive inert world) in favor of a relational ethics/aesthetics of matter that attends to material complexity and embodied ways of knowing materials, and a critique of war and the nuclear industry not from a position of distance or through traditional written criticism but by irreverently pursuing an "extradisciplinary" genealogy of material arts and sciences—including nuclear science.[136]

There are numerous and notable hazards here, already aired in some of the public's reaction to Acord's work. Such a reading could lend undeserved credibility to the nuclear industry. It also might be perceived as celebrating what Haraway refers to as "productionism and its corollary, humanism . . . the story line that 'man makes everything, including himself, out of the world that can only be resource and potency to his project and active agency.' This productionism is about man the tool-maker and -user, whose highest technical production is himself."[137] The nuclear facilities at Hanford and Rocky Flats, as featured in chapter 2, are not simply residues of an earlier era but ongoing manifestations of objectifying paradigms—namely, a deep belief in the separation of humans and the material world.[138] This is resoundingly evidenced by the ongoing accumulation of nuclear waste and its constant denial and distancing from the everyday concerns of most of the U.S. population. And yet Acord's transmutation vision persistently sought to wrest crafts, tools, and material knowledges away from the deterministic views of arrogant productionism, situating them within preindustrial alchemy or "applied sciences" such as metallurgy.[139] He provocatively placed the nuclear age within a materials-oriented genealogy that, while including the military–industrial complex, did not affirm war as the necessary history of the present.

Acord had admired the wonderful metalwork that was being done as a way to support the nuclear age: "The beautiful metallic elements used in fuel assemblies appealed to me as a sculptor."[140] Acord saw that the best technology and use of metallurgy, including titanium and zirconium, were monopolized by the nuclear industry. This indicated that among the many negative impacts of the nuclear age, it also harbored a legacy of material crafts. If the nuclear age had become residual, still materially all around us in the form of nuclear waste, there were also more potentially positive residues. This utopian hope was crystallized in Acord's admiration for the construction materials and work that went into the FFTF facility itself and his determination to use FFTF to transmute long-lived nuclear waste into material for his sculptures. Reviewing Acord's actions raises the pragmatic question: What are we going to do with nuclear waste and nuclear technologies? Going further, could the actual FFTF facility ever be converted, through collaborative creative research practices, into something other than what some have called "rotten perfection"?[141] Acord dared to show one potential way: forging a material arts and sciences—with an ethical basis in sensuous skillful knowledge and ways of recognizing and working with nonhuman material agencies—from within the nuclear industry, using nuclear facilities, technologies, and processes.

The resistance he experienced went beyond the nuclear industry and the Cold War. Humanist institutions and the historical organization of disciplines and knowledge have profoundly disregarded materials and material complexity. The development of early science, split off from everyday "applied science," the trades, and home/table-top pursuits of alchemy, followed a somewhat dematerializing or antimaterial trajectory; philosophy, too, historically marginalized materiality:

> Through most of history, matter has been a concern of metaphysics more than physics, and materials of neither. Classical physics at its best turned matter only into mass, while chemistry discovered the atom and lost interest in properties . . . [In both metaphysical speculation and scientific research] sensitivity to the wonderful diversity of real materials was lost, at first because philosophical thought despised the senses, later because . . . the new science could only deal with one thing at a time. It was atomistic, or at least simplistic in its very essence.[142]

Consideration of material agencies was also stifled in philosophy and the humanities by investments in an organic/inorganic divide, as well as human-centered language/matter and culture/nature divisions: "As the behavior of metals and other mineral materials became routine, and hence, unremarkable, philosophical attention became redirected to the more interesting behavior of living creatures, as in early twenty-century *[sic]* forms of vitalism, and later on, to the behavior of symbols, discourses, and texts, in which any consideration of material or energetic factors was completely lost."[143] Regardless of its accomplishments, industrial production severely dissipated material craft and knowledge; mass production and economies of scale resulted in countless losses, among them the dehumanizing and deskilling of labor, a radical reduction in the complexity and diversity of materials (such as the homogenizing of metallic behavior), and the spectacular separation and fetishizing of the machine, meaning that the transfer of skills from human to machine no longer produced or responded to sensuous knowledge between human and nonhuman agencies (i.e., wood, stone, metal) in the same way.

Metallurgy and other skilled trades were recognized, valued, and drawn into U.S. bomb production within the national "Defense Material Complex" (another DOE name for the nuclear complex), even though, as chapter 3 made clear, many of these laborers were later found to be repeatedly misinformed or deceived, not adequately protected in their work environments, and rejected when compensation for illness and death was sought. World War II and the Cold War set a firm agenda on investigating material complexity. The emergence of the transdiciplinary and dynamic field of material science today can be directly tied to the nuclear age:

> The contemporary science of materials is an offspring of World War II and the Manhattan Project. While prior to the war the field constituted a collage of minor sciences, engineers and metallurgists who had participated in wartime government projects finally unified and gave this discipline the prestige that it deserved. The study of material complexity is now the rule, and a new awareness of the self-organizing capacities of matter is beginning to emerge in this field.[144]

Acord affirmed this aspect of the nuclear era, but tactically, within an abiding interest in material sensibility that confronted nuclear institutions, nuclear waste regulators, and dominant nuclear affects with a captivatingly different understanding of those nuclear tools, technologies, and social investments. In other words, this was a tactical genealogy of materialism/material knowledge—made manifest through the embodied methodology of sculpture—that refused human and nature, organic and inorganic, divisions. In collaboration with nonhuman material agencies and with alchemy as ethical "touchstone," Acord labored to transmute military Cold War–era institutions and humans as much as the materials themselves.

Drawing on a format that channels Acord's interest in medieval art, the following triptych plays with nuclear matters—the materials and issues—and extends a reenchantment story of nuclear waste. The three panels individually draw out some of the intertwined ways that waste and social life are administered. The developmental arch of the triptych explores how Acord's sculptural counterspectacles of nuclear economy, moralism, and containment, from home to nation, substantiate the ambiguities of proliferation and containment. The triptych exhibits the different ways Acord sought to *proliferate access to nuclear processes and transmute military stockpiling, national/natural purity, nuclear technologies, and waste burial into more artful forms of endurance.*

This is a transnatural art of existence that plays with negative material legacies of war via sculpture to create other material and social arrangements equally persistent.

Left—Artheads, not Warheads

In a project conceived but never executed, Acord proposed replacing the nuclear warheads in U.S. arsenals with granite sculptures called "artheads." Transforming the stockpile would proceed in such a way that artheads would cost as much as the warheads they replaced. Through shrewd calculations of cost overruns and kickbacks, the military–industrial complex—the economic base of the nation—would be allowed to continue, but with significantly less-toxic materials in circulation and art as its product. The scheme sarcastically concedes that a permanent war economy might be the condition for life itself in the United States; it implicitly acknowledges that understanding war not as a disruption of order but as the condition upon which order is constructed *makes society intelligible.*[145] However, it refuses the warhead end product and, thus, the status quo by, ironically, preserving the productive economic capacities of the military–industrial complex in the service of producing a nondestructive end product. By punning "warhead" and "arthead," Acord even allows the same phallic sculptural form to persist, as well as the hefty costs, irrational management, and errors that are seen as fundamental to the military–industrial complex. The scheme's purpose is simply to convert military stockpiling and the whole underlying military basis of the economy to more constructive art-making purposes.

The contemporary "greening of nuclear weapons" provides interesting comparison. Under the aegis of security, stewardship, and even sustainability, nuclear weapons in the national stockpile are being refurbished, swapping some warhead materials, which are extremely toxic to handle, with more "environmental-friendly" ingredients.[146] This remediation of weapons responds to contemporary environmental concerns in order to maintain business as usual; it preserves the status quo by tweaking some of the material elements in order to make weapons somehow more appealing, spectacularly deferring the destruction and toxicity of the weapons from producer to target. Toxic waste has in many U.S. popular venues replaced nuclear weapons as a live threat. The greening of nuclear weapons shows how the fallout of World War II and the Cold War—in the form of toxic wastes and landscapes—now serves as the basis for the persistence of economic and political modes of organization via a "green war" model. This essentially proliferates the stockpiling of weapons through the greening of materials, whereas Acord proposed to alter not only the materials but the fetishized end product's purpose—from mass destruction and ruinous aftermath to massive art exposure.

Center—Transuranic Kitchen

After Acord was denied access to a nuclear reactor and his uranium was confiscated, he did not give up on his vision of making a sculpture that embodied the concept of transmutation. He simply had to resort to his own means: Fiestaware pottery and the construction of his own kitchen table-top nuclear reactor.[147] The technical preparations for the project involved stockpiling an excessive amount of orange/red-glazed tableware known as Fiestaware purchased from thrift stores. This pottery line, which started in the 1930s, utilized uranium in its glaze to give off a bright (some say glowing) orange glaze, before uranium was requisitioned by the government and no longer permitted on the market for decorative use.[148] Acord crushed up pottery and then separated the uranium from the clay using a technique similar to panning for gold: washing off the crushed clay and allowing the uranium ore to sink and stick to the tray.[149] He would also need an alpha source in order to generate the neutrons required to transmute uranium into plutonium in a miniscule amount. For this, he acquired everyday home-use smoke detectors, which each contain a radioactive dot of americium. Americium is a transuranic material that is produced in nuclear reactors as an alpha source. The rules requiring a license for the handling, producing, and selling of anything with radioactive materials were changed to allow for these everyday smoke detectors. Finally, Acord purchased emeralds—a crystal gem made largely of beryllium that, when hit by an alpha particle, would give off a neutron; he also needed beeswax to slow down the neutron.

In contradistinction to the assumed complexity of nuclear processes, the kitchen table-top nuclear reactor operated according to the following basic formula: the smoke-detector-acquired dot of americium strikes the emerald (beryllium), which gives off a neutron that is then slowed down through the beeswax to "breed" plutonium 239 from the Fiestaware-acquired uranium 238 ore. Because the creation or ownership of plutonium is strictly forbidden in the United States, Acord publicly announced that he had successfully performed the transmutation during a talk at a foreign arts conference in 2010.[150] When asked whether such work would be shown, he wryly replied that his sculpture would be confiscated; furthermore, any sculptures of his containing radioactive materials would never make it past customs for international art exhibition, and anyone interested in purchasing one of his sculptures would have to obtain a license to handle radioactive materials, even though his granite sculptures were designed to safely hold nuclear waste for thousands of years. His art conflicted with transuranic material controls, national and international nonproliferation rules, and historical efforts to make nuclear-free domestic zones. His inability to ship and show his work made tangible how the art world is subject to nuclear proliferation regulations and radioactive materials licensing, effectively implicating the artist and the international art world of exhibition and collecting in the national and international regulatory politics of nuclear power.

Before transmutation was his sculptural goal, Acord had simply extracted the uranium from the pottery and incorporated it directly inside granite sculpture, using a stainless steel vessel. In the early stages, after taking some Fiestaware glaze into a local office of Washington State's radiation control for testing, he was told that what he was doing was classified as uranium "mining and milling," an activity that necessitated a license and a particular setup. There were no clear guidelines for dealing with artists trying to manufacture their own nuclear waste, and his excessive Fiestaware collection was confiscated. He was restricted to possessing no more than fifteen pounds of the pottery at a time. However, the federal Nuclear Regulatory Commission later ruled that the fifteen-pound limit applied to source material—the uranium mineral—not the dinnerware itself. He was allowed to complete his nuclear sculpture as long as it contained no more than 20 percent of the glaze. Acord's Fiestaware fiasco reveals the ambiguous "value" of uranium's radioactivity, utilized for purposes ranging from decorative arts to bomb parts. Furthermore, it exposes the military natures of everyday home-use products, that is, the way military demands for and production of certain materials have organized, impacted, and manufactured the very materials constituting the home and everyday life.

In a very surreal way, it also elicits reflection on lingering social attachments to "the nuclear gaze" of the Cold War: "The atomic age mandated a gaze—a nuclear gaze—that defined the difference between dangerous and nondangerous activity, universal and specific jurisdiction, containment and proliferation."[151] A way of seeing that strictly emphasized binaries, the nuclear gaze worked to contain surpluses of knowledge production, artistic expression, atomic secrets, nuclear energy, sexual license and gender roles, and opposition to American liberal capitalist paradigms. Here this social organization of life is partly thrown into relief and challenged by Acord's two nuclear home projects. His excessive collection of radioactive orange dinnerware and his homemade table-top transuranic reactor reveal how the classification, placement, and containment of nuclear materials and waste are arbitrary, uncertain, and therefore potentially open to reorganization. The Fiestaware collection and table-top reactor negate the taken-for-granted arrangements of materials and spaces, prompting hilarious confusion between kitchen and mining operations, home and nuclear reactor, and in doing so foreground the rules to contest their necessity, inevitability, and even rationality. More bizarre, while seemingly stringent controls over nuclear materials underpin everything, Acord's home reactor shows how exceptions have been made for the military-based transuranic materials of domestic products, such as smoke detectors.

Right—Monstrance Meditations

Acord designed sculptures for the purposes of marking sites of radioactive waste, nuclear graveyards, and contaminated lands in ways that he felt would communicate nonverbally, through the affects generated by sculptural forms, textures, and placement. He also envisioned radioactive sculptures that would be buried beneath the earth and viewed through radiation-shielded telescopes on the earth's surface. He is reported to have spent nearly ten years working on a single sculpture, *Monstrance for a Grey Horse*.[152] It is carved from a single block of granite and depicts a skull sitting atop a five-foot trapezoidal column, an archetypal totem of death. A triumph of craft, displaying a rich vocabulary of stonework techniques all done by hand and eye, the monolith is itself a waste depository; embedded in it is a canister of "live" nuclear material, which Acord generated from his Fiestaware-dish home "mining" operation. Incorporating Catholic church imagery into his nuclear sculpture ideas, nuclear materials were treated as "sacred substance," referencing the way the Eucharist is contained in the traditional Catholic monstrance during Easter. A series of reliquaries, too, were modeled after the medieval reliquaries used to preserve sacred Christian objects and relics, such as a shard of the true cross or a saint's lock of hair. Acord's ornamental containers, however, housed symbolic items from the nuclear age, including mathematical formulas and scientific information. In an exhibition titled "Atomic," organized by Arts Catalyst at the Yard Gallery in Wollaton Park, Nottingham, in 1999, Acord showed three wooden reliquaries shaped like church windows and fastened to the wall.[153] While these boxes did not contain transmuted materials or nuclear waste, as Acord had intended, they still played with the idea of the sacred as a way to frame and reflect on waste as the living relics of the nuclear age.

Relics of the Catholic church were historically considered living matter; they provided points of contact between mundane existence and the divine world, life and death. Drawing on the association, Acord's reliquaries encourage serious contemplation of the materiality of waste in the context of a playful irreverence toward nuclear institutions. He compared the remote special communities, secrecy, and nuclear bureaucracy to the medieval Catholic church: "The Department of Energy is like the Catholic Church in the 12th century. It's a fully mature bureaucracy. It's committed to the dogma that waste should be buried in the ground. It has thirty years of studies worth billions of dollars saying that waste burial is good. Getting it to backtrack is hard."[154] Reconceiving burial as "devout belief" (backed by significant investments) estranges this taken-for-granted way of handling waste. Such medieval associations with nuclear waste also remove this material (symbolically, at least) from the current business opportunities related to disposal, which are rewriting Cold War containment as a lucrative way to extract further value from the contaminated residues of war.

Monstrance for a Grey Horse, by James Acord.
Photograph courtesy of Lucas Adams, Southwestern University.

Instead of recuperating cultural artifacts as signs of American triumphalism, Acord's "nuclear medieval" reliquaries and monstrances encourage meditation on the nation not as a pristine landscape or therapeutic story of innocence but as repository of the abjected remains of war. They also anticipate an "archaeology of war" that would dig into the military nature of different material histories, and potentially excavate the complicity of burial strategies with concealment of exposures.[155] Perhaps the sculpture *Monstrance for a Grey Horse* will, at the very least, transform studies of nuclear matters into more edifying experiences at Southwestern University, where the sculpture, previously purchased by an alumnus, was donated and installed in the summer of 2011. Stationed near a library entrance, the both beautiful and monstrous monolith was referred to in university press materials as not only public art but a significant teaching tool.[156]

James Acord's nuclear waste reliquaries, kitchen table-top reactor, artheads, and transmutation follies serve as counterspectacles to the pervasive consumer culture of exposure. These projects make manifest the material linkages between war and everyday life and tactically respond by materially arranging for other possibilities. They play with the residues of the Cold War—especially nuclear waste—through sculpture, as part of a new "transmuted" material arts and sciences, an effort that would take as its mandate the discovery and proliferation of ways to convert the residues of war into forms of artful endurance: a way of persisting through making knowable/visible/relatable the interrelations between bodies, materials, and aesthetic orders. Whereas the figure of Nuclia Waste reorients nuclear waste's excess and mutation—through the irreverence of camp—toward new responses to military-to-wildlife conversions and new coalitions of depleted citizens, the nuclear waste sculptor, as dramatized by the life and times of James Acord, demonstrates the administration of our social relations with nuclear waste and nuclear technologies—and the ambiguities, material complexities, and residual matters that the supposedly postnuclear era would negate even as exposures continue. How do we learn to live otherwise with waste and endure the residues of war, when their organization, governing, and proliferation underwrite our daily lives? Both figures enact a certain subjectivity in waste that tactically affirms other ways of living than denial, repudiation, austerity, depletion, or doing nothing. They both offer a transnatural arts of existence that ethically demands *persistent irreverence.*

CONCLUSION / HOT SPOTTING

The western United States has served as a "proving ground" for testing the nation's strategies and technologies.[1] Throughout the U.S. West, state of emergency powers have taken spatial form through the creation of countless arsenals and training grounds, resource extraction projects, atomic bomb facilities and hundreds of nuclear detonations, permanent waste disposal sites, and internment camps for populations cast out from the nation.[2] While the beauty and expansiveness of the western landscape represents American openness and symbolic liberty, its "remoteness" has in part been produced through legally sanctioned constraints on political freedom and public access, often by federal land withdrawals for war purposes. The intensive militarization of the West has meant that it is always at war, linked irrevocably to American geopolitics; it has served as the material grounds for military–industrial power and a permanent war economy, and as a social-cultural model of emergency preparedness. The material effects of war within the nation are also intensely concentrated in the West's poisoned populations, fallout, and contaminated terrain. While discussions of war continue to be externalized by American exceptionalism and national imaginations of the West, the reformulation of warfare under post–Cold War conditions and the restructuring of the United States' twentieth-century military–industrial foundations have further expanded the "collateral casualties" of state power to include massive former workforces, now economically redundant and sick.[3]

The military–industrial infrastructures, colonial projects of resource extraction, and logic of national sacrifice that have settled in the U.S. West show that sovereign power has waged war not only externally but internally to the nation. The historical geography of the U.S. West presents war as an internal institution of the nation, not only the raw event of battle; death and violence are part of domestic affairs, even fundamental to the regulation and administration of the population.[4] Drawing on the analytic perspective offered by Michel Foucault, the "biopolitics of war" maintains that *life is managed through war; war is paradoxically the condition for life of the nation itself.*[5] Being capable of killing is sustained as a principle integral to state strategies of promoting life; war is not waged on behalf of a sovereign who must be defended but "on behalf of the existence of everyone; . . . this is not because of a recent return of the ancient right to kill; it is because power is situated and exercised at the level of life, the species, the race, and the large-scale phenomena of population."[6] Such power relations are key to understanding how sites of exclusion in the U.S. West—for instance, secret weapons facilities, internment camps, military bases, and toxic mines—are accepted by the vast majority of the population as part of the "natural environment" of the nation or the means of protection and security, and, further, how innumerable deaths go unnoticed or are consumed as a kind of "inevitable violence," part of the increasing risks to exposure to toxic chemicals and radiation required by American power and modernity. Threats persist as opportunities to promise protection, and such threats are in fact the residues of war, such as lingering contamination or blank spots of information resulting from state secrecy. The threat of violence and exposure to death is "written into the regulation of life, normalizing the exceptional conditions of war as part of everyday affairs."[7]

Foucault traces allowable, rationalized, even necessitated forms of death—what he calls practices of "letting die" that inhere within institutions and laws of liberal society, which seek to maximize the relationship between wealth, territory, population, and resources with a minimum of force or direct state intervention.[8] By this account, law is not the pacification of war; rather, "war is the motor behind institutions and order. In the smallest of its cogs, peace is waging a secret war."[9] Safety and general welfare, domination and oppression, are produced simultaneously by the organization and deployment of political and economic capacities; even ordinary material existence is affected by and saturated with powerful logics that direct ways of living.[10] Foucault contends, "Something fragile and superficial will be built on top of this web of bodies, accidents, and passions, this seething mass which is sometimes murky and sometimes bloody: a growing rationality. The rationality of calculations, strategies and ruses; the rationality of technical procedures that are used to perpetuate the victory, to silence, or so it would seem, the war, and to preserve or invert the relationship of force."[11] In short, *the nation's institutions, regulations, and administration of various logics, such as cost-benefit budgetary decisions or risk assessments, perpetuate permanent war, sustain threats, and rationalize death in order to "enhance society."*

This book has explored several sites that evidence new proving grounds of war in the United States. These emerging sites refer to the concessions that have been made to populations and lands impacted by military activity focused in the U.S. West. The new proving grounds include postmilitary and postnuclear nature refuges, nuclear compensation programs and occupational-illness epidemiology, "legacy management" and long-term stewardship, and other efforts to secure the health and safety of the population and environment. Acknowledgments of human harm and environmental devastation wrought by military activities are hard-won triumphs that have involved the efforts of many and should not go unnoticed. However, military-to-wildlife conversions are also utopian projects of the nation-state, which propagate the idea that such lands are somehow demilitarized and can be returned to the public for recreation and observation of nature. Concessions and liquidation, here, go hand in hand; compensation and recognition of environmental responsibility can be seen as strategic acts that maintain consent, annex counterhistories, shift capacities and research infrastructures to ensure the survival of war, make profits on wastes, and further abandon the collateral damage of war in order to support military realignment.

As declared elsewhere in this volume, the partial reconstitution and management of the military to pursue sustainability and environmental security is not merely "greenwashing." Criticizing "military natures" as green packaging and ideological mirage does not adequately capture what is going on; for example, postmilitary nature refuges are both natural *and still* industrial, technological, and "human." Furthermore, ideological greenwashing offers a limited understanding of truth as something that is hidden and that must be exposed; this ontology relies on binary modes of thinking and relegates matter/materialism to mere medium of a deceptive image. By contrast, biopolitical analysis investigates the material arrangements of truth as practices of governing; it is the administration of truth or reality that is the focus (e.g., the organization of secrecy that makes the secret possible). Applied to current military restructuring and concessions, the closure and decommissioning of landholdings of the Department of Defense (DOD) and Department of Energy (DOE) and their pursuit of environmental compliance, cleanup actions, and stewardship are responses to the threats generated by the military–industrial projects they sponsored. Such responses have shifted resources, remaindered land, and reduced responsibility under the auspices of ecological improvement, genetic preservation, or science. Efforts to create a more efficacious military apparatus and establish consistency between discourses of security/defense and good economic and environmental management do not indicate the erosion of war but, rather, the ushering in of new principles of biopolitical management, new forms of rationality, and denial under new names.[12] "Green" sustains war.

The new proving grounds of war have developed new capacities and forms of value—whether through local real-estate development, waste management opportunities, or new scientific methods that organize new industries—just as they coordinate, allow for, and justify ongoing environmental harm and death. Chapter 1 detailed how certain animal bodies, those of the American bald eagles, have become signs of survival and adaptation, rather than guides to military practices of environmental ruin. Chapter 1 ended by discussing the "deadstock" of animals hidden by the spectacle of nature and purity at the DOD's Rocky Mountain Arsenal, and the way the environmental harm to the area has been scientized/rationalized through the biomonitoring of particular animals as neutral collateral damage. The radioactive and chemical legacies of World War II and the Cold War have been largely treated by the DOE as a bureaucratic management problem. Chapter 2 explored how the DOE contained the cleanup of a former plutonium plant by predetermining the site's future use, and initiated the new function of legacy management to manage radioactive and contaminated land and ensure safety in perpetuity—an impossible mandate that "ends" with the organization disposing of itself through a spiral of efficiency. In the instance of nuclear worker compensation, the science of "dose reconstruction" was inaugurated to provide information for the epidemiological assessment of exposure to hazardous and radioactive materials at facilities integral to the U.S. nuclear complex. Chapter 3 excavated the way this compensation law operates to deny claims even as it was intended to celebrate Cold War–era labor. Further consideration of the qualification and channeling of claimants reveals the ways biopolitical administration of cancers and diseases that are the living traces of nationalist projects extend the United States' economic and political influence.[13] These are three examples not of greenwashing so much as the biopolitics of war and administration of life through "green" arrangements of salvation, efficiency, and abandonment.

As toxicity, wastes, and the material fallout of war persist, risk and other modes of calculative rationality provide techniques of governance through which such threats are managed. As the Rocky Flats example of chapter 2 illustrated, risk analysis was used to determine the future of the former nuclear facility. The "end state" of nature refuge mitigated uncertainty about the hazards and material-ecological transformations that were taking place by narrowing the cleanup options to specific short-term tasks, and by asserting that safety would be achieved through limited human contact and nature spectacle. The Rocky Flats case demonstrated that risk is a complex discourse that does not merely describe the world nor assess arbitrary chance; it produces knowledge about uncertainty and defines what can be achieved, what the future should be, and what is acceptable—for example, the setting of safe levels, standards, and statistics.[14] Even beyond DOE environmental remediation and military policy making, risk assessment has become a budgetary requirement for the enforcement of legislation related to health and safety. As a result, ethical considerations are subsumed under budgetary considerations, and economic scarcity is continually wheeled in to justify the orderings of reality through risk.[15] Risk calculations, furthermore, are employed to transform uncertainty into new baselines with their associated futures, and to regulate—to regularize the environment and moderate dissent. Risk is not an epiphenomenon of society; it is central to the continuous governmental management of uncertainty that now characterizes the biopolitics of war.

In the crisis-prone environment of the United States and "the on-all-the-time, everywhere-at-the-ready of military response" amplified since September 11, 2001, certain risks are increasingly and heavily militarized; a widespread condition of permanently "at war" potential becomes "as ubiquitously irruptible as the indiscriminate threats it seeks to counter."[16] Preemptive strikes and permanently critical conditions are not the only responses to the figure of a nonstandard environment possessed with a threatening autonomy of indiscriminate danger. Other responses have been detailed in this book. The handling of the numerous compensation claims of nuclear workers in chapter 3 presented another strategy: denial and obscurantism. Drawing on the legacy of secrecy, sovereign immunity, and misinformation characterizing the public's interface with America's atomic bomb project, truth productions about occupational exposure to hazardous materials can arrange for "not knowing" in order to *not be responsible* and to maintain the status quo. What this essentially entails is the purposeful production of ignorance and the maintenance of uncertainty (or, as chapter 3 argued, the adamancy of "certain uncertainty") in order to take no action, silence any claims "to know," and discipline subjects to document—confess—evidence otherwise. The rigorous pursuit of indeterminacy in the finding of links between radiation exposure and cancer continues to defer the obligation to do something for former nuclear workers.

The military-to-wildlife transformations of the Rocky Mountain Arsenal examined in chapter 1 and Rocky Flats in chapter 2 show a second way that present modes of doing nothing have been justified, even celebrated. Instrumentalizing the on-site appearances of the bald eagle, the conversion of the Arsenal, in particular, co-opted the concept of ecosystemic integrity to bury controversial history and lingering material dangers as things of the past. The decommissioned DOD site provides an early model of "nature spectacle" as a way to reduce responsibility for and abandon the material remains of World War II and the Cold War. Essentially, the ontology of spectacle—meaning, the separation of nature and human, human and waste—disposes of the site; the site is experienced through visual consumption of nature's salvation, limiting any understanding of the complex ecology of the industrial past or at present.[17] Binaristic thinking and the treating of human bodies as separate from the environment are part of the military legacy. Such spectacle produces its own remains: humans and nonhumans within the U.S. nuclear project that pay invisible "toxic debts" so that society and the military–industrial complex may live on. This abandonment—in the form of the collateral damage of war—is effectively obscured by the spectacle of nature. Compared to the documented wildlife at the Rocky Flats National Wildlife Refuge (the former plutonium production facility), ex–Rocky Flats workers who now seek compensation for occupational illnesses appear to be unadaptable subjects and subsequently an abnormal burden on society. At best, workers are made hypervisible by media reports that drum up sympathy; however, little is done for them beyond compassionate consumption of their imminent death.

As explored in chapter 4, the biophysical spectacle of the contained ecology—and purportedly contained waste—of the postmilitary and postnuclear nature refuge is an ongoing system of exposure to toxicity. Attempts by regulation to keep everything in place facilitate this; environmental harm is constricted to binary modes of thought rather than transgenerational, interecosystemic, and transcorporeal scenarios.[18] Spectacle, as understood here, arranges the consumption of exposure and, as a result, extends "slow death" and banal endemic forms of violence as the "background conditions" of everyday life.[19] Military restructuring and concessions made to the public for the casualties of war strategically utilize this ontology to abandon certain responsibilities for past injustices as well as literal material residues—for example, wastes—in plain sight. The state's investment in maintaining spectacular divisions and discrete categories orchestrates refusal of the remainders: the perforated land, depleting former workers, and subjugated knowledges. Spectacle also forestalls ethical consideration of the way the past persists in conditioning the present. This intensifies the tactical necessity of responding to state rationalities that have normalized "quiet deaths," cancers, and environmental exposures to toxicity as inevitable, even natural.[20] Resistant to accounts of causality and subjectivity, such deaths and exposures are sometimes administered as the threat to be guarded against but often strategically left as ambiguous, acceptable, and indistinguishable from the everyday negation and abandonment of the outmoded, redundant, and already vulnerable.

Contrary to the new proving grounds, concessions, and stated environmental aims of American military organizing, "The reverberations of past injustice are not calmed by moments of official contrition but persist in shaping the conditions of possibility within which place, identity, and agency, however abjected, are constructed and contested."[21] Furthermore, this legacy of injustices continues; governmental practices and rationalities that pride themselves on being concerned with the welfare of the governed continually allow for suffering and death as the "silent residue of policy."[22] The case studies in this book have argued for—and attempted to model—tactical ethical responses to the new proving grounds of the military. The overall approach has considered government to be "ways of doing things" (rather than a political philosophy or ideology) and has irreverently sought to use governmental technologies and logics against themselves.[23] Moving away from a normative understanding of rights to a more performative stance, the entire project calls on the "right to question" governmental regimes of truth, regardless of whether this practice requires a tactical recourse to juridical rights. The method is all too familiar to those who, in their efforts to survive, continually confront policy choices and governmental strategies; nuclear workers, for example, have had to scrutinize the compensation law justified in their name. There are also those exempted from the right to question government—the bare life of animals, the social death of racism, and the abject life-forms excepted out of "agonistic games of truth" between governors and governed. My effort has been to reflect on the ways subjects are made through governing—made into objects needing further government intervention—and on a subject's ethical relation to itself and others, including responses to such governmental actions and attempts to carve out other forms of existence: *in a nutshell, the relations between politics and ethics, between forms of governing and counterresponses/counterconducts.*

Instead of premising criticism and dissent on the excesses or illegitimacy of government (although these might be employed as tactics), the project has sought out the logics and conditions of acceptability of governmental systems. To that end, it has largely eschewed general universal terms in favor of tactical analysis and detailed problematizing, in order to open up possible and practical reversals. "In order for governmental systems to be considered fragile and contingent achievements, and to avoid any recourse to transcendental or empirical universals—such as, by asking how we might resist modern governmentalities or how entities such as the state shift over time—Foucault *must* begin with the question, ignited by governmental counter-conducts, of 'how not to be governed *like* that, by that, [and] in the name of those principles.'"[24] This is an irreverent demand for a different regulatory terrain, laws, even concepts, as the basis for challenging how we envisage environmental harm and might ameliorate the damages of war. *The irreverence must also be directed at what we are*—the ways we are engaged as objects and subjects of government. The challenge requires ethical self-transformations that do not serve the current governmental regime of truth, that "risk themselves" in order to decolonize the war basis of institutions. It's not simply a matter of being against institutions—"we" are the institutions—but, rather, to ask what other capacities and competencies, agencies and potentials, could be cultivated besides the organization of war. "Maybe the target nowadays is not to discover what we are [or what we will become], but to refuse what we are [now]."[25]

Tactical ethical responses develop as immanent forms of criticism. In the context of this project, they have insisted on indeterminacy or uncertainty to deontologize present ways of doing things. Insisting on indeterminacy is necessary "because it operates to undercut the confidence that anything can be fully and finally known; it rejects the notion of a fixed and stable narrative subject, and it acknowledges that the introduction of previously unforeseen or unknown information or events can reshape shorelines, family histories, and narrative accounts of them."[26] Moreover, it rejects the governing of uncertainty through closed systems, end states, or risk-based futures, thereby creating opportunities to develop a more material-relational ethics. This is realized by genealogical excavations that explore the way unknowns have come to be governed, known, calculated, and routinized, for example, techniques of containment such as audits, regulatory documents, the so-called transparency of numbers, and secrecy. The effort does not celebrate indeterminacy; the mutability of truth can be painful and difficult to live. However, the indeterminate natures and vulnerabilities produced by the biopolitical organization of green war necessitate responses that do not revert to truth claims about nature but, instead, keep open the possibility that more can be identified, that more can be said, and that more can be done. Tactical refusals of the finality of truth render the present more ambiguous and controversial and, therefore, allow for different coalitions. Genealogical ethical tactics can also make relations of spectacle and systemic exposure manifest and, thereby, encourage other arrangements—ways of "living otherwise" in the present.

A typical response to contaminated lands and abandoned workers of former military operations is to expose them, to bring them to the public's attention. While this draws public attention and opens debate, pedagogical "exposure" does not ameliorate forms of violence that are hypervisible from the start, such as the terminal survival of the bald eagle of the postmilitary nature refuge. "While there is plenty of hidden violence that requires exposure there is also, and increasingly, an ethos where forms of violence that are hypervisible from the start may be offered as an exemplary spectacle rather than remain to be unveiled as a scandalous secret."[27] Nor does transparency or visibility guarantee that anything different will be done; in spite of the media blitz and congressional wrangling over sick nuclear workers, a social order/affect of sympathy does not abate worker suffering. Another example: the relentless exposure of the militarized West to tourist maneuvers and nostalgic imaginations has actually served to obscure the violences still at work. Different methods are also needed to understand the natures of indeterminate and banal forms of exposure and death, which are nonlocalizable (displaced in relation to victim/perpetrator), excessive, ambiguous, and without closure. A forensics is called for that subtly shows such "forms of violence distributed in arbitrary and not so arbitrary ways and as a by-product of systems that require consumers and exposures."[28]

Through genealogical analysis and tactical ethics, each chapter of this book made visible and contested biopolitical truth productions, such as genetic preservation of bison, the risk-based end state of the nature refuge, or the determination of "no known link" between exposures and illnesses in the nuclear workers compensation program, to cite but three examples. The chapters dealt with institutions in such a way as to "bury" truth production back into hitherto unseen power relations, in order to lend more ambiguity to the present. Underlying them all was an ethical imperative: the indeterminacy of the effects of war—the contamination and social ruination—and the shutting down of options in the name of that indeterminacy *should compel creative responses*. "The culture that sustains the procedural rigor that perpetuates the finding of indeterminacy stands accused *de facto* of perpetually deferring the obligation to *do* something beyond critical interrogation of the problem."[29] The tactical ethics developed in this book, therefore, acknowledged the need to continue to engage in truth-telling exercises, but as the means for opening up debate and proliferating controversy—not to deliver a new ontology as the end game. Chapter 4's pursuit of "transnatural aesthetics" in response to nuclear waste and toxicity insisted that there be no recourse to pure nature because this would *not* address the relations of harm and degradation that continue to condition the present, *nor* the unevenness of the vibrancy of life, of uncertainty, or of the residues of war. The desire to affirm life can install new binarisms, for example, between organic and inorganic, and further repudiate what is considered to be negative: pain, suffering, half-lives, abandonment, death. This volume has taken a different tack; it has developed accounts of the mutability of life and death of humans and nonhumans, and scrutinized institutions and practices that manage the negative (management that often involves repression, denial, or disposal). Furthermore, the chapters showed ways of "playing with the negative"—playing with the style of response to deepen resources for regeneration and affirmation. Addressing past injustices and their conditioning of the present does not mean that responses remain tethered to recognition, reparations, modernist definitions of harm, and binaristic modes of thinking that separate humans from the environment and humans from nonhumans. Instead, this volume has endeavored to practice criticism as a kind of artful forensics, as "hot spotting."

Hot spots, in the most general sense, are areas of intense activity, dense relations, and/or focused attention, embedded within larger areas of stable or calm relations. The term can be applied in a range of contexts, from regions of significant biodiversity to popular entertainment attractions, from geologic areas of high volcanic activity to wireless Internet access points or the areas of the computer screen where users hover. The concept also refers to sites of concern, such as areas of potential defects in metal casting, or of potential disaster. This volume draws more on the conceptual schemata of radiation hot spots. Hot spots are areas where something remains unassimilated and nagging; they are reminders of the "stickiness" of radiation and the impossibility of pure spaces, pure categories, or a pure self.[30] Hot spots are unidirectional, irreversible, and contextually unique. As the state has invested in maintaining spectacular divisions and discrete categories, hot spots manifest the transgressions of those boundaries and containment mechanisms. They are sites of stubborn ambiguity.

Hot spotting, then, demands attention to hot spots—ambiguous bodies of land or embodied subjects, such as former nuclear workers, abandoned through military projects of environmental purity, postnuclear spectacle, and discrete categories of waste and human, or human and nature, maintained by the state. In different theaters of operation, hot spotting might draw attention to the limits of sovereignty of these terms, contesting what is epistemologically or legally determined "inside" or "outside," "valuable" or "ruined." The places or sites where power and its limits are inscribed might also be drawn out through hot spotting. Hot spots thus foment awareness of the limits, contradictions, and inward spirals of green affirmations, military concessions, and environmental reconstructions that are underpinned by logics that perpetuate war. Hot spotting essentially involves the operations of identifying, making visible, and keeping open the possibility that more can be identified—and that all will be experienced differently by different people.

As a diagnostic, hot spotting problematizes the visual; hot spotting references the operation "to spot" in order to question the veracity of what is seen—not to denigrate visuality wholesale but to explore the relations that make something visible and the technologies of detection that are needed to produce knowledge and grant the status of material existence to the unseen, such as toxicity or radiation. In contrast to an ethics grounded in exposing the truth as the end point of critique, hot spotting tries to arrange for other relations than exposure between hot spotter and hot spot. In response to the everyday conditions of toxic stealth and the ambiguity of death and life owing to environmental exposures, hot spotting acknowledges the interdependence of the framework of observation and the vulnerability, complicity, and embeddedness of the viewer with the materials under observation. However, this environmental forensics is not mere reflexivity or spectacular positioning, but an exercise in practicing "good ambiguity" in relation to others, whether human or nonhuman, and experimenting with different diagnostic arrangements.[31] Empiricism here moves from a depth model (or representation-based model) to a diagnostic art that encourages play and invests in practices, but not finished form. Hot spotting trespasses bureaucracies and sanitized histories to irreverently question truths about nature or uncertainty, and to rearrange conventional boundaries of bodies and categories. The hot spotter is a historically shaped agent who focuses on limits to existing thought and material conditions with an eye to conceiving historically possible alternatives.[32]

Hot spotting not only bears witness to but seeks to cultivate ethical relations with the remains of war. A methodology of the residual, hot spotting excavates the conditions of military spectacle by drawing on material remainders—nuclear waste, bald eagles, nuclear workers, and so on—as material reminders that war is both external to the U.S. nation and internal, not merely spectral but geographical and material. This is not "deconstruction" for the sake of a philosophy of loss or a fascination with margins, lack, or the abject; it is a way of thinking about biopolitics as radical positivity, as expressive capacities and ways of "living on" that exceed the relations of power.[33] Under conditions of toxicity, uncertainty, partial knowledge, and secrecy, hot spotting seeks to disrupt routinized denial, disposability, and war. "War sustains its practices through acting on the senses, crafting them to apprehend the world selectively, deadening affect in response to certain images and sounds, and enlivening affective responses to others."[34] By proliferating and constantly shifting critique, hot spotting can "*draw attention to the possible by showing the contingent dimension of the actual*."[35] The task is not to refine a representation in relation to a preexisting truth but to produce representation upon representation with the aim of rearranging affect and building collective responses; "any stabilities that are produced need to be provisional, working categories, that enable rather than disable further learning and another reconstitution of nature."[36] As a tactical ethics, it is not a "plan or blueprint of how social relationships will be structured . . . [but rather] a continual process of composition and decomposition through social encounters, on an immanent field of forces."[37] It is a call to experiment with different aesthetics.[38] Hot spotting attempts to practice critique as a skillful way of putting things together, in order "to cultivate an attention to the conditions under which things become 'evident,' . . . ceasing to be objects of our attention and therefore seemingly fixed, necessary, and unchangeable."[39] It is the enactment of empirical research as a process of artful arranging and rearranging.[40]

The chapters in this book utilized various technologies of form, such as performative techniques, material figures, documentary play, humor, and the absurd. The chapters frequently combined empirical and theoretical writing with satirical elements, wedding social-science case-study methods to performance art as part of a collaborative effort to bring together some of the sensibilities of environmentalists, artists, scientists, journalists, and various activists. The cases attempted to develop modes of expression that resonated with the materials used. Inspired by a wide array of activist theory writing, including the philosophy-poetry of Walter Benjamin, the creative historical geographies of the present of Allan Pred, and material figuration strategies of Donna Haraway, the project sought "to show" as much as tell the reader how to assess a problem, using theatrical techniques such as "being in character" as a particular figure, concentrating things to the point that they became defamiliarized, or mimicking and overidentifying with dominant discourses and institutions.[41] Essentially, the performative case studies attempted to morphologically adapt criticism to the materials they addressed, often amplifying those implications to satirize the performances in the process of doing them. They also developed and employed various figures to re-present visual-textual discourses and rhetorical contrivances *and* to playfully explore material links between bodies of land and language, animal and human bodies, the laboring body to the environment and law. The figures provided the author a way to pursue the material conditions of criticism.[42] Based on ethnographic fieldwork and combining academic research with popular journalism and personal narrative, the figures employed in the chapters are grounded in ordinary reality and mark thresholds of transformative political potential. In the case of chapter 3, the "Hole in the Head Gang" was not only a collective figure that channeled commentary by nuclear workers, but also captured how nuclear workers are simultaneously hot spot and hot spotter: they cannot intellectually exit the scene of criticism, for they are the watch and witness of state denial of their embodied waste.

Other figures, such as the E.A.G.L.E. collective of chapter 1 and EGO organization of chapter 2 served as fictitious "front companies" for the author's forensics activity. They sought points that radiate potential for critical disclosure of truth as a practice. All of the chapters developed different expressions of documentary play, cultivating a documentary sensibility while considering documentary realism, documentary agencies within biopolitics, and, in the case of nuclear workers, the disciplinary function of documentation and the burden of proof. Each chapter featured a different kind of document or some kind of fragile intervention consisting of documents, including government reports, PowerPoints, charts, workbooks, even waste's common function as document. The documents performed important aspects of the analysis, providing illustration, contesting official or popular accounts, playing with affect in relation to the topic, mimicking institutions and the evidentiary qualities and reality effects of the documentary format of "the report." Clustering real and fictive elements, chapters included faux documents that "tell the truth" as well as real documents that are barely credible. They endeavored to expose the cultural system of reports and foreground the realism of text and visual documentation as objects of consideration and sources of fiction and error in their own right; realism is shown to be something that is arranged and made. This reflects not only on the conduct of academic writing, which is usually about truth acquisition and report writing, but also, more generally, documentary ethics and the material agencies of documents.[43]

Documentation is imbricated in ethics and politics. Many activist and academic practices involve documenting injustices as well as tracking and sharing government documents. Documentation is more than mere description or statement, and documents more than semiological product: "Writing and documents are material, thereby invoking technological apparatuses and institutional embeddings."[44] Official writing devices are management tools; they manage complexity and define the world through the power to prescribe compliance with them.[45] As such, they are ontologically generative and performative and can bring different entities into being, producing public-forming or nation-building effects or conjuring the collective ethical subject of self-referential audits. The hot spotter endeavors to contest official documentary worlds; to do so, the hot spotter must draw on "fiction" and illegitimate or repressed knowledges to induce effects of truth that engender other understandings of the world. "It seems to me that the possibility exists for fiction to function in truth, for a fictional discourse to induce effects of truth, and for bringing it about that a true discourse engenders or 'manufactures' something that does not as yet exist, that is, 'fictions' it."[46] Playing with documentation can be the material grounds for serious institutional reinvention. The entire *Hot Spotter's Report* participates in forms of documentary play that continually shift registers, sometimes by circling a point with different perspectives, or by telling the same story using different rhetorical or visual conventions. Hot spotting attempts to recite truth in different arenas in order to avert the way truths become sedimented in particular ways, and to encourage new understandings and—attendant to that—multiple publics. The "Hole in the Head Gang" figure of chapter 3, in particular, demonstrates the importance of this aspect of hot spotting: the state demands documentation from former nuclear workers; workers or their relations must enter the field of documentation as part of the compensation claims process, conferring truths about themselves, their work and relations. Hot spotting responds to this endless disciplining of workers through the field of documentation—the routinizing of "confessional" illness and the treatment of workers as the remainders of the Cold War—with even more fervent efforts to share documentation in different ways, so that the information is encountered and heard differently. Operating with the assumption that freedom is about the minimum of domination, hot spotting is dedicated to documentary openness and play as serious responses to official documentary routines and policies.

All four chapters and their featured figures also performatively enacted hot spotting to explore the materiality of bodies and bureaucracy, contributing to geographical and environmental criticism that has shifted registers from a focus on language and text, and on the land, nature, or environment "out there," to "the intimate fabric of corporeality that includes and redistributes the 'in here' of the human being."[47] Several of the chapters put bodies into critical relief, whether bald eagle or nuclear worker. From a biopolitical perspective, the presence and effects of toxins on any body—human or nonhuman—remain epistemologically uncertain and deeply contested; interpretations of biological processes are a means of social regulation. The concessions of disaster-responsive social institutions and the vulnerabilities of different populations exposed to militarized environments have fostered bureaucratic dependencies (water supply programs, etc.), modes of objectification (dosages, biomonitoring), and distributions of benefits, entitlements, and blame.[48] Bodies are projects in the making, and conduits for commodification, toxicity, and militarization that are constantly changing. The politics of injury, victimhood, and recognition tends to retroactively—and continually—recite the violence that led to injury, limiting the definition of self and community to that of the state. The retroactive desire to return to a truth before—a clean and proper body, unblemished land, and so forth—risks closing down ways to live otherwise in the present and future. While the demand for reparations is important and essential to many humans—and nonhumans—that live in or as the remains of war, it should not forestall the demand to be organized and governed differently. The recourse to injury is but one way to understand the self. *Hot Spotter's Report*, therefore, attempts to intervene subtly in the biopolitical organization of green war by reworking "agency" as something other than the militarized, heroic, and/or masculine subject, as recourse to injury.

Chapter 4 ruminated on the potential of "transnatural aesthetics," which tactically denies the human subject as the sovereign, disembodied central position or creator of the universe to instead consider the "human body" to be always already part of an active material world.[49] The figures of the radioactive drag queen and nuclear waste sculptor facilitated awareness of the impasse imposed by the sovereign subject of liberal humanism, bequeathed from modern Enlightenment philosophical ideas, and at the other end of the spectrum, a postmodern nihilism that seemingly destroys the ethic of responsibility that is traditionally aligned with sovereign agency.[50] Sovereignty as the foundation of individual autonomy, politics, and ethics overidentifies the similarity of self-control to sovereign performativity and state control over geographic boundaries.[51] This has encouraged a militaristic and melodramatic view of agency that does not—cannot—address the barely discernible forms of slow violence and deaths in the spaces of everyday life and the existence of diverse and subtle forms of living on, enduring, and surviving. The emphasis on heroic agency and bodily sovereignty makes out such compromised tactical life-forms to be failures by comparison, and risks denying the possibilities for shared political life that are not based on pure positions or autonomous natures. By contrast, hot spotting considers agency in more material, corporeal, contingent, and "impure" forms, and life and death in terms of composition, ambiguity, banality, and playful negotiation. Ambiguity and vulnerability are opportunities to proliferate different coalitions, experiment with social and political options, and sustain indeterminacy and controversy where they have been shut down.

As imaginatively practiced in the four preceding chapters, hot spotting is concerned with critiquing institutions and rendering ambiguity through irreverence, developing relationships that are theoretically committed to contingency and permanent critique but are also politically motivated to material and social justice.[52] Hot spotting does not merely point out contradictions through empirical evidence; the book models hot spotting as a practice of rendering the absurdities that are part of the practices of governing—but that are presented as rational. Essentially, *Hot Spotter's Report* investigates governmental absurdism, but rather than recuperate truth or rationality, the book proposes hot spotting as a counterpractice that meets governmental absurdism with absurdism, because the rationalities targeted in the book are revealed to be in fact absurd. This absurdist counterpractice works on multiple registers—through different fields and audiences, different power relations, and different endgames—*to question the rationality of the absurdism immanent to governing, and to refuse to be governed that way.*

Absurdism is put to work throughout the volume as a methodology of making things more ambiguous, strange, and curious. Along these lines, Foucault discussed "curiosity" as a vital attitude that informed his understanding of ethics:

> I like the word [curiosity]; it suggests something quite different to me. It evokes "care"; it evokes the care one takes of what exists and what might exist; a sharpened sense of reality, but one that is never immobilized before it; a readiness to find what surrounds us strange and odd; a certain determination to throw off familiar ways of thought and to look at the same things in a different way; a passion for seizing what is happening now and what is disappearing; a lack of respect for the traditional hierarchies of what is important and fundamental.[53]

This irreverence is a material engagement with the world that emboldens creative action. Hot spotting employs tactical deontologizing play that dramatizes material agencies and reports on, even amplifies, "ridiculousness" as a way of reconceiving the present—particularly institutions and laws—as contingent, uncertain, and ultimately undecided. "The serious figure of the detached critic as neutral, expert, committed, progressive or connected judge is markedly out of place in this process—in its place is the vitally engaged and changing subject; a searching, meandering spirit that refuses to abide aspects of the present's lived realities. . . . not the moralizing cynic, but the ironist, the laughing spirit, who moves on and on, ever-searching for signs of new life."[54] Absurdism, while it may appear to be a postmodern game of representation or extravagant fantasy or nihilistic irony, is essentially realistic and optimistic; it grapples with the empirical to produce the slippages of strange laughter, unhinging the taken-for-granted and engaging material contradictions, the residual, waste—the negative in general—*in order to affirm,* to rearrange, and to generate something new.[55] Friedrich Nietzsche described the effort as follows: "When we criticize something, this is no arbitrary and impersonal event; it is, at least very often, evidence of vital energies in us that are growing and shedding a skin. We negate and must negate because something in us wants to live and affirm—something that we perhaps do not know or see as yet."[56]

Coda: Hot Spotting as Decolonizing the War Imagination

Hot spotting has been used as the name for more geographically deterministic practices. For example, in the U.S. health-care system, hot spotting locates neighborhoods, even buildings, that generate high costs in hospital care, targeting the residents of such areas in order to develop other conveyances of care and, thereby, alleviate expenses.[57] Although there are many admirable qualities to this diagnostic when based in community efforts to develop alternative relations and institutions of health, it can be seen to target/fix racial and genetic susceptibility in individual bodies and their residences, on a grid of data that denies social relations. This enlists epidemiology as a new governing strategy, and geography as calculating rationality (essentially, a map of costs). It surveys the population and imposes race, class, disease, and citizenship in place, in order carve out singularities—single bodies—from larger events of toxicity and structural-institutional forms of racism; these are hot spots annexed to and instrumentalized by a system-wide cost-benefit analysis. The danger of this operation is that "hot spots" essentially serve a colonial function: advocacy of domestic targeting and the geographic marketing of social death through the imperative to live.[58] Hot spotting in this book, by comparison, has worked for something of the reverse: decolonizing the U.S. war imagination. It has endeavored to show domestic geographies of toxic occupation of the nation. Furthermore, it has excavated the new proving grounds of military restructuring, institutional accountability, and concessions as *the colonial grounds for an emerging biopolitical project of the U.S. nation: "green war."* Discussion of a "hot spot" design contest provides further insight.

In 2001, the *Bulletin of the Atomic Scientists* hosted a competition that addressed the problem of plutonium disposal—the material used in nuclear weapons and with a half-life of 24,110 years. The "Plutonium Memorial" competition directed talent toward the important issue of nuclear waste via a hypothetical memorial to the excesses of the U.S. nuclear enterprise. Although the exercise was tongue in cheek, artists, architects, and other visionaries were invited to send in schemes with critical design impact but also serious consideration of the safety and security requirements that such a structure would necessitate. Among the many entries, Michael Simonian's *24110*—named after the half-life of plutonium—was the winner.[59] His design proposed a prominent plutonium storage facility south of the White House on the National Mall in Washington, D.C., under a partially lifted circular lawn cover. In D.C., the memorial could be visited and policed, and would silently reproach lawmakers for shortsighted nuclear policies. A concrete and steel tub would be sunk into the ground holding a five-hundred-ton stash of plutonium casks. Beneath a circle of peeled-up lawn, the concrete lip of the memorial would be embossed with portraits of politicians and scientists, as well as logos of nuclear-industry corporations. A capillary layer of gravel and volcanic turf above the casks would theoretically expose visitors to just 0.01 millirem of radiation annually, to be incorporated into the 360 millirem that the average American is said to receive annually.[60] A series of flared steel tabs—241 "clock totems"—would splay out from the walkway; one would be bolted down to the ground every century to mark the passing of plutonium's half-life. The memorial would serve as a reminder that there is still no solution to nuclear waste, despite claims that nuclear energy is "green." Moreover, it would turn inside out two prominent strategies of waste disposal: hiding nuclear waste "out of sight" in western deserts and/or "the poor's backyard," and "the great American lawn cover-up."[61] Through a process of reverse colonization, the memorial would symbolically and materially transform the National Mall into proving ground, bringing the U.S. militarized West to the heart of the nation and forcing the president, Congress, the DOE, and the polis in general to face the domestic effects of American war and energy policies. The critique relies on a clever juxtapositioning of "outside" and "inside." If American exceptionalism always externalizes war, obscuring militarism domestically, then this memorial would internalize it in the symbolic space of the nation and compel a reckoning with domestic toxicity and nuclear colonialism. The imagined memorial asks: What would it mean to bring excessively deadly residues of war to the front door of the president and to contaminate the geographic imagination of the nation?

Within the broader terrain of the colonial present, this intervention demonstrates the pedagogical importance of hot spots—including purposely designed memorial hot spots.[62] A plutonium memorial in

D.C. would surely draw attention to the legacy of waste from U.S. nation projects, as well as draw out interesting and unexpected symbolic relations with other memorials to American wars through juxtaposition.[63] Aside from the uncanny affinity between long-lasting nuclear materials and memory projects, the temporality and location of this particular memorial project reveals productive contradictions: Have we surpassed the nuclear, so far as to memorialize it? In the context of the National Mall, would a plutonium memorial merely become more kitsch of the nation, with nuclear waste as destination, so that "clean-tech" wars may be waged elsewhere?[64] How might the national imagination *still* assimilate a plutonium container within the progress of the nation in spite of its promise of critical-symbolic contamination of the nation?

These are difficult questions that by no means make a critical plutonium memorial in the nation's capital less worthy; it raises the stakes, particularly with respect to the perpetually colonial conditions of Washington, D.C. Without formal statehood, Washington, D.C., is the territorial seat of the U.S. government that props up the national imagination, *and* it is a landfill of material remains of that nation, evidenced by mass incarceration, one of the worst AIDS rate in the country, a beleaguered educational system, and exploitational land development, to name but a few colonial legacies. There is no domestic peacetime in the nation's center; the violences of spectacle—the relentless erasure of the domestic grounds of war and its casualties—examined throughout this book are also conditions of existence of the "last colony" of the United States. Given this context, the critique of nuclear colonialism exhibited by a plutonium memorial in D.C. would not necessarily produce the radical effect intended, even reproducing the colonial basis of U.S. war making or of the national imagination more generally. However, as this book has endeavored to show, such creative consideration of "waste" is crucial to the project of decolonizing the war imagination of the nation—an increasingly "green" colonial imagination.

As the largest energy user in the country, the DOD has been investing more than a billion dollars a year in alternative energy, making the U.S. military one of the major defenders of green policies, such as lowering carbon emissions and greenhouse gases, in Washington, D.C., legislative circles. The DOE, too, has been "greening" the nation's stockpile, making weapons of mass destruction less toxic to handle and organizing energy farms on land that was formerly part of bomb production. Here is potential for a hot spotter's report that would, via a yet-to-be-determined form, explore the perpetuity of war in/by the United States, drawing together the domestic colonialism of nuclear waste (the nation's waste) *and* the colonialism of national spectacle (the nation's capital) to show clean or green war to be materially impossible yet absurdly inevitable.

NOTES

Introduction

1. Dwight D. Eisenhower, Farewell Radio and Television Address to the American People (January 17, 1961), in *Public Papers of the Presidents of the United States: Dwight D. Eisenhower. 1960–61: Containing the Public Messages, Speeches, and Statements of the President January 1, 1960, to January 20, 1961* (Washington, D.C.: U.S. Government Printing Office, 1961), 1035–40.

2. Dwight D. Eisenhower, address "Chance for Peace" delivered before the American Society of Newspaper Editors (April 16, 1953), in *Public Papers of the Presidents of the United States: Dwight D. Eisenhower. 1953: Containing the Public Messages, Speeches, and Statements of the President January 20 to December 31, 1953* (Washington, D.C.: U.S. Government Printing Office, 1960), 179–88.

3. Dwight D. Eisenhower, Speech before the NATO Council (November 27, 1951), http://www.eisenhowermemorial.org (accessed March 20, 2012).

4. Martin L. Calhoun, "Cleaning up the Military's Toxic Legacy," *USA Today Magazine* 124, no. 2604 (Society for the Advancement of Education, September 1995), http://FindArticles.com (accessed March 3, 2012).

5. Ann Laura Stoler, "Imperial Debris: Reflections on Ruins and Ruination," *Cultural Anthropology* 23, no. 2 (2008): 205.

6. Catherine Lutz, "Empire Is in the Details," *American Ethnologist* 33, no. 4 (2006): 593–611; Ruth Wilson Gilmore, *The Golden Gulag: Prisons, Surplus, Crisis, and Opposition in Globalizing California* (Berkeley: University of California Press, 2007); Michael Dillon and Julian Reid, *The Liberal Way of War: Killing to Make Life Live* (London and New York: Routledge, 2009).

7. Rachel Woodward, "From Military Geographies to Militarism's Geographies: Disciplinary Engagements with the Geographies of Militarism and Military Activities," *Progress in Human Geography* 29, no. 6 (2005): 720; Rachel Woodward, *Military Geographies* (Malden, Mass.: Blackwell, 2004).

8. National security normalized military preparations of the home, via bomb shelters, domestic waste management, home technologies, and attendant designations of gender. For more discussion, refer to Ann Markusen, Peter Hall, Sabina Deitrick, and Scott Campbell, *The Rise of the Gunbelt: The Military Remapping of Industrial America* (New York: Oxford University Press, 1991); Ann Markusen, "The Military Industrial Divide: Cold War Transformation of the Economy and the Rise of New Industrial Complexes," *Environment and Planning D: Society and*

Space 9, no. 4 (1991): 391–416; Ann Markusen and Joel Yudken, *Dismantling the Cold War Economy* (New York: Basic Books, 1992).

9. Barney Warf, "The Geopolitics/Geoeconomics of Military Base Closures in the USA," *Political Geography* 16, no. 7 (1997): 541–63.

10. John Lowering, "The Production and Consumption of the 'Means of Violence': Implications of the Reconfiguration of the State, Economic Internationalisation, and the End of the Cold War," *Geoforum* 25, no. 4 (1994): 481.

11. Ibid., 471–86.

12. Ibid., 479.

13. Kenneth N. Hansen, *The Greening of Pentagon Brownfields: Using Environmental Discourse to Redevelop Former Military Bases* (Lanham, Md., and Oxford: Lexington Books, 2004).

14. Amy Standen, "U.S. Military Boosts Clean Energy, with Startup Help," KQED, October 24, 2011.

15. Andrew Ross, "The Future Is a Risky Business," in *FutureNatural: Nature, Science, Culture,* ed. George Robertson, Melinda Mash, Lisa Tickner, Jon Bird, Barry Curtis, and Tim Putnam (London and New York: Routledge, 1996), 7–20; Robert F. Durant, *The Greening of the U.S. Military: Environmental Policy, National Security, and Organizational Change* (Washington, D.C.: Georgetown University Press, 2007).

16. David Biello, "A Need for New Warheads?" *Scientific American* 297 (November 2007): 80–85, esp. 83–84. Additionally, powerful transnational corporations have consolidated control over nuclear materials and nuclear technologies in global energy markets.

17. Éric Darier, "Foucault and the Environment: An Introduction," in *Discourses of the Environment,* ed. Éric Darier (Oxford and Malden, Mass.: Blackwell, 1999), 22.

18. Michel Foucault, "24 January 1979," in *The Birth of Biopolitics: Lectures at the College de France, 1978–1979,* ed. Michel Senellart, trans. Graham Burchell (New York: Palgrave Macmillan, 2008), 51–73; Michel Foucault, "Part Five: Right of Death and Power over Life," in *The History of Sexuality, Volume 1: An Introduction,* trans. Robert Hurley (New York: Vintage Books, 1990), 135–45; Thomas Lemke, *Biopolitics: An Advanced Introduction* (New York and London: New York University Press, 2011).

19. Not surprisingly, defining threats often mobilizes or becomes aligned with discriminatory differences around disabilities, race, class, sexuality, and so on.

20. Lauren Berlant, "Slow Death (Sovereignty, Obesity, Lateral Agency)," *Critical Inquiry* 33, no. 4 (2007): 756.

21. Diana Coole and Samantha Frost, "Introducing the New Materialisms," in *New Materialisms: Ontology, Agency, and Politics,* ed. Diana Coole and Samantha Frost (Durham, N.C., and London: Duke University Press, 2010), 32–33.

22. Leerom Medovoi, "A Contribution to the Critique of Political Ecology: Sustainability as Disavowal," *New Formations: A Journal of Culture/Theory/Politics,* no. 69 (2010): 141–42.

23. Stuart J. Murray, "Care and the Self: Biotechnology, Reproduction, and the

Good Life," *Philosophy, Ethics, and Humanities in Medicine* 2, no. 6 (2007), http://www.peh-med.com/content/2/1/6 (accessed January 14, 2011).

24. John Beck, *Dirty Wars: Landscape, Power, and Waste in Western American Literature* (Lincoln and London: University of Nebraska Press, 2009), 37.

25. Cultural critic Walter Benjamin saw his task to be that of sifting through the pile of debris trailing progress and modernity, to redeem objects and people that had been cast aside as worthless by dominant modes of organizing value. See Walter Benjamin, *The Arcades Project*, trans. Howard Eiland and Kevin McLaughlin (Cambridge, Mass., and London: Belknap Press of Harvard University Press, 1999); Raymond Williams, "Dominant, Residual, and Emergent," in *Marxism and Literature* (Oxford and New York: Oxford University Press, 1977), 121–27; Jean Baudrillard, "The Remainder," in *Simulacra and Simulation*, trans. Sheila Faria Glaser, http://www9.georgetown.edu/faculty/irvinem/theory/Baudrillard-Simulacra_and_Simulation.pdf (accessed July 17, 2011), 95.

26. The category "waste" is used in this volume to refer to specific categorical distinctions and divisions of matter/materials, e.g., nuclear waste.

27. According to this paradigm, the environment is considered to be a physical setting, with negligible conditions that do not significantly impact the economic and military activities on site. The capitalist-industrial understanding of the environment also treats natural resources as free to be appropriated and available as a dumping ground for excess outputs from economic production. Essentially, the environment only makes appearances as a blank slate or as abstract parameters. The spectacle of the postindustrial/postmilitary/postnuclear nature refuge maintains this binary view of humans and nature. See Nicole Shukin, *Animal Capital: Rendering Life in Biopolitical Times* (Minneapolis and London: University of Minnesota Press, 2009), 184–85; Martin O'Connor, "Codependency and Indeterminacy: A Critique of the Theory of Production," in *Is Capitalism Sustainable? Political Economy and the Politics of Ecology*, ed. Martin O'Connor (New York and London: Guilford Press, 1994), 63–64.

28. Nigel Thrift, *Spatial Formations* (London: SAGE Publications, 1996), 36.

29. Michel Foucault, "Nietzsche, Genealogy, History," in *Language, Counter-memory, Practice: Selected Essays and Interviews*, ed. Donald F. Bouchard, trans. Donald F. Bouchard and Sherry Simon (Ithaca, N.Y.: Cornell University Press, 1977), 139–64.

30. Avery F. Gordon, *Keeping Good Time: Reflections on Knowledge, Power, and People* (Boulder, Colo., and London: Paradigm Publishers, 2004), viii.

31. Valerie Kuletz, personal communication (January 3, 2012).

32. Gay Hawkins, *The Ethics of Waste: How We Relate to Rubbish* (Lanham, Md., and Oxford: Rowman and Littlefield, 2006).

33. See chapter 4.

34. George Pavlich, "Nietzsche, Critique and the Promise of Not Being Thus . . . ," *International Journal for the Semiotics of Law* 13, no. 4 (2000): 357–75.

35. Rosalyn Diprose, "Toward an Ethico-Politics of the Posthuman: Foucault and Merleau-Ponty," *Parrhesia* 8 (2009): 7–19.

36. Hawkins, *The Ethics of Waste*, 7.

37. Bernd Frohmann, "Documentary Ethics, Ontology, Politics," *Journal of Archival Science* 8, no. 3 (2008): 165–80; Bruno Latour, *Reassembling the Social: An Introduction to Actor-Network-Theory* (New York: Oxford University Press, 2005).

38. See, for example, Donna J. Haraway, *Modest_Witness@Second_Millennium.FemaleMan©_Meets_OncoMouse™* (New York and London: Routledge, 1997).

39. While the efforts of deep ecology resonate with my statements, the relational environmental ethics advanced here is different and argues for interconnectivity without sameness or full incorporation of human and nonhuman. "Hot spotting" encourages a more postcolonial sensibility. See Noel Castree, "A Post-environmental Ethics?" *Ethics, Place and Environment* 6, no. 1 (2003): 3–12.

40. The U.S. Fish and Wildlife Service of the Department of Interior administers the National Wildlife Refuge System, the nation's network of lands and waters managed for the conservation of wildlife and habitats. Rocky Flats is now administered as part of the Rocky Mountain Arsenal National Wildlife Refuge Complex.

41. "Album: Unnatural Nature," in *Uncommon Ground: Rethinking the Human Place in Nature,* ed. William Cronon (New York and London: W. W. Norton, 1996), 62; Alan Berger, *Drosscape: Wasting Land in Urban America* (New York: Princeton Architectural Press, 2006); David Havlick, "Bombs Away: New Geographies of Military-to-Wildlife Conversions in the United States," Ph.D. dissertation, University of North Carolina, Chapel Hill, 2006); Peter Coates, Tim Cole, Marianna Dudley, and Chris Pearson, "Defending Nation, Defending Nature? Militarized Landscapes and Military Environmentalism in Britain, France, and the United States," *Environmental History* 16, no. 3 (2011): 456–91.

42. Berger, *Drosscape,* 67–70.

43. Brownfield development projects allow developers to benefit from federal subsidies and reduced liability to offset short-term cleanup costs of contaminated land. They provide new commercial uses for contaminated sites, such as parking garages, malls, or basketball courts, even residential subdivisions.

44. Kathryn Yusoff, "Biopolitical Economies and the Political Aesthetics of Climate Change," *Theory, Culture and Society* 27, nos. 2–3 (2010): 74.

45. The production of the bomb, here, refers to all processes associated with making, testing, and storing of nuclear weapons. Simultaneously, it references the bomb as a crucial figure organizing cultural production during the Cold War. The concept of a "nuclear complex" is employed to refer to the large-scale, multi-sited effort of nuclear weapons production.

46. Michel Foucault, "17 March 1976," in *Society Must Be Defended: Lectures at the Collège de France, 1975–1976,* ed. Mauro Bertani and Alessandro Fontana, trans. David Macey (New York: Picador, 2003), 239–63; Stuart J. Murray, "Thanatopolitics: On the Use of Death for Mobilizing Political Life," *Polygraph* 18 (2006): 191–215.

47. Gwendolyn Blue and Melanie Rock, "Trans-biopolitics: Compexity in Interspecies Relations," *Health* 15, no. 4 (2011): 353–68; Stacy Alaimo, *Bodily Natures: Science, Environment, and the Material Self* (Bloomington and Indianapolis: Indiana University Press, 2010).

48. Michel Foucault, "The Masked Philosopher," *Foucault Live: Collected Interviews, 1961–1984*, ed. Sylvère Lotringer, trans. Lysa Hochroth and John Johnston (New York: Semiotext[e], 1996), 305.

1. Where Eagles Dare

1. Adriana Petryna, *Life Exposed: Biological Citizens after Chernobyl* (Princeton, N.J., and Oxford: Princeton University Press, 2002).

2. Bryan C. Taylor, "'Our Bruised Arms Hung Up as Monuments': Nuclear Iconography in Post–Cold War Culture," *Critical Studies in Media Communication* 20, no. 1 (2003): 1–34; Jeffrey Sasha Davis, "Introduction. Military Natures: Militarism and the Environment," *GeoJournal* 69, no. 3 (2007): 131–34; Jeffrey Sasha Davis, "Scales of Eden: Conservation and Pristine Devastation on Bikini Atoll," *Environment and Planning D: Society and Space* 25, no. 2 (2007): 213–35; David Havlick, "Bombs Away: New Geographies of Military-to-Wildlife Conversions in the United States," Ph.D. dissertation, University of North Carolina, Chapel Hill, 2006; David Havlick, "Logics of Change for Military-to-Wildlife Conversions in the United States," *GeoJournal* 69, no. 3 (2007): 151–64; Nick Brown and Sarah Kanouse, "Fishing for Uranium: Race, Nature and the Military at the Crab Orchard National Wildlife Refuge" (paper presented at the "Visible Memories" conference, Syracuse, New York, October 4, 2008); Peter Coates, Tim Cole, Marianna Dudley, and Chris Pearson, "Defending Nation, Defending Nature? Militarized Landscapes and Military Environmentalism in Britain, France, and the United States," *Environmental History* 16, no. 3 (July 2011): 456–91.

3. Davis, "Introduction. Military Natures"; Davis, "Scales of Eden"; Valerie Kuletz, "Invisible Spaces, Violent Places: Cold War Nuclear and Militarized Landscapes," in *Violent Environments*, ed. Nancy L. Peluso and Michael Watts (Ithaca, N.Y., and London: Cornell University Press, 2001), 237–60.

4. Dinesh Joseph Wadiwel, "Cows and Sovereignty: Biopower and Animal Life," *Borderlands* 1, no. 2 (2002), http://www.borderlands.net.au/vol1no2_2002/wadiwel_cows.htm (accessed July 17, 2001); Dinesh Wadiwel, "Animal by Any Other Name? Patterson and Agamben Discuss Animal (and Human) Life," *Borderlands* 3, no. 1 (2004), http://www.borderlands.net.au/vol3no1_2004/wadiwel_animal.html (accessed July 17, 2001).

5. Joseph Masco, "Mutant Ecologies: Radioactive Life in Post–Cold War New Mexico," *Cultural Anthropology* 19, no. 4 (2004): 517–50; Joseph Masco, *The Nuclear Borderlands: The Manhattan Project in Post–Cold War New Mexico* (Princeton, N.J., and Oxford: Princeton University Press, 2006); Jake Kosek, *Understories: The Political Life of Forests in Northern New Mexico* (Durham, N.C., and London: Duke University Press, 2006).

6. Giorgio Agamben discusses the animal's "zone of indeterminacy" in *The Open: Man and Animal*, trans. Ken Attell (Stanford, Calif.: Stanford University Press, 2004), 37. Also refer to Cary Wolfe, "Before the Law: Animals in a Biopolitical Context," *Law, Culture and the Humanities* 6, no. 1 (2010): 8–23; Michael Dillon and Luis Lobo-Guerrero, "The Biopolitical Imaginary of Species-being,"

Theory, Culture and Society 26, no. 1 (2009): 1–23; Gwendolyn Blue and Melanie Rock, "Trans-biopolitics: Complexity in Interspecies Relations," *Health* 15, no. 4 (2011): 353–68. Gwendolyn Blue and Melanie Rock explore how value is assigned to life—the way species are demarcated, differentiated, arranged, and regulated, and the consequences of this, which are sometimes lethal; they attend to animal-to-animal relations mediated by human actions and inequalities, honing the idea of "trans-biopolitics" to emphasize the arrangements of value, of life.

7. John F. Hoffecker, *Twenty-seven Square Miles: Landscape and History at the Rocky Mountain Arsenal National Wildlife Refuge* (U.S. Fish and Wildlife Service, 2001), 84; Camille Colatosti, "A 'Toxic Tour' of Denver: Working for Environmental Justice at the Grassroots," *The Witness* (July–August 2000), http://www.thewitness.org/archive/julyaug00/toxictour.html (accessed July 17, 2011).

8. The vignette "Site Background: Eagle Vision" draws on numerous documents, among them Hoffecker, *Twenty-seven Square Miles;* U.S. Environmental Protection Agency (EPA), "Colorado Site Locator," http://www.epa.gov/region8/superfund/co/ (accessed February 5, 2012); EPA, "Superfund Program: Rocky Mountain Arsenal," http://www.epa.gov/region8/superfund/co/rkymtnarsenal/index.html (accessed February 5, 2012); Rocky Mountain Arsenal (RMA) Information Center, "History of Rocky Mountain Arsenal: Commerce City, Colorado" No. 84324RO4 (May 1980), http:///www.dtic.mil/cgi-bin/GetTRDoc?AD=ADA288453 (accessed February 12, 2012); RMA Remediation Venture Office, "Rocky Mountain Arsenal History," http://www.rma.army.mil/site/sitefrm.html (accessed February 12, 2012); U.S. Department of Defense (DOD), "Final Agreement on Rocky Mountain Arsenal Cleanup Signed" No. 354-96 (June 11, 1996), http://www.defense.gov/releases/release.aspx?releaseid=931 (accessed February 5, 2012); Colorado Department of Public Health and Environment (CDPHE), "Preassessment Screen Determination" (January 12, 2007), http://www.colorado.gov (accessed February 5, 2012); EPA, "Partial Deletion of the Rocky Mountain Arsenal Federal Facility," 40 CFR Part 300 [EPA-HQ-SFUND-1987-0002; FRL-9199-5], *Federal Register* 75, no. 176 (September 13, 2010/ FR Doc No. 2010-22747); RMA Remediation Venture Office, "Bison Reintroduction," http://www.rma.army.mil/refuge/bisonfrm.html (accessed February 5, 2012); "Nerve Gas Bomb Discovered during Colorado Arsenal Conversion," *Environment News Service* (October 23, 2000), http://ens-newswire.com/ens/oct2000/2000-10-23-15.asp (accessed February 4, 2012); Raquel Fuentes, "Rocky Mountain Arsenal Communications Program Named One of the Top Five PR Campaigns of the Decade," *Denver Post* (April 8, 2010); U.S. Department of Interior Recovery Investments, "Rocky Mountain Arsenal National Wildlife Refuge" (September 29, 2011), http://recovery.doi.gov/press/us-fish-and-wildlife-service/rocky-mountain-arsenal-national-wildlife-refuge/ (accessed February 4, 2012).

9. Edmund Russell, *War and Nature: Fighting Humans and Insects with Chemicals from World War I to Silent Spring* (Cambridge: Cambridge University Press, 2001); Richard P. Tucker and Edmund Russell, eds., *Natural Enemy, Natural Ally: Toward an Environmental History of War* (Corvallis: Oregon State University

Press, 2004); U.S. Army, "Rocky Mountain Arsenal: Turning Vision into Action," http://www.rma.army.mil/ (accessed July 17, 2011).

10. Gayle Worland, "Toxic Wait: Some Residents Say the Rocky Mountain Arsenal Still Isn't Clean Enough to Polish Commerce City's Image," *Westword* (February 11, 1999), http://www.westword.com/1999-02-11/news/toxic-wait/ (accessed July 17, 2011); Julie Jargon, "This Is a Job for Superfund!" *Westword* (July 5, 2001), http://www.westword.com/2001-07-05/news/this-is-a-job-for-superfund/ (accessed February 2, 2005); Gayle Worland, "Shy, but Not Retiring," *Westword* (October 29, 1998), http://www.westword.com/issues/1998-10-29/news/news4.html (accessed February 2, 2005).

11. John C. Ensslin, "Rocky Mountain Arsenal Wildlife Refuge Closed: Traces of Chemical Weapon from '43 Found at Arsenal," *Rocky Mountain News* (November 2, 2007); Jeff Kass, "Wildlife Refuge Programs Canceled," *Rocky Mountain News* (November 3, 2007); John C. Ensslin, "Lewisite Tests to Begin Today," *Rocky Mountain News* (November 13, 2007); "Arsenal Cleanup Work Suspended," *Rocky Mountain News* (May 19, 2008), http://www.rockymountainnews.com/news/2008/may/19/arsenal-cleanup-work-suspended/ (accessed May 20, 2008); CDPHE, "Rocky Mountain Arsenal Wildlife Refuge Temporarily Closed to the Public" (November 1, 2007), http://www.cdphe.state.co.us/release/2007/110107.html (accessed July 17, 2011).

12. RMA Remediation Venture Office, "Experts Safely Dispose of Laboratory Jar at Rocky Mountain Arsenal," http://www.rma.army.mil/involve/PressReleases08/DisposalofLabJar.html (accessed July 24, 2011).

13. Petryna, *Life Exposed*, 216–17.

14. On the court cases, refer to Matthew Greene, "The Rocky Mountain Arsenal: States' Rights and the Cleanup of Hazardous Waste" (working paper 94-58, Conflict Resolution Consortium, University of Colorado, Boulder, 1994), http://www.colorado.edu/conflict/full_text_search/AllCRCDocs/94-58.htm (accessed July 17, 2011); *State of Colorado v. United States of America, Shell Oil Company, et al.*, 83-C-2386 (D.C. Cir. 2009); John W. Suthers, Attorney General, Colorado Department of Law, "Colorado Settles Rocky Mountain Arsenal Suit" (May 29, 2008), http://www.coloradoattorneygeneral.gov/press/news/2008/05/29/colorado_settles_rocky_mountain_arsenal_suit (accessed July 17, 2011); Todd Hartman, "Arsenal Cleanup Suit Settled," *Rocky Mountain News* (May 30, 2008), 4; John Ingold, "Arsenal Deal Opens Tap for Cleanup," *Denver Post* (May 30, 2008), 1A and 19A.

15. Thomas J. Knudson, "Rocky Mountain Journal," *New York Times* (March 2, 1987), http://www.nytimes.com/1987/03/02/us/rocky-mountain-journal-bad-idea-bad-art-bad-dog.html (accessed February 10, 2011).

16. Jamie Lorimer, "Nonhuman Charisma," *Environment and Planning D: Society and Space* 25, no. 5 (2007): 911–32; Steve Hinchliffe, "Reconstituting Nature Conservation: Towards a Careful Political Ecology," *Geoforum* 39 (2008): 88–97; Suzanne M. Michel, "Golden Eagles and the Environmental Politics of Care," in *Animal Geographies: Place, Politics, and Identity in the Nature-Culture Borderlands*, ed. Jennifer Wolch and Jody Emel (London and New York: Verso, 1998), 162–87.

17. Stephen Gascoyne, "From Toxic Site to Wildlife Refuge," *Christian Science Monitor* (September 12, 1991), 10; Franklin Hoke, "Nature Reclaims Toxic Arsenal," *Environment* 33, no. 8 (1991): 22; Stephen Gascoyne, "Toxic Refuge," *Environmental Action* 23, no. 3 (1991): 9; Bill Breen, "From Superfund Site to Wildlife Refuge," *Garbage* 4, no. 3 (1992): 22; Stephen Gascoyne, "Slipcovering a Superfund Site," *Bulletin of the Atomic Scientists* 49, no. 7 (1993): 33; Karen B. Wiley and Steven L. Rhodes, "From Weapons to Wildlife: The Transformation of the Rocky Mountain Arsenal," *Environment* 40, no. 5 (1998), 4–11; Jeffrey P. Cohn, "A Makeover for Rocky Mountain Arsenal," *BioScience* 49, no. 4 (1999): 273–77; Kirk Johnson, "Weapons Moving Out, Wildlife Moving In," *New York Times* (April 19, 2004), http://www.nytimes.com/2004/04/19/national/19ROCK.html (accessed July 20, 2011); Bruce Finley, "Arsenal of Greenery," *Denver Post* (October 16, 2010), 1–2B; U.S. Fish and Wildlife Service (FWS) Rocky Mountain Arsenal National Wildlife Refuge (RMANWR) brochure (Commerce City, Colo., 2003).

18. Nicole Shukin, *Animal Capital: Rendering Life in Biopolitical Times* (Minneapolis and London: University of Minnesota Press, 2009); Kathryn Yusoff, "Biopolitical Economies and the Political Aesthetics of Climate Change," *Theory Culture Society* 27, nos. 2–3 (2010): 73–99.

19. FWS, "Draft Reintroduction Plan for Bison at RMA-NWR" (2007), http://www.fws.gov/rockymountainarsenal/Public%20Comment/BisonRe.pdf (accessed July 21, 2011).

20. The inventory of quotations about the bison was developed from the following sources: FWS, "Draft Reintroduction Plan for Bison"; RMA Remediation Venture Office, "Bison Pilot Program," *Milestones* (spring 2007), 1; FWS, "Comprehensive Management Plan" (1996), 52, quoted in FWS, "Draft Reintroduction Plan," 9; FWS, "News Release: U.S. Fish and Wildlife Service to Establish Pilot Bison Project at Rocky Mountain Arsenal National Wildlife Refuge" (January 10, 2007), http://www.fws.gov/mountain-prairie/pressrel/07-01.htm (accessed July 21, 2011). Footage of the bison release can be found on YouTube, such as "Bison Reintroduction to the RMA National Wildlife Refuge," http://www.youtube.com/watch?v=rPpjoAhrlcc (accessed July 21, 2011).

21. Geoffrey C. Bowker, "Time, Money, and Biodiversity," in *Global Assemblages: Technology, Politics, and Ethics as Anthropological Problems*, ed. Aihwa Ong and Stephen J. Collier (Malden, Mass., and Oxford: Blackwell, 2005), 107–23.

22. Scott Lauria Morgensen, "Settler Homonationalism: Theorizing Settler Colonialism within Queer Modernities," *GLQ: A Journal of Lesbian and Gay Studies* 16, nos. 1–2 (2010): 105–31.

23. Martin O'Connor, "On the Misadventures of Capitalist Nature," in *Is Capitalism Sustainable? Political Economy and the Politics of Ecology*, ed. Martin O'Connor (New York and London: Guilford Press, 1994), 125–51, esp. 130–31.

24. As per John Beck: The landscape of representation "naturalizes (and thereby de-historicizes) violence as a prehistory of national becoming." Furthermore, the landscape itself *is* the "ordering of space in the aftermath of territorial wars . . . the granulated remains of the exterminated, turned into landscape by the violence

left unchecked by an expansionist politics that articulates itself as a force of nature" (John Beck, *Dirty Wars: Landscape, Power, and Waste in Western American Literature* [Lincoln and London: University of Nebraska Press, 2009], 86 and 72, respectively).

25. Donna Haraway, *When Species Meet* (Minneapolis and London: University of Minnesota Press, 2008).

26. Donald S. Moore, Jake Kosek, and Anand Pandian, eds., *Race, Nature, and the Politics of Difference* (Durham, N.C.: Duke University Press, 2003).

27. Ghassan Hage, *White Nation: Fantasies of White Supremacy in a Multicultural Society* (New York: Routledge, 2000), 166–73.

28. James D. Proctor, "The Spotted Owl and the Contested Moral Landscape of the Pacific Northwest," in *Animal Geographies: Place, Politics, and Identity in the Nature–Culture Borderlands*, ed. Jennifer Wolch and Jody Emel (London and New York: Verso, 1998), 191–217, esp. 193.

29. Matthew Chrulew, "Managing Love and Death at the Zoo: The Biopolitics of Endangered Species Preservation," *Australian Humanities Review*, no. 50 (May 2011), http://www.australianhumanitiesreview.org/archive/Issue-May-2011/chrulew.html (accessed November 28, 2011). Chrulew asserts that "the wild" is eventually transformed into an extension of the zoo, such as when strategies of "soft release" provide animals with intense support after reintroduction.

30. Jean Baudrillard, "The Animals: Territory and Metamorphoses," in *Simulacra and Simulation*, trans. Sheila Faria Glaser, http://www9.georgetown.edu/faculty/irvinem/theory/Baudrillard-Simulacra_and_Simulation.pdf (accessed July 17, 2011), 89; emphasis added.

31. Catherine Waldby develops the term "biovalue" in her work *The Visible Human Project: Informatic Bodies and Posthuman Medicine* (London: Routledge, 2000). See also Stefan Helmreich, "Species of Biocapital," *Science as Culture* 17, no. 4 (2008): 463–78. Biovalue is generated whenever the productivity of life/living entities can be instrumentalized in ways that are useful to human projects.

32. Shukin, *Animal Capital*, 20.

33. Chrulew, "Managing Love and Death at the Zoo."

34. This vignette draws heavily on Timothy Luke's discussion of spectacular nature photography: Timothy Luke, "Nature Protection or Nature Projection: A Cultural Critique of the Sierra Club," *Capitalism, Nature, Socialism* 8, no. 1 (1997): 37–63, esp. 51.

35. Gary Gerhardt, "Ready to Reel 'Em Back In," *Rocky Mountain News* (April 12, 2004).

36. Cindi Katz, "Whose Nature, Whose Culture? Private Productions of Space and the 'Preservation' of Nature," in *Remaking Reality: Nature at the Millennium*, ed. Bruce Braun and Noel Castree (New York: Routledge, 1998), 46–63, esp. 52. Also refer to the case of the Vandenberg Air Force Base in Leslie Friedman and Ken Wiley, "Protecting Military Secrets: A Wealth of Rare Species on U.S. Department of Defense Lands Sparks Innovative Partnership," *Nature Conservancy* 39, no. 4 (1989): 4–9.

37. Andrew Ross, "The Future Is a Risky Business," in *FutureNatural: Nature, Science, Culture,* ed. George Robertson, Melinda Mash, Lisa Tickner, Jon Bird, Barry Curtis, and Tim Putnam (London and New York: Routledge, 1996), 7–20.

38. Tetra Tech FW, Inc., http://www.ttfwi.com (accessed September 2, 2005); current Web site is Tetra Tech EC, Inc., http://www.tteci.com (accessed February 2, 2012).

39. Ross, "The Future Is a Risky Business, 14–15.

40. Ibid., 8.

41. Havlick, "Bombs Away."

42. John Wills, "'Welcome to the Atomic Park': American Nuclear Landscapes and the 'Unnaturally Natural,'" *Environment and History* 7, no. 4 (2001): 462.

43. See, for example, the FWS RMANWR brochures, especially September 2003.

44. Katz, "Whose Nature, Whose Culture?" 54–55.

45. Worland, "Toxic Wait."

46. The mixed-use development was originally named "Prairie Gateway"; however, a rebranding campaign shifted the title to "Victory Crossing." See Joey Kirchmer, "Commerce City Mixed-Use Project Renamed Victory Crossing," *Denver Post* (December 6, 2010), http://www.denverpost.com/business/ci_16785053 (accessed July 24, 2011); Commerce City Economic Development Office, "New Colorado Rapids Soccer Stadium to be Located in Commerce City," http://www.ci.commerce-city.co.us/discovery/soccerStadium.html (accessed March 27, 2005); John Ingold, "Commerce City Pins Hopes on Developing Land Arsenals," *Denver Post* (September 21, 2004), http://www.cpeo.org/lists/military/2004/msg00799.html (accessed July 17, 2011); Commerce City Economic Development Office, "Prairie Gateway: Discover the Wildlife Refuge through the Prairie Gateway!" http://www.ci.commerce-city.co.us/discovery/prairieGateway.html (accessed March 26, 2005); Adams County Economic Development, "Locating Your Business in Adams County, Colorado," http://www.adamscountyed.com/ez.html (accessed October 1, 2006); Todd Hartman and Burt Hubbard, "Home Ownership Growing: Census Finds Increase among Coloradans Enjoying the Benefits of Buying Residences," *Rocky Mountain News* (July 3, 2001), http://www.rockymountainnews.com/news/2001/Jul/03/home-ownership-growing/ (accessed June 21, 2007). On the new visitor center at the refuge, see "New Visitor Center Grand Opening Celebration at Rocky Mountain Arsenal National Wildlife Refuge," *Wild News* (February–May, 2011), 1; "A New Visitor Center for Rocky Mountain Arsenal National Wildlife Refuge," *Wild News* (July–September, 2010).

47. "The Mascots," http://www.coloradorapids.com/supporters/rapids-mascots (accessed February 10, 2011).

48. Jennifer Wolch, "Zoopolis," *Capitalism, Nature, Socialism* 7, no. 2 (1996): 21–47. Wolch explains, "The distanciation of wild animals has simultaneously stimulated the elaboration of a romanticized wildness used as a means to peddle consumer goods, sell real estate, and sustain the capital accumulation process, reinforcing urban expansion and environmental degradation" (34).

49. See the Shea Homes Reunion housing development Web site, http://www.reunionco.com (accessed February 1, 2005); and Amber Homes' numerous residential development projects in the area, such as http://www.amberhomes.com/eaglecreek.html (accessed October 1, 2006).

50. Wolch, "Zoopolis."

51. EPA, "Rocky Mountain Arsenal: From Weapons to Wildlife," http://www.epa.gov/superfund/accomp/success/rma.htm (accessed February 1, 2005).

52. Patricia Calhoun, "And Not a Drop to Drink," *Westword* (April 3, 1997), http://www.westword.com/1997-04-03/news/and-not-a-drop-to-drink/ (accessed July 17, 2011).

53. The Colorado People's Environmental and Economic Network former Web site, http://www.copeen.org/toxictours.htm (accessed September 5, 2005).

54. For examples of child-focused visitor activities, see the "Go!wild" activity books funded by Shell and FWS; FWS, *Fish and Wildlife News,* Special Issue "Children and Nature" (summer/fall 2007), http://www.fws.gov/news/pdf/News_SuFall07.pdf (accessed February 17, 2011); and a range of activities posted on the FWS RMANWR Web site, including "Refuge Time Capsules," "Wheels to Wildlife," and "Project WILD Colorado" workshops, http://www.fws.gov/rockymountainarsenal/teachers/teachertraining.htm (accessed February 17, 2011).

55. Glascoyne, "Slipcovering a Superfund Site," 33; Gayle Worland, "Mr. Clean: Superfund Ombudsman Robert Martin Is Either a White Knight—or an EPA Whitewash," *Westword* (April 1, 1999), http://www.westword.com/1999-04-01/news/mr-clean/ (accessed July 17, 2011).

56. From the "June, Education" section of the scenic calendar "Denver's Urban Wildlife Refuge," produced by the RMA in 1991; partly reprinted in "Album: Unnatural Nature," in *Uncommon Ground: Rethinking the Human Place in Nature,* ed. William Cronon (New York and London: W. W. Norton, 1996), 62.

57. Clayton Pierce, "The Promissory Future(s) of Education: Rethinking Scientific Literacy in the Era of Biocapitalism," *Educational Philosophy and Theory* 44, no. 7 (2011): 721–45.

58. U.S. Army, "Visitor Safety Fact Sheet," http://www.pmrma.army.mil/refuge/Export1.htm (accessed July 17, 2011); FWS, "Visitor Services," http://rockymountainarsenal.fws.gov/visitors/vc.htm and http://rockymountainarsenal.fws.gov/visitors/trails.htm (accessed July 17, 2011).

59. FWS Office of Law Enforcement, http://www.fws.gov/le/Natives/EagleRepository.htm and http://www.fws.gov/mountain-prairie/law/eagle/ (accessed May 3, 2009); Shauna Stephenson, "Where Evidence from Wildlife Crimes Goes to Be Destroyed," *Wyoming News* (June 14, 2010), http://www.wyomingnews.com/articles/2010/06/19/outdoors/10ut_06-13-10.txt (accessed July 21, 2011); Electa Draper, "National Repository: Where Dead Eagles Land," *Denver Post* (September 1, 2009), http://www.denverpost.com/ci_13242945 (accessed July 21, 2011); Gary Gerhardt, "Feathers in Big Demand," *Rocky Mountain News* (August 4, 2005), 33A; Gayle Worland, "The Wild Life," *Westword* (May 19, 1999), http://www.westword.com/1999-05-20/news/the-wild-life/ (accessed February 1, 2005).

60. Jody Baker, "Modeling Industrial Thresholds: Waste at the Confluence of Social and Ecological Turbulence," *Cultronix,* no. 1, http://eserver.org/cultronix/baker (accessed July 17, 2011).

61. Ross, "The Future Is a Risky Business," 12.

62. Valerie Kuletz powerfully suggests that the U.S. Southwest might be understood as an intense grounds of environmental racism in *The Tainted Desert: Environmental and Social Ruin in the American West* (New York and London: Routledge, 1998), esp. 13–14; Mike Davis, "Dead West: Ecocide in Marlboro Country," *New Left Review* 200 (1993): 49–73. See also Mike Davis, "Utah's Toxic Heaven," *Capitalism, Nature, Socialism* 9, no. 2 (1998): 35–39; Peter C. van Wyck, *Signs of Danger: Waste, Trauma, and Nuclear Threat* (Minneapolis and London: University of Minnesota Press, 2005).

63. Giorgio Agamben, *Homo Sacer: Sovereign Power and Bare Life* (Stanford, Calif.: Stanford University Press, 1998).

64. Dick Foster, "The Bald and the Beautiful: Soap Opera of Regal Eagles' Life-or-Death Struggle Nears Happy Conclusion," *Rocky Mountain News* (February 24, 2007), 1 and 6–7.

65. Wadiwel, "Cows and Sovereignty"; Wadiwel, "Animal by Any Other Name?"; Shukin, *Animal Capital.*

66. Beck reports that "gothic epistemology" is a product of the military's activity in the U.S. militarized West. Tropes of concealment and exposure permeate everyday life in this region, and the despoliation of the landscape has resulted in an intensely gothic organization of material life and affect: the threat of the unseen, hidden histories, shadow kingdoms, blank spots, secret crypts, and—as imagined here via the crypt keeper's account of the bald eagle—subaltern histories of the materially unknown (Beck, *Dirty Wars,* 179–82).

67. Colorado Departments of Law and Public Health and Environment, News Release "Report Details Environmental Damage at Rocky Mountain Arsenal" (October 29, 2007), http://www.cdphe.state.co.us/release/2007/RMA-NRDSrelease_10292007.pdf (accessed July 17, 2011); CDPHE, "Hazardous Materials and Waste Management Division Natural Resource Damage Assessment for the Rocky Mountain Arsenal, Commerce City, Colorado," http://www.cdphe.state.co.us/hm/rmaplan.htm (accessed July 17, 2011). "Deadstock" indirectly refers to the "mortality effect" articulated by Sarah Lochlann Jain, "The Mortality Effect: Counting the Dead in the Cancer Trial," *Public Culture* 22, no. 1 (2010): 89–117. Jain refers to the spectral presence of the numbers of the dead that enable others to survive; such deaths are required in order to make diagnoses that predict lives. The mortality effect is this ghosting of lives that undergo the cancer trials, where deaths "maintain an everywhere and nowhere quality" yet anchor the whole process.

68. John W. Suthers, Attorney General, Colorado Department of Law, "Natural Resource Trustees Announce Conservation Easement in Rocky Mountain Arsenal Natural Resource Damages Case" (October 22, 2010), http://www.coloradoattorneygeneral.gov/press/news/2010/10/22/natural_resource_trustees_announce_conservation_easement_rocky_mountain_arsena (accessed July 24, 2011).

69. Shukin, *Animal Capital;* Jacques Derrida, *The Animal That Therefore I Am,* ed. Marie-Louis Mallet, trans. David Wills (New York: Fordham University Press, 2008); Jacques Derrida, "And Say the Animal Responded?" in *Zoontologies: The Question of the Animal,* ed. Cary Wolfe (Minneapolis and London: University of Minnesota Press, 2003), 121–46.

70. "The Use of Wildlife Biomonitoring at Hazardous Waste Sites," *Research Brief* 19 (1998) from the Superfund Basic Research Program, http://www.tiehh.ttu.edu/mhooper/rocky.htm (accessed July 17, 2011) and http://www.niehs.nih.gov/research/supported/srp/ (accessed July 17, 2011).

71. Stacy Alaimo discusses the use of biomonitoring to detect exposures, its post hoc nature, and the danger of individualizing and biomedicalizing humans to document toxicity rather than prevent it—"shifting the responsibility away from those who produce, disseminate, or fail to regulate toxic substances and to the individuals whose bodies harbor those substances" (Stacy Alaimo, *Bodily Natures: Science, Environment, and the Material Self* [Bloomington and Indianapolis: Indiana University Press, 2010], 82).

72. Jean Baudrillard, "Darwin's Artificial Ancestors and the Terroristic Dream of Transparency of the Good," *International Journal of Baudrillard Studies* 4, no. 2 (2007), http://www.ubishops.ca/baudrillardstudies/vol4_2/v4-2-baudrillard.html (accessed July 17, 2011).

73. "Bald Eagle Tired of Everyone Assuming It Just Supports War," *The Onion* (February 4, 2010), 1 and 6.

2. Alien Still Life

The congressional testimonial PowerPoint, comprising the first part of this chapter, was originally performed as an allegorical drama at the "Practice, Power, Politics, and Performance" symposium in honor of Allan Pred at the Department of Geography, University of California at Berkeley, on April 26, 2006.

1. "EGO" stands for "Endgame of Government Oversight" and is loosely based on the Government Accountability Office (GAO); the EGO report aesthetically mimics GAO congressional reporting. In this chapter, truth involves the rules and practices according to which true and false are delineated and specific effects of power are attached to the "true." Game refers to the ensemble of procedures, which designate statements as valid (or not) and arrange truths in relation to each other, and which lead to a certain result. See Michel Foucault, "Truth and Power," in *Power/Knowledge: Selected Interviews and Other Writings, 1972–1977,* ed. Colin Gordon, trans. Colin Gordon, Leo Marshall, John Mepham, and Kate Soper (New York: Pantheon Books, 1980), 132.

2. Nikolas Rose and Peter Miller, "Political Power beyond the State: Problematics of Government," *British Journal of Sociology* 43, no. 2 (1992): 173–205.

3. Michel Foucault, "Governmentality," in *The Foucault Effect: Studies in Governmentality,* ed. Graham Burchell, Colin Gordon, and Peter Miller (Chicago: University of Chicago Press, 1991), 95.

4. Paul Rutherford, "Ecological Modernization and Environmental Risk," in *Discourses of the Environment,* ed. Éric Darier (Oxford and Malden, Mass.: Blackwell, 1999), 116–17.

5. Elizabeth Dunn, "*Escherichia coli,* Corporate Discipline and the Failure of the Sewer State," *Space and Polity* 11, no. 1 (2007): 37.

6. For a critique of psychoanalytic readings of waste, see Gay Hawkins, *The Ethics of Waste: How We Relate to Rubbish* (Lanham, Md., and Oxford: Rowman and Littlefield Publishers, 2006), esp. 3–7. See also Elizabeth A. Povinelli, *Economies of Abandonment: Social Belonging and Endurance in Late Liberalism* (Durham, N.C., and London: Duke University Press, 2011), 108–9.

7. U.S. Department of Energy (DOE), Memorandum "Report on Audit of the U.S. Department of Energy's Identification and Disposal of Nonessential Land" (January 8, 1997), http://www.aforr.org/IG-0399.html (accessed June 18, 2011).

8. Marc Fioravanti and Arjun Makhijani, "Containing the Cold War Mess: Restructuring the Environmental Management of the U.S. Nuclear Weapons Complex" (1997), http://www.ieer.org/reports/cleanup/ccwm.pdf (accessed July 22, 2011).

9. DOE Memorandum "Report on Audit of the U.S. Department of Energy's Identification and Disposal of Nonessential Land."

10. Ibid.

11. DOE Office of Environmental Management (EM), "Report to Congress: Status of Environmental Management Initiatives to Accelerate the Reduction of Environmental Risks and Challenges Posed by the Legacy of the Cold War," DOE/EM-0004 (Washington, D.C.: DOE, 2009).

12. The withdrawal of land "for nature" and the withdrawal "for war" share the same history of executive power in spatial form. See Darrin Hostetler, "The Wrong War, with the Wrong Enemy, at the Wrong Time: The Coming Battle over the Military Land Withdrawal Act and an Experiment in Privatizing the Regulation of Public Lands," *Environmental Law* 29, no. 2 (summer 1999): 303–38.

13. Elmar Altvater, "Ecological and Economic Modalities of Time and Space," in *Is Capitalism Sustainable? Political Economy and the Politics of Ecology,* ed. Martin O'Connor (New York and London: Guilford Press, 1994), 89.

14. Jody Baker, "Modeling Industrial Thresholds: Waste at the Confluence of Social and Ecological Turbulence," *Cultronix,* no. 1 (1994), http://eserver.org/cultronix/baker (accessed July 22, 2011).

15. Ibid.

16. Joseph Masco, "Mutant Ecologies: Radioactive Life in Post–Cold War New Mexico," *Cultural Anthropology* 19, no. 4 (2004): 535; John Wills, "'Welcome to the Atomic Park': American Nuclear Landscapes and the 'Unnaturally Natural,'" *Environment and History* 7 (2001): 449–72. Also refer to Peter Galison's exploration of nuclear wastelands and wilderness, and the removal of parts of the earth in perpetuity for reasons of sanctification or despoilment, as reviewed by Corydon Ireland, "Wasteland and Wilderness," *Harvard Science* (October 8, 2009), http://news.harvard.edu/gazette/story/2009/10/wasteland-and-wilderness/ (accessed October 29, 2011); Peter Galison, interview by Jamie Kruse, *Friends of the*

Pleistocene blog (March 31, 2011), http://fopnews.wordpress.com/2011/03/31/galison/ (accessed October 29, 2011).

17. Joe Fiorill, "Rocky Flats Called Cleanup Model at Senate Hearing," *Global Security Newswire* (November 15, 2005), http://www.nti.org/gsn/article/rocky-flats-called-cleanup-model-at-senate-hearing (accessed August 21, 2011).

18. GAO, "Recovery Act: Most DOE Cleanup Projects Appear to Be Meeting Cost and Schedule Targets, but Assessing Impact of Spending Remains a Challenge," GAO-10-784 (Washington, D.C.: DOE, 2010).

19. Rob Nixon, *Slow Violence and the Environmentalism of the Poor* (Cambridge, Mass., and London: Harvard University Press, 2011), 18–19.

20. Ibid., 9, 60, and 185.

21. Ann Laura Stoler, "Imperial Debris: Reflections on Ruins and Ruination," *Cultural Anthropology* 23, no. 2 (2008): 193.

22. Nixon, *Slow Violence,* 57.

23. Stoler, "Imperial Debris," 191–219.

24. Ibid.; Povinelli, *Economies of Abandonment,* 108–9.

25. *Westword,* "Up and Atom! Our Rocky Flats Contest Did a Booming Business" (January 5, 1994), http://www.westword.com/1994-01-05/news/up-and-atom (accessed July 24, 2011).

26. Len Ackland, *Making a Real Killing: Rocky Flats and the Nuclear West* (Albuquerque: University of New Mexico Press, 1999), 56.

27. Zsuzsa Gille, *From the Cult of Waste to the Trash Heap of History: The Politics of Waste in Socialist and Postsocialist Hungary* (Bloomington and Indianapolis: Indiana University Press, 2007), 27; Tim Cooper, "Recycling Modernity: Waste and Environmental History," *History Compass,* nos. 8/9 (2010): 1120.

28. Jennifer Gabrys, "Sink: The Dirt of Systems," *Environment and Planning D: Society and Space* 27, no. 4 (2009): 669.

29. Keith Schneider, "Weapons Plant Pressed for Accounting of Toll on Environment and Health," *New York Times* (February 15, 1990), http://www.nytimes.com/1990/02/15/us/weapons-plant-pressed-for-accounting-of-toll-on-environment-and-health.html (accessed July 25, 2011); Tom Ruwitch, "Rockwell International: Gouging the Government," *Multinational Monitor* 11, no. 3 (1990), http://multinationalmonitor.org/hyper/issues/1990/03/ruwitch.html (accessed July 25, 2011); "Plutonium as a Liability/Dangers of Plutonium" *Science for Democratic Action* 3, no. 2 (spring 1994), http://www.ieer.org/sdafiles/vol_3/3-2/index.html (accessed July 25, 2011); Richard Fleming, "Glowing Reports: There's Plenty of Good News about Rocky Flats, and You're Paying for It," *Westword* (March 15, 1995), http://www.westword.com/1995-03-15/news/glowing-reports (accessed July 25, 2011).

30. Environmental investigators found that operations at the plant resulted in releases of hazardous substances, as defined by the Comprehensive Environmental Response, Compensation, and Liability Act (CERCLA), the Resource Conservation and Recovery Act (RCRA), and the Colorado Hazardous Waste Act (CHWA). As a result, Rocky Flats was listed on the U.S. Environmental Protection Agency's "Superfund" National Priorities List, the federal government's program to clean

up the nation's waste sites. When the DOE surveyed its nuclear sites to identify where the weapons complex was vulnerable in terms of threats to health and safety, Rocky Flats was found responsible for several of the top ten most vulnerable buildings. Refer to DOE, "Plutonium ES&H Vulnerability Assessment Project Plan" (Washington, D.C.: DOE, 1994); DOE EM, "Closing the Circle on the Splitting of the Atom: The Environmental Legacy of Nuclear Weapons Production in the United States and What the Department of Energy Is Doing about It," DOE/EM-0266 (Washington, D.C., 1996). Among the trenchant criticisms of the Rocky Flats cleanup are Jacque Brever, "An Analysis of the Department of Energy's Cleanup Plans for Four Areas at Rocky Flats: The Coverup Continues" (2004), http://www.utwatch.org/war/jacquebrever_rockyflatscleanup.html (accessed June 17, 2011); the Rocky Mountain Peace and Justice Center, http://rmpjc.org/; and Dr. LeRoy Moore's blog, http://leroymoore.wordpress.com/ (accessed August 23, 2011).

31. "Cleanup" here refers to the process of addressing contaminated land, facilities, and materials in accordance with applicable requirements; it does not mean that all hazards have been removed. See DOE EM, "Linking Legacies: Connecting the Cold War Nuclear Weapons Production Processes to Their Environmental Consequences (Washington, D.C., 1997); DOE, "Accelerating Cleanup: Paths to Closure," DOE/EM-0362 (Washington, D.C., 1998). The challenges to cleanup of Rocky Flats were vast. When production activities were halted in 1989, the plant anticipated that production would restart shortly and did not prepare for total shutdown. Materials remained in configurations and conditions unsuitable for extended storage; plutonium, in particular, was found in an array of forms. See the U.S. Department of Health and Human Services (HHS), Agency for Toxic Substances and Disease Registry, "Rocky Flats Environmental Technology Site" (May 13, 2005), http://www.atsdr.cdc.gov/hac/pha/RockyFlats%28DOE%29/RockyFlatsPHA051305a.pdf (accessed July 25, 2011).

32. DOE Office of Public Affairs, "DOE Certifies Rocky Flats Cleanup 'Complete'" (December 8, 2005), http://energy.gov/articles/doe-certifies-rocky-flats-cleanup-complete (accessed July 25, 2011); Hillary Rosner, "At the Foot of the Rockies, Cleaning a Radioactive Wasteland," *New York Times* (June 7, 2005), http://www.nytimes.com/2005/06/07/science/earth/07flat.html?pagewanted=print (accessed July 25, 2011); Todd Hartman and Kevin Vaughan, "End of Road at Flats; Inspections under Way; Weapons Plant to Revert to Prairie," *Rocky Mountain News* (October 13, 2005), front page and 5A.

33. Mary Buckner Powers, "First DOE Weapon Site Meets Cleanup Target," *Engineering News-Record* (October 24, 2005), 10–11; Ed Bodey, "Closure Is in Sight," *Rocky Flats Endvision* 11, no. 1 (January 19, 2004), 1 and 3.

34. "Final Rocky Flats Cleanup Agreement," State of Colorado Docket #96-07-19-01 (July 19, 1996), http://rockyflats.apps.em.doe.gov/references/003%20-RFCA%20Doc-FNLRFCA-All.pdf (accessed July 25, 2011).

35. DOE Policy 455.1 "Use of Risk-Based End States" (approved July 15, 2003), https://www.directives.doe.gov/directives/0455.1-APolicy/view (accessed August 23, 2011).

36. Safety here is determined by risk's calculative rationality; risk "measures" safety according to scenarios that seek to master time and discipline future land relations in order to produce cost efficiencies in the present.

37. Dan Melamed, Bryan Skokan, Mathew Zenkowich, and Dan Kocher, "Estimating Uncertainty for a Life-Cycle Cost Management Program," *Cost Engineering* 50, no. 11 (November 2008), http://www.ppc.com/Documents/EstimatingUncertaintyforaLifeCycleCostManagementProgram.pdf (accessed July 25, 2011).

38. DOE Rocky Flats Project Office, "Closure Legacy: From Weapons to Wildlife" (2006), http://rockyflats.apps.em.doe.gov/references/Closure_Legacy_Document.pdf (accessed August 23, 2011), 2–6.

39. The Fernald site in Ohio also underwent accelerated cleanup.

40. John Abbotts, "Remediation, Land Use, and Risk at Rocky Flats, and a Comparison with Hanford," *Remediation* 21, no. 3 (summer 2011), 145–62; DOE, "Rocky Flats Closure Legacy: Accelerated Closure Concept" (2006), http://rockyflats.apps.em.doe.gov/TOC.aspx (accessed July 29, 2011), 1-1–1-20.

41. DOE EM, "Report to Congress: Status of Environmental Management Initiatives to Accelerate the Reduction of Environmental Risks and Challenges Posed by the Legacy of the Cold War," DOE/EM-0004 (Washington, D.C.: DOE, 2009), http://www.em.doe.gov/pdfs/NDAA%20Report-%2801-15-09%29a.pdf (accessed August 23, 2011).

42. DOE, "Rocky Flats Closure Legacy: Accelerated Closure Concept," 1–6.

43. Ibid., 1–10.

44. Michel Callon, "An Essay on Framing and Overflowing: Economic Externalities Revisited by Sociology," in *The Laws of the Markets*, ed. Michel Callon (Oxford and Malden, Mass.: Blackwell, 1998), 244–69.

45. Refer to Appendix 9, "The Rocky Flats Vision," of the "Final Rocky Flats Cleanup Agreement."

46. DOE, "Rocky Flats Closure Legacy: Accelerated Closure Concept," 1–5.

47. Under CERCLA and in accordance with Executive Order 12580, the DOE is responsible for the response action to hazardous substance releases at Rocky Flats, with the EPA and CDPHE serving as support agencies. Executive Order 12580, signed into effect by President Reagan in 1987, removed the EPA from its position as administrator of Superfund regulations for federal facilities. The order places agencies directly responsible for causing contamination in charge of their own cleanups. In 1993, the Clinton administration directed federal agencies to reuse cleaned-up federal facilities, making land parcels available for public recreational use. In response, the EPA, Department of Defense (DOD), and DOE developed accelerated Superfund models that fast-tracked remediation. The DOD implemented the president's "Fast Track Cleanup Program" to facilitate the Base Realignment and Closure (BRAC I, II, III, and IV) program; the DOE operationalized the guiding document *Accelerating Cleanup: Paths to Closure;* and the EPA drafted the "Superfund Redevelopment Initiative." All three prioritize putting federal Superfund sites into reuse as expeditiously as possible.

48. GAO, "Nuclear Cleanup: Preliminary Results of the Review of the Department of Energy's Rocky Flats Closure Project," GAO-05-1044R (Washington,

D.C., 2005), http://www.gao.gov/new.items/d051044r.pdf (accessed July 25, 2011); GAO, "Nuclear Cleanup of Rocky Flats: DOE Can Use Lessons Learned to Improve Oversight of Other Sites' Cleanup Activities," GAO-06-352 (Washington, D.C., 2006), http://www.gao.gov/htext/d06352.html (accessed July 25, 2011); Seth Kirshenberg, Paul Kalomiris, David Abelson, and Sara Szynwelski, "The Politics of Cleanup: Lessons Learned from Complex Federal Environmental Cleanups," under cooperative agreement DE-FC02-02AL67852 with the DOE EM (Energy Communities Alliance, Inc., 2007), http://www.energyca.org/PDF/ECACleanupforPosting.pdf (accessed July 25, 2011).

49. This allowed for the sequencing of work by the cleanup manager, not DOE; Kaiser-Hill set its own cost and schedule boundaries by treating the contract as a project with a discrete end.

50. Nancy Tuor, President and Chief Executive Officer of Kaiser-Hill Company, "The Rocky Flats Closure Project: Making the Impossible Possible" (October 9, 2007), http://www.rc2007.org/Download/Presentations/Nancy_Tuor.pdf (accessed July 29, 2011).

51. The first Accelerated Site Action Project (ASAP) proposed demolishing the buildings in place and leaving existing radioactivity on-site after closure. The 1995 revised agreement entailed substantial removal of building radioactivity and waste, and led to ASAP II in 1996, a working draft intended to guide future plutonium consolidation, safety, and land use.

52. DOE, "Rocky Flats Closure Legacy: Accelerated Closure Concept," 1–16.

53. "Rocky Flats Workers Stiffed for Efficiency," *Denver Post* (November 17, 2005), B6. Todd Hartman, "New Probe at Flats: Contractor Accused of Pitching Gear to Speed Site Cleanup," *Rocky Mountain News* (May 18, 2006), 4A and 18A.

54. Leroy Moore, "Rocky Flats: The Bait-and-Switch Cleanup," *Bulletin of the Atomic Scientists* 61, no. 1 (January/February 2005): 50–57.

55. Under the Rocky Flats Cleanup Agreement, "action level" refers to the level of environmental contamination that is used to decide whether remedial action, such as removal of soil, is necessary. For plutonium, Cleanup Agreement parties proposed an action level of 651 picocuries per gram of soil, based on an estimated 1 in 10,000 risk of a wildlife refuge worker developing cancer in the Rocky Flats open space. By contrast, the Rocky Flats Future Site Use Working Group, which was convened to involve the public and local stakeholders in developing the future use of the site, had recommended that the area be returned to conservative "background levels." Owing to the disagreement, the action levels were reexamined by the Radionuclide Soil Action Level (RSAL) Oversight panel, which developed a new RSAL, that of 35 picocuries per gram of soil, which was expected to enable a hypothetical rancher and family to live on the land without interruption, drinking water and eating food grown on the property. Regardless, it was revealed that the cleanup would be predetermined by fiscal restraints, and therefore closed to meaningful public input; the site's future use would be based on the acceptable risk for a wildlife refuge worker. Refer to Abbotts, "Remediation, Land Use, and Risk at Rocky Flats"; Theresa Satterfield and Joshua Levin, "From Cold War Complex to Nature Preserve: Diagnosing the Breakdown of a Multi-Stakeholder

Decision Process and Its Consequences for Rocky Flats," in *Half-Lives and Half-Truths: Confronting the Radioactive Legacies of the Cold War,* ed. Barbara Rose Johnston (Santa Fe, N.Mex.: School for Advanced Research Press, 2007), 165–91.

56. Information in this section was assembled from Bodey, "Closure Is in Sight," 1 and 3; Hartman and Vaughan, "End of Road at Flats"; Todd Hartman, "Rocky Flats Wraps Up Radioactive Cleanup," *Rocky Mountain News* (October 8, 2005), 6A; DOE and Kaiser-Hill Company, "Rocky Flats: A Proud Legacy, a New Beginning," http://rockyflats.apps.em.doe.gov/references/Weapons to Wildlife Brochure.pdf (accessed August, 23, 2011). For a map of the redistribution of Rocky Flats waste, see GAO, "Department of Energy/Accelerated Closure of Rocky Flats: Status and Obstacles," GAO/RCED-99-100 (Washington, D.C., 1999), 39; DOE, "Rocky Flats Closure Legacy: Waste Disposition" (2006), http://rockyflats.apps.em.doe.gov/chapters/09%20-%20Waste%20Disposition.pdf (accessed July 29, 2011).

57. For the statement made by former Rocky Flats Manager Frazer Lockhart, see Martin Schneider, "Rocky Flats: R.I.P.," *Weapons Complex Monitor* 16, no. 42 (2005), 13. Additionally, refer to DOE, "Final Rocky Flats Cleanup Agreement"; Arjun Makhijani and Sriram Gopal, "Setting Cleanup Standards to Protect Future Generations," report prepared for the Rocky Mountain Peace and Justice Center by the Institute for Energy and Environmental Research (December 2001), http://www.ieer.org/reports/rocky/toc.html (accessed January 15, 2007); LeRoy Moore, "Plutonium and People Don't Mix: Why the Rocky Flats National Wildlife Refuge Should Remain Closed to the Public" (November 30, 2010), http://rmpjc.org/2010/12/01/plutonium-people-dont-mix/ (accessed July 18, 2010).

58. Environmental remediation was budgeted at approximately $470 million: Alliance for Nuclear Accountability, "Danger Lurks Below: The Threat to Major Water Supplies from U.S. Department of Energy Nuclear Weapons Plants," report prepared by Radioactive Waste Management Associates (April 2004), http://www.ananuclear.org/Portals/0/documents/Water%20Report/waterreportrockyflats.pdf (accessed August 23, 2011), 187.

59. The idea is to avoid environmentally destructive remediation and to allow for natural attenuation of contaminants in situ. On this strategy, see F. Ward Whicker, Thomas G. Hinton, Margaret MacDonnell, John Pinder III, and L. J. Habegger, "Avoiding Destructive Remediation at DOE Sites," *Science* 303, no. 5664 (March 12, 2004): 1615–16; Margaret MacDonell, "Scientists Advocate New DOE Policy to Avoid Environmentally Destructive Remediation" (March 22, 2004), http://www.ead.anl.gov/new/dsp_news.cfm?id=68 (accessed July 15, 2011); Robert Nelson, "From Waste to Wilderness: Maintaining Biodiversity on Nuclear-Bomb-Building Sites" (Washington, D.C.: Competitive Enterprise Institute, 2001).

60. S. Surovchak, of DOE Office of Legacy Management, and L. Kaiser, R. DiSalvo, J. Boylan, G. Squibb, J. Nelson, B. Darr, and M. Hanson, of S. M. Stoller Corporation, "Long-Term Surveillance and Maintenance at Rocky Flats: Early Experiences and Lessons Learned" (paper presented at the Waste Management Conference "WM2008," Phoenix, Arizona, February 24–28, 2008).

61. Rosner, "At the Foot of the Rockies."

62. Kelley Hunsberger, "Finding Closure: The Rocky Flats Plant Is Transformed from a Dangerous Nuclear Wasteland to a Community Asset," Project Management Institute 2006 Project of the Year Award (2007), http://www.pmi.org/About-Us/Our-Professional-Awards/~/media/PDF/Awards/ PMN0107_Rockyflats.ashx (accessed August 20, 2011).

63. DOE and Kaiser-Hill Company, "Rocky Flats."

64. The Department of Interior commissioned the writing of its own department history. See Robert M. Utley and Barry Mackintosh, *The Department of Everything Else: Highlights of Interior History* (Washington, D.C.: DOI, 1989). The administrative miscellany of the DOI grew as Congress historically streamlined the definitions of the other preexisting departments; because of this, the DOI was often referred to as "the government's dumping ground."

65. After celebrating its 160th anniversary, the DOI received $3 billion from President Obama's signing of the American Recovery and Reinvestment Act. Colorado's wildlife refuges were earmarked to receive a boost in their funding, particularly since Interior Secretary Ken Salazar formerly served as a Colorado senator and helped spearhead the reuse of Rocky Flats land as a wildlife refuge.

66. U.S. Fish and Wildlife Service (FWS), "Rocky Flats National Wildlife Refuge Vision and Issues Workshop Summary Report" (2002), http://www.fws.gov/rockyflats/Documents/vision_report.pdf (accessed August 23, 2011); FWS, "Summary of the Comprehensive Conservation Plan" (2005), http://www.fws.gov/rockyflats/Documents/SummaryBrochure.pdf (accessed August 23, 2011).

67. FWS, "Record of Decision" for the Rocky Flats National Wildlife Refuge Final Comprehensive Conservation Plan (Rocky Mountain Arsenal, Commerce City, Colo., 2005); DOE EM and DOI FWS, "Memorandum of Understanding," *Federal Register* 70, no. 54, FR Doc 05-5597 (March 22, 2005), 14452–57.

68. DOI Office of Inspector General, "Status of the Rocky Flats National Wildlife Refuge," Report No. C-IS-FWS-0017-2010 (July 2011). Further complicating the future of the refuge is the proposed selling or trading of a strip of property on the refuge's eastern edge for transportation service. One bid is for the Jefferson Parkway, a ten-mile toll road that would connect Colorado 128 with Colorado 93 in an effort to complete a beltway around the Denver metro area. Others would like to see open space and a bike path; and many prefer that the site remain closed to the public permanently.

69. Ulrich Beck, *Ecological Politics in an Age of Risk,* trans. Amos Weisz (Cambridge and Malden, Mass.: Polity Press, 1995), 79–86.

70. Len Ackland, *Making a Real Killing: Rocky Flats and the Nuclear West* (Albuquerque: University of New Mexico Press, 1999), 203–27; Ann Imse, "Lawyer Fees Mount in Legal Wrangling over Flats, Taxpayers Pay Bulk of Tab in Sweetheart Deal," *Rocky Mountain News* (April 21, 2007), http://www.therocky.com/news/2007/apr/21/lawyer-fees-mount-in-legal-wrangling-over-flats/ (accessed July 27, 2011).

71. Boxes of sealed court records about the Rocky Flats grand jury case were nearly released to the public in 2009 after the death of U.S. District Judge

Sherman Finesilver, who had impaneled the grand jury in 1989. However, these files were quickly sealed again, and there is no schedule for their release at this time. See Joey Bunch, "Some Flats Data Public," *Denver Post* (May 6, 2008), http://www.denverpost.com/news/ci_9164267 (accessed August 20, 2011); Patricia Calhoun, "Boxed In," *Westword* (August 13, 2009), 9.

72. This claim is particularly suspect in light of an ongoing class-action lawsuit, filed in 1993, by property owners living downwind of the Rocky Flats Plant, against former plant operators, who are indemnified by the government.

73. Prior to the establishment of the DOE Office of Legacy Management (LM), the DOE EM had developed an "Office of Long-Term Stewardship" for the purposes of coordinating stewardship activities and sites, and protecting human health and the environment; see DOE EM, Office of Long-Term Stewardship, "A Report to Congress on Long-Term Stewardship," Volume I—Summary Report, DOE/EM-0563 (Washington, D.C.: DOE, 2001); DOE EM, "From Cleanup to Stewardship: A Companion Report to *Accelerating Cleanup: Paths to Closure,*" DOE/EM-0466 (Washington, D.C.: DOE, 1999).

74. DOE LM, "Strategic Plan: Managing Today's Change, Protecting Tomorrow's Future" (Washington, D.C., 2004).

75. LM, "LM Site Management Guide aka the 'Blue Book,'" Revision 11 (June 2011), http://www.lm.doe.gov/WorkArea/linkit.aspx?LinkIdentifier=id&ItemID=1834 (accessed August 23, 2011). Several types of facilities are under the purview of the LM; in addition to departmental sites, the LM is responsible for sites under the Formerly Utilized Sites Remedial Action Program, as well as that of Title II of the Uranium Mill Tailings Radiation Control Act.

76. "Office of Legacy Management Designated as a High-Performing Organization" (July 6, 2009), http://www.lm.doe.gov/default.aspx?id=2102 (accessed August 24, 2011).

77. Michael W. Owen, LM, "Letter from the Director," http://www.lm.doe.gov/WorkArea/linkit.aspx?LinkIdentifier=id&ItemID=417 (accessed August 23, 2011).

78. DOE, "FY 2004 Congressional Budget Request/Budget Highlights," DOE/ME-0023 (Washington, D.C.: DOE, 2003), 84.

79. DOE, "Five Year Plan FY 2007–FY 2011," vol. 1 (March 15, 2006), http://www.cfo.doe.gov/cf20/doefypvol1.pdf (accessed August 23, 2011), 170.

80. Energy Communities Alliance, "Environmental Remediation and Long-term Stewardship" (2004), http://www.energyca.org/policies_erlts.htm (accessed August 23, 2011); GAO, "Nuclear Cleanup of Rocky Flats: DOE Can Use Lessons Learned to Improve Oversight of Other Sites' Cleanup Activities," GAO-06-352 (U.S. GAO: Washington, D.C., 2006), http://www.gao.gov/new.items/d06352.pdf (accessed August 21, 2011).

81. Tony Carter, "Office of Legacy Management Organized to Ensure Effective and Efficient Management of Department of Energy Legacy Responsibilities" (paper presented at the Waste Management Conference "WM2007," Tucson, Arizona, February 25–March 1, 2007).

82. The Rocky Flats Legacy Management Agreement established the regulatory

framework for implementing the final response action, and for ensuring that it remains protective of human health and the environment. This agreement coordinates all DOE obligations under CERCLA, RCRA, CHWA; see EPA Region VIII, the State of Colorado, and DOE, "Rocky Flats Legacy Management Agreement," CERCLA 8-96-21, RCRA (3008(h)) 8-96-01, State of CO Docket 96-07-19-01 (February 2007).

83. Beck, *Ecological Politics in an Age of Risk*, 98–99.

84. Lynda Crowley-Cyr, "Contractualism, Exclusion, and 'Madness' in Australia's 'Outsourced Wastelands,'" *Macquarie Law Journal* 5 (2005), http://www.austlii.edu.au/au/journals/MqLJ/2005/5.html (accessed July 25, 2011).

85. "Office of Legacy Management Receives Management Award" (January 24, 2011), http://www.lm.doe.gov/default.aspx?id=7563 (accessed August 24, 2011).

86. For alternative approaches to legacy waste sites, see International Institute for Indigenous Resource Management, "Taking Control: Opportunities for and Impediments to the Use of Socio-Cultural Controls for Long-Term Stewardship of U.S. Department of Energy Legacy Waste Sites" Final Report (Roundtable, Denver, 2004), http://www.iiirm.org/publications/Articles%20Reports%20Papers/Environmental%20Restoration/resolve.htm (accessed August 23, 2011); "Rocky Flats: A Call to Guardianship" lecture series and workshops (2011), http://www.rockyflatsnuclearguardianship.org/ (accessed August 23, 2011).

87. DOI National Park Service, "National Register of Historic Places" (May 19, 1997), http://www.nps.gov/nr/listings/971212.htm (accessed July 25, 2011).

88. Stuart Steers, "Chain Reaction: The Cold War Has Ended, but the Fight over Where to Put a Rocky Flats Museum Has Just Begun," *Westword* (September 28, 2000), http://www.westword.com/2000-09-28/news/chain-reaction/ (accessed July 25, 2011).

89. For a more detailed account of the Rocky Flats Cold War Museum struggle with the DOE LM, see Jason Krupar and Stephen Depoe, "Cold War Triumphant: The Rhetorical Uses of History, Memory, and Heritage Preservation within the Department of Energy's Nuclear Weapons Complex," in *Nuclear Legacies: Communication, Controversy, and the U.S. Nuclear Weapons Complex*, ed. Bryan C. Taylor, William J. Kinsella, Stephen P. Depoe, and Maribeth S. Metzler (Lanham, Md.: Lexington Books, 2008), 135–66. For further research on long-term stewardship, preservation, and the LM, see Jason Krupar, "The Challenges of Preserving America's Nuclear Weapons Complex," in *The Atomic Bomb and American Society*, ed. Rosemary B. Mariner and G. Kurt Piehler (Knoxville: University of Tennessee Press, 2009), 381–405; Bryan C. Taylor, "(Forever) at Work in the Fields of the Bomb: Images of Long-Term Stewardship in Post–Cold War Nuclear Discourse," in Taylor et al., *Nuclear Legacies*, 199–236; Jason Krupar, "Burying Atomic History: The Mound Builders of Fernald and Weldon Spring," *Public Historian* 29, no. 1 (2007): 31–58.

90. Joey Bunch, "Museum Sought for Rocky Flats," *Denver Post* (August 25, 2003), 1B and 4B.

91. This echoes a general trend of excluding the Cold War landscape from U.S. heritage. See Tom Vanderbilt, *Survival City: Adventures among the Ruins of Atomic*

America (New York: Princeton Architectural Press, 2002); Jon Wiener, *How We Forgot the Cold War* (Berkeley and Los Angeles: University of California Press, 2012).

92. In April 2011, the Rocky Flats Cold War Museum established a lease on the historic Arvada Post Office in Olde Town Arvada. The museum's opening, originally set for September 2012, was delayed until 2013 in order to continue fundraising and complete the exhibit development process. Refer to the museum's Web site for press releases and information on exhibitions: http://www.rockyflatscoldwarmuseum.org (accessed March 10, 2012).

93. Peggy Lowe, "Udall Opposes Idea of Radiation Warning Signs, Consent Forms," *Rocky Mountain News* (January 6, 2005), http://rockymountainnews.com/drmn/legislature/article/0,1299,DRMN_37_3448839,00.html (accessed June 12, 2005).

94. Whether signs and brochures that address these issues will be provided by the state, and whether informed consent should be established for site visitations, have been the subject of heated debate. The FWS developed a draft plan for interpretive signage about the history and safety of the site and opened it to public comment; see http://www.fws.gov/rockyflats/Signage/Sign.htm (accessed July 25, 2011). Colorado State Representative Wes McKinley advocated and reintroduced a bill that would require the posting of on-site signs, and the distribution of brochures to schoolchildren before field trips to the site, that would explain the history of environmental violations and pollution from nuclear weapons production at Rocky Flats.

95. Ann Schrader, "Agency's Purge of Flats Documents Triggers Outcry," *Denver Post* (May 11, 2008), 1B and 12B; Ann Schrader, "Protests Stir Energy Agency to Preserve Flats Papers," *Denver Post* (June 8, 2008), B28. Letters of concern from various agencies can be found in the packet of meeting minutes and correspondences at http://www.rockyflatssc.org/RFSC_agendas/RFSC_Bd_mtg_packet_8_08.pdf (accessed August 23, 2011).

96. Dunn, "*Escherichia coli,* Corporate Discipline and the Failure of the Sewer State," 49. Also see Peter C. van Wyck, *Signs of Danger: Waste, Trauma, and Nuclear Threat* (Minneapolis and London: University of Minnesota Press, 2005), 82.

97. Dunn, "*Escherichia coli,* Corporate Discipline and the Failure of the Sewer State," 41–43.

98. Ibid., 50.

99. Ibid., 43.

100. Alexander Wendt and Raymond Duvall, "Sovereignty and the UFO," *Political Theory* 36, no. 4 (2008): 626.

101. Marilyn Strathern, "'Improving Ratings': Audit in the British University System," *European Review* 5, no. 3 (1997): 318.

102. Michael Power, *The Audit Explosion* (London: Demos, 1994), 15; Michael Power, *The Audit Society: Rituals of Verification* (Oxford: Oxford University, 1997).

103. Marilyn Strathern, "Robust Knowledge and Fragile Futures," in *Global Assemblages: Technology, Politics, and Ethics as Anthropological Problems,* ed. Aihwa Ong and Stephen J. Collier (Malden, Mass., and Oxford: Blackwell, 2005), 464–81.

104. Katarzyna Kosmala MacLullich, "The Emperor's 'New' Clothes? New Audit Regimes: Insights from Foucault's Technologies of the Self," *Critical Perspectives on Accounting* 14, no. 8 (2003): 791.

105. DOE LM, "2011–2020 Strategic Plan," DOE/LM-0512 (Washington, D.C.: DOE, 2011), 19; DOE LM, "Legacy Management Goals and Performance Measures" (January 29, 2009), http://www.lm.doe.gov/WorkArea/linkit.aspx?LinkIdentifier=id&ItemID=498 (accessed August 23, 2011); DOE LM, "Joint Environmental Management System Sustainability Programs," http://www.lm.doe.gov/default.aspx?id=4283 (accessed August 23, 2011).

106. DOE LM, "2011–2020 Strategic Plan"; Tracy Ribeiro, Environmental Program Manager, DOE LM, "Legacy Management's EMS and Sustainability Plans" (no date), http://www.google.com (accessed February 10, 2012).

107. "Alien still life" refers to but differs significantly from Neil Evernden's ideas about the "natural alien." Instead of Evernden's ontology of the human as "natural alien," I use the concept of "alien" to explore the alienation of humans, especially workers, from the land, *and,* as a figure related to waste, to trouble taken-for-granted understandings of human sovereignty and separation from the environment. See Neil Evernden, *The Natural Alien: Humankind and Environment* (Toronto, Buffalo, and London: University of Toronto Press, 1999).

108. Jean Pierre La Porte, "Variations on Waste," essay posted on the Johannesburg Workshop in Theory and Criticism blog (July 27, 2009), http://jhbwtc.blogspot.com/2009/07/variations-on-waste.html (accessed July 25, 2011).

109. Crowley-Cyr, "Contractualism, Exclusion, and 'Madness.'"

110. In 2005, the author conducted a series of interviews by e-mail with former DOE Rocky Flats employees. The interview questions focused on their impressions of the site after cleanup. Among those shared was the remarkable statement about "alien still life" explored in this section and included in the PowerPoint script.

111. Jean Baudrillard, "Darwin's Artificial Ancestors and the Terroristic Dream of the Transparency of the Good," *International Journal of Baudrillard Studies* 4, no. 2 (2007), http://www.ubishops.ca/baudrillardstudies/vol4_2/v4-2-baudrillard.html (accessed July 29, 2011).

112. Lauren Berlant developed the analytic concept of "slow death" in her article "Slow Death (Sovereignty, Obesity, Lateral Agency)," *Critical Inquiry* 33, no. 4 (2007): 754–80.

113. Refer to chapter 3 for further summary of the EEOICPA program and history.

114. Kimberly McGuire, "No Clean Slate for Rocky Flats," *Denver Post* (November 28, 2005), 1B and 6B.

115. Beck, *Ecological Politics in an Age of Risk,* 88.

116. Mick Smith, "Suspended Animation: Radical Ecology, Sovereign Powers, and Saving the (Natural) World," *Journal for the Study of Radicalism* 2, no. 1 (2008): 1–25.

117. Masco, "Mutant Ecologies."

118. John Scanlan conceptualizes the temporality of waste as a kind of baroque

"lasting on" in his article "In Deadly Time: The Lasting On of Waste in Mayhew's London," *Time and Society* 16, nos. 2–3 (2007): 189–206.

119. Walter Moser, "The Acculturation of Waste," in *Waste-Site Stories: The Recycling of Memory*, ed. Brian Neville and Johanne Villeneuve (Albany: State University of New York Press, 2002), 85–105.

120. The idea of "itching" nuclear legacies references Ulrich Beck, "All Aboard the Nuclear Power Superjet, Just Don't Ask about the Landing Strip," *Guardian* (July 16, 2008), http://www.guardian.co.uk/commentisfree/2008/jul/17/nuclear power.climatechange (accessed July 25, 2011).

121. Zoë Sofia, "Exterminating Fetuses: Abortion, Disarmament, and the Sexo-semiotics of Extraterrestrialism," *Diacritics* 14, no. 2 (1984): 48.

122. Couze Venn, "Rubbish, the Remnant, Etcetera," *Theory, Culture and Society* 23, nos. 2–3 (2006): 44. Also see John Scanlan, *On Garbage* (London: Reaktion Books, 2005), 41.

123. Jean Baudrillard, "The Remainder," in *Simulacra and Simulation*, trans. Sheila Faria Glaser, http://www9.georgetown.edu/faculty/irvinem/theory/Baudrillard-Simulacra_and_Simulation.pdf (accessed July 17, 2011), 95.

124. Ibid., 96.

125. Ann Imse, "Flats Still Has Hot Spots: Low-level Radiation Found in Some Areas Thought to Be Clean," *Rocky Mountain News* (September 2, 2005), 4A and 29A; Manny Gonzales, "Rocky Flats to Get Aerial 'Hot Spot' Inspection," *Denver Post* (February 11, 2005), B3.

126. Kim McGuire, "'Hot Spots' Discovered at Flats: Areas May Not be Cleaned," *Denver Post* (September 2, 2005), B1 and B4; Kim McGuire, "DOE Agrees to Clean 13 Flats 'Hot Spots,'" *Denver Post* (September 13, 2005), B2 and B5.

127. Kim McGuire, "Crews Begin Cleaning Up Radioactive Flats Ponds," *Denver Post* (February 4, 2005), A1.

128. Americium was detected in select lung, muscle, and kidney tissues; plutonium was found in bone samples; and uranium was documented in select liver and muscle tissues. The FWS calculations of risk from ingesting deer tissue with americium—determined to be the greatest risk to humans—fell within the EPA's acceptable risk range. See Andrew S. Todd and R. Mark Sattelberg, "Actinides in Deer Tissues at Rocky Flats Environmental Technology Site," http://www.fws.gov/rockyflats/Documents/DeerTissue_ExSummary.pdf (accessed February 13, 2012).

129. Gabrys, "Sink," 679.

130. Ibid., 678.

131. Baker, "Modeling Industrial Thresholds."

132. Ibid.

133. Brian Neville and Johanne Villeneuve, "Introduction: In Lieu of Waste," in Neville and Villeneuve, *Waste-Site Stories*, 20.

134. Paul Rutherford, "The Entry of Life into History," in *Discourses of the Environment*, ed. Éric Darier (Oxford and Malden, Mass.: Blackwell, 1999), 41; Martin Hewitt, "Biopolitics and Social Policy: Foucault's Account of Welfare," *Theory,*

Culture and Society 2, no. 1 (1983): 69; Michel Foucault, *The History of Sexuality, Volume 1: An Introduction*, trans. Robert Hurley (New York: Vintage Books, 1990), 144.

135. Michel Foucault, "24 January 1979," in *The Birth of Biopolitics: Lectures at the Collège de France, 1978–1979*, ed. Michel Senellart, trans. Graham Burchell (New York: Palgrave Macmillan, 2008), 67.

136. Foucault, "Governmentality," 92.

137. Éric Darier, "Foucault and the Environment: An Introduction," in Darier, *Discourses of the Environment*, 28.

138. Rutherford, "The Entry of Life into History," 45.

139. Leerom Medovoi, "A Contribution to the Critique of Political Ecology: Sustainability as Disavowal," *New Formations: A Journal of Culture/Theory/Politics*, no. 69 (2010): 142.

140. Foucault, *History of Sexuality*, 144.

3. Hole in the Head Gang

1. *Energy Employees Occupational Illness Compensation Program Act*, Public Law 106-398, 42 U.S.C. 7384-7385. The Act passed on October 30, 2000, and became effective July 31, 2001. See Michael Flynn, "A Debt Long Overdue: Nuclear Weapons Work Made People Sick—At Last, Workers May Be Compensated," *Bulletin of the Atomic Scientists* 57, no. 4 (July/August 2001): 38–48; Arjun Makhijani, "The Burden of Proof," *Bulletin of the Atomic Scientists* 57, no. 4 (July/August 2001): 49–54; Robert Alvarez, "Making It Work: Will the Legislation Do the Job?" *Bulletin of the Atomic Scientists* 57, no. 4 (July/August 2001): 55–60; Mark J. Parascandola, "Compensating for Cold War Cancers," *Environmental Health Perspectives* 110, no. 7 (July 2002): A404–A407.

2. Ken Silver, "The Energy Employees Occupational Illness Compensation Program Act: New Legislation to Compensate Affected Employees," *American Association of Occupational Health Nurses Journal* 53, no. 6 (June 2005): 267.

3. Ibid. For an overview of the problems associated with the former DOE-run Part D, refer to U.S. Senate, Committee on Energy and Natural Resources, *To Conduct Oversight of the Implementation of the Energy Employees Occupational Illness Compensation Program Hearings*, 108th Cong., 1st sess. 108-334, November 21, 2003 (Washington, D.C.: U.S. Government Printing Office, 2004); U.S. House of Representatives, Subcommittee on Workforce Protection of the Committee on Education and the Workforce, *Energy Employees Workers' Compensation: Examining the Department of Labor's Role in Helping Workers with Energy-Related Occupational Illnesses and Diseases Hearings*, 108th Cong., 1st sess. 108-41, October 30, 2003 (Washington, D.C.: U.S. Government Printing Office, 2004).

4. U.S. Department of Energy (DOE) Office of Environmental Management (EM), "Closing the Circle on the Splitting of the Atom: The Environmental Legacy of Nuclear Weapons Production in the United States and What the Department of Energy Is Doing About It," DOE/EM-0266 (Washington, D.C., U.S. Government Printing Office, 1995).

5. Leslie R. Groves, *Now It Can Be Told: The Story of the Manhattan Project* (New York: Harper, 1962); Gordon E. Dean, *Report on the Atom: What You Should Know about the Atomic Energy Program of the United States* (New York: Knopf, 1953). For recent interdisciplinary resources, see Stephen I. Schwartz, ed., *Atomic Audit: The Costs and Consequences of U.S. Nuclear Weapons since 1940* (Washington, D.C.: Brookings Institution Press, 1998); Cynthia C. Kelley, ed., *The Manhattan Project: The Birth of the Atomic Bomb in the Words of Its Creators, Eyewitnesses, and Historians* (New York: Black Dog and Leventhal Publishers, 2007); Bryan C. Taylor, William J. Kinsella, Stephen P. Depoe, and Maribeth S. Metzler, eds., *Nuclear Legacies: Communication, Controversy, and the U.S. Nuclear Weapons Complex* (Lanham, Md.: Lexington Books, 2008); Joseph Masco, *The Nuclear Borderlands: The Manhattan Project in Post–Cold War New Mexico* (Princeton, N.J., and Oxford: Princeton University Press, 2006); Hugh Gusterson, *People of the Bomb: Portraits of America's Nuclear Complex* (Minneapolis and London: University of Minnesota Press, 2004); Barbara Rose Johnston, ed., *Half-Lives and Half-Truths: Confronting the Radioactive Legacies of the Cold War* (Santa Fe, N.Mex.: School for Advanced Research Press, 2007); Valerie Kuletz, *The Tainted Desert: Environmental and Social Ruin in the American West* (New York and London: Routledge, 1998); Matthew Farish, *The Contours of America's Cold War* (Minneapolis and London: University of Minnesota Press, 2010).

6. Parascandola, "Compensating for Cold War Cancers"; U.S. Government Accountability Office (GAO), "Energy Employees Compensation: Additional Independent Oversight and Transparency Would Improve Program's Credibility," GAO-10-302 (Washington, D.C., 2010), 3.

7. David Michaels, *Doubt Is Their Product: How Industry's Assault on Science Threatens Your Health* (Oxford and New York: Oxford University Press, 2008), 212–31. Also see Arjun Makhijani, Stephen I. Schwartz, and William J. Weida, "Nuclear Waste Management and Environmental Remediation," in Schwartz, *Atomic Audit*, 353–94.

8. Silver, "The Energy Employees Occupational Illness Compensation Program Act."

9. Ibid.

10. See Makhijani, "The Burden of Proof"; Eileen Welsome, *The Plutonium Files: America's Secret Medical Experiments in the Cold War* (New York: Delta Press, 2000).

11. Silver, "The Energy Employees Occupational Illness Compensation Program Act," 267. This was not the first time Congress tried to address radiation exposure claims. Earlier attempts included the Veterans Dioxin and Radiation Exposure Compensation Standards Act in 1984, the Radiation-Exposed Veterans Compensation Act of 1988, and the Radiation Exposure Compensation Act (RECA), initially enacted in 1990 and subsequently amended twice. Refer to the Advisory Committee on Human Radiation Experiments, "Overview of Radiation Exposure Compensation Programs" (October 4, 1994), http://www.gwu.edu/~nsarchiv/radiation/dir/mstreet/commeet/meet7/brief7/br7j1.txt (accessed August 4, 2011).

12. Eileen Welsome, "A Cure for the Common Cold Warrior," *Westword* (September 28, 2000), http://www.westword.com/2000-09-28/news/a-cure-for-the-common-cold-warrior/5/ (accessed August 4, 2011); "Ill Workers May Be Paid $100,000," Evansville Courier & Press (November 18, 1999), listed on *NucNews* (November 19, 1999), http://nucnews.net/nucnews/1999nn/9911nn/991119nn.htm (accessed August 4, 2011).

13. Subcommittee on Research and Development of the Joint Committee on Atomic Energy, *Employee Radiation Hazards and Workmen's Compensation Hearings*, 86th Cong., 1st sess., March 10–19, 1959 (Washington, D.C.: U.S. Government Printing Office, 1959).

14. U.S. House of Representatives, Subcommittee on Immigration, Border Security and Claims of the Committee on the Judiciary, *Energy Employees Occupational Illness Compensation Program: Are We Fulfilling the Promise We Made to These Cold War Veterans When We Created This Program? (Part V) Hearings*, 109th Cong., 2nd sess. 109-59, December 5, 2006 (Washington, D.C.: U.S. Government Printing Office, 2007).

15. Great emphasis was placed on the legislation's intent to restore the "wholeness" of workers, meaning the combined compensation for their occupational illnesses with commendation for their service.

16. Michaels, *Doubt Is Their Product*, 230.

17. The Energy Employees Claimant Assistance Project (EECAP) breaks down the complicated channeling of claims on its Web site: http://www.eecap.org (accessed July 28, 2011).

18. GAO, "Department of Energy, Office of Worker Advocacy: Deficient Controls Led to Millions of Dollars in Improper and Questionable Payments to Contractors," GAO-06-547 (June 27, 2006), http://www.gao.gov/htext/d06547.html (accessed August 4, 2011). The former assistant secretary of energy for environment, safety, and health candidly summates the DOE's implementation of Part D as "farcically incompetent at best and intentionally subversive at worst" (Michaels, *Doubt Is Their Product*, 230).

19. The Government Accountability Office (GAO) was formerly called the General Accounting Office. Refer to the following reports: GAO, "Department of Energy: Contractor Litigation Costs," GAO-02-418R (March 8, 2002), http://www.gao.gov/new.items/d02418r.pdf (accessed July 15, 2011); GAO, "Energy Employees Compensation: Case-Processing Bottlenecks Delay Payment of Claims," GAO-04-298T (December 6, 2003), http://www.gao.gov/new.items/d04298t.pdf (accessed August 4, 2011); GAO, "Energy Employees Compensation: Even with Needed Improvements in Case Processing, Program Structure May Result in Inconsistent Benefit Outcomes," GAO-04-516 (May 2004), http://www.gao.gov/new.items/d04516.pdf (accessed August 4, 2011); GAO, "Energy Employees Compensation: Obstacles Remain in Processing Cases Efficiently and Ensuring a Source of Benefit Payments," GAO-04-571T (March 30, 2004), http://www.gao.gov/new.items/d04571t.pdf (accessed August 4, 2011).

20. Senate Committee on Energy and Natural Resources, *To Conduct Oversight.*

21. Several amendments have been made to EEOICPA.

22. The facilities and workplaces covered by EEOICPA include Atomic Weapons Employers (AWE), which are privately owned plants that processed/produced radioactive materials for nuclear weapons, DOE facilities, and any buildings, structures, or premises where radioactive materials and beryllium were handled or over which the DOE had proprietary interest, including certain universities and private companies.

23. The three original SEC sites were Paducah, Oak Ridge, and Portsmouth. Petitions to expand SEC status are made to the Advisory Board on Radiation and Worker Health, which reviews the petitions and gives NIOSH supposedly independent advice on dose reconstruction and probability of causation.

24. EEOICPA Bulletin No. 08-01, "Rocky Flats SEC Designations" (issued October 15, 2007/effective September 5, 2007), http://www.dol.gov/owcp/energy/regs/compliance/PolicyandProcedures/finalbulletinshtml/EEOICPABulletin08-01.htm (accessed August 4, 2011).

25. The Denver-area now-defunct newspaper *Rocky Mountain News* dedicated reporting to the struggle of former Rocky Flats workers seeking SEC status. Although their petition was denied, reportage for the *Rocky Mountain News*, the *Denver Post*, and *Westword* rallied public and legislative support behind the workers. See, for example, Laura Frank's "Deadly Denial" three-day series in the *Rocky Mountain News* (July 21–23, 2008), http://www.rockymountainnews.com/special-reports/deadly-denial/ (accessed August 30, 2011).

26. U.S. Department of Health and Human Services (HHS), 42 CFR Parts 81 and 82, "Guidelines for Determining the Probability of Causation and Methods for Radiation Dose Reconstruction under the Energy Employees Occupational Illness Compensation Program Act of 2000; Final Rules," *Federal Register* 67, no. 85 (May 2, 2002): 22296–336; EEOICPA Bulletin No. 03-23, "Review of Dose Reconstruction in the Final Adjudication Branch Hearing Process" (May 2, 2003); NIOSH, "Radiation Dose Reconstruction/Probability of Causation—NIOSH-IREP" http://www.cdc.gov/niosh/ocas/ocasirep.html (accessed July 19, 2010).

27. Refer to NIOSH's IREP Web site and user's guide: http://www.niosh-irep.com/irep_niosh/ (accessed July 18, 2010). Also see NIOSH, "Frequently Asked Questions (FAQs) Probability of Causation," http://www.cdc.gov/niosh/ocas/faqspoc.html (accessed July 19, 2010); D. C. Kocher et al., "Interactive RadioEpidemiological Program (IREP): A Web-based Tool for Estimating the Probability of Causation/Assigned Share of Radiogenic Cancers," *Health Physics* 95, no. 1 (2008): 119–47.

28. Mark Parascandola, "Uncertain Science and a Failure of Trust: The NIH Radioepidemiologic Tables and Compensation for Radiation-Induced Cancer," *Isis* 93, no. 4 (2002): 559–84; HHS, "Report of the NCI-CDC Working Group to Revise the 1985 NIH Radioepidemiological Tables," NIH Publication no. 03-5387 (Washington, D.C., 2003).

29. Parascandola, "Uncertain Science and a Failure of Trust," 560.

30. Makhijani, "The Burden of Proof," 54.

31. For example, Labor is allowed to change final decisions at any time, and NIOSH can repeatedly reconstruct doses that have already been completed in order to execute new methods.

32. These questions directly echo that of Michelle Murphy's research on historical knowledge production of exposure and more specifically "sick building syndrome." See Michelle Murphy, *Sick Building Syndrome and the Problem of Uncertainty* (Durham, N.C., and London: Duke University Press, 2006), esp. 7, 82–83. Also see Linda Nash, *Inescapable Ecologies: A History of Environment, Disease, and Knowledge* (Berkeley and Los Angeles: University of California Press, 2006).

33. Adriana Petryna, *Life Exposed: Biological Citizens after Chernobyl* (Princeton, N.J., and Oxford: Princeton University Press, 2002), 13.

34. Ibid., 9–16.

35. Stuart J. Murray, "Thanatopolitics: On the Use of Death for Mobilizing Political Life," *Polygraph* 18 (2006): 191–215; John Protevi, "The Terri Schiavo Case: Biopolitics, Biopower, and Privacy as Singularity," in *Deleuze and Law: Forensic Futures,* ed. Rosi Braidotti, Claire Colebrook, and Patrick Hanafin (New York: Palgrave Macmillan, 2009), 59–72.

36. Petryna, *Life Exposed,* 13.

37. See Murray, "Thanatopolitics," 193 and 198; Stuart J. Murray and David Holmes, eds., *Critical Interventions in the Ethics of Healthcare* (London: Ashgate, 2009); Adele E. Clarke, Laura Mamo, Jennifer Ruth Fosket, Jennifer R. Fishman, and Janet K. Shim, eds., *Biomedicalization: Technoscience, Health, and Illness in the U.S.* (Durham, N.C., and London: Duke University Press, 2010); Bruce Braun, "Biopolitics and the Molecularization of Life," *Cultural Geographies* 14, no. 1 (2007): 6–28; Nikolas Rose, *The Politics of Life Itself: Biomedicine, Power, and Subjectivity in the Twenty-first Century* (Princeton, N.J., and Oxford: Princeton University Press, 2007); Deborah Lupton, *The Imperative of Health: Public Health and the Regulated Body* (Thousand Oaks, Calif.: Sage, 1995).

38. See EECAP's 2010 survey results: "Survey on Claimants' Experience with EEOICPA," http://www.eecap.org/PDF/Surveys/Final_EECAP_Survey.pdf (accessed August 15, 2011); Alliance of Nuclear Worker Advocacy Groups (ANWAG) survey on claim denials and dose reconstruction (updated January 2011/ hosted on the EECAP Web site), http://www.eecap.org/PDF/Surveys/DR%20Survey%201-28-11.pdf (accessed August 15, 2011); DOL, Office of Inspector General—Office of Audit, "Energy Employees Occupational Illness Compensation Program—DOL Could Do More to Assist Claimants and Further Improve Timeliness," Report no. 04-09-002-04-437 (November 12, 2008).

39. Murphy, *Sick Building Syndrome,* 9–10.

40. HHS, "Report of the NCI-CDC Working Group to Revise the 1985 NIH Radioepidemiological Tables."

41. Mariachiari Tallacchini, "Before and beyond the Precautionary Principle: Epistemology of Uncertainty in Science and Law," *Toxicology and Applied Pharmacology* 207 (2005): S645–51; Helen M. Regan, Mark Colyvan, and Mark A.

Burgman, "A Taxonomy and Treatment of Uncertainty for Ecology and Conservation Biology," *Ecological Applications* 12, no. 2 (2002): 618–28.

42. C. Greig Crysler and Shiloh Krupar, "Designing (In)Security: A Message from Jayne S. Phace" (paper presented at the Association of American Geographers Annual Meeting, Las Vegas, Nevada, March 27, 2009).

43. Robert N. Proctor, *Cancer Wars: How Politics Shapes What We Know and Don't Know about Cancer* (New York: Basic Books, 1995), 8; Robert N. Proctor and Londa Schiebinger, eds., *Agnotology: The Making and Unmaking of Ignorance* (Stanford, Calif.: Stanford University Press, 2008). Also see James D. Brown and Sarah L. Damery, "Uncertainty and Risk," in *A Companion to Environmental Geography*, ed. Noel Castree, David Demeritt, Diana Liverman, and Bruce Rhoads (Malden, Mass., and Oxford: Wiley-Blackwell, 2009), 81–94.

44. Michel Foucault, *The History of Sexuality, Volume 1: An Introduction*, trans. Robert Hurley (New York: Vintage Books, 1990), 144.

45. Ibid.

46. This phrase was related to the author in an informal interview with a former Rocky Flats worker (June 19, 2009). See Lauren Berlant, "Slow Death (Sovereignty, Obesity, Lateral Agency)," *Critical Inquiry* 33, no. 4 (2007): 754–80, esp. 769. Rob Nixon takes further steps to conceptualize the connections between "disposable people" and "disposable ecosystems" in his book *Slow Violence and the Environmentalism of the Poor* (Cambridge, Mass., and London: Harvard University Press, 2011).

47. For the purposes of this chapter, note that while cancer can be considered a kind of perforation in relation to the body, this is not an assertion that a whole or natural body existed beforehand.

48. Murphy, *Sick Building Syndrome*, 7–10.

49. Proctor, *Cancer Wars*, 6–7.

50. DOL, Office of the Ombudsman for EEOICP, *2010 Annual Report to Congress* (Washington, D.C., 2010), 57. "Claim numbers represent each employee or survivor who filed for benefits. Case numbers [by contrast] represent employees (living or deceased) whose employment and illness are the basis for the claim." For the latest statistics, see DOL-OWCP EEOICP Program Statistics, http://www.dol.gov/owcp/energy/regs/compliance/weeklystats.htm (accessed August 31, 2011).

51. Charlie Wolf (former project manager at Rocky Flats), Laura Frank (former reporter with the *Rocky Mountain News*), and Peter Turcic (former director of EEOICP), interview by Tamara Banks, *Studio 12*, KBDI-TV, July 30, 2008. In the exchange between Wolf and Turcic regarding Wolf's condition of brain cancer (and that of other workers), Turcic states that the legislation was not set up to recognize such disease clusters; "the law ties our hands."

52. See, for example, Joseph L. Lyon, "Nuclear Weapons Testing and Research Efforts to Evaluate Health Effects on Exposed Populations in the United States," *Epidemiology* 10, no. 5 (1999): 557–60.

53. Steve Wing, "Objectivity and Ethics in Environmental Health Science," *Environmental Health Perspectives* 111, no. 14 (November 2003): 1809–18.

54. Lyon, "Nuclear Weapons Testing"; Mark Parascandola, "Chances, Individuals and Toxic Torts," *Journal of Applied Philosophy* 14, no. 2 (1997): 147–57.

55. Representative Tom Udall of New Mexico, testifying for the *Energy Employees Occupational Illness Compensation Program: Are We Fulfilling the Promise We Made to These Cold War Veterans When We Created This Program? (Part II) Hearings,* on May 4, 2006, to the Subcommittee on Immigration, Border Security, and Claims, of the Committee on the Judiciary, U.S. House of Representatives, 109th Cong., 2nd sess. (Washington, D.C.: U.S. Government Printing Office, 2006), 8.

56. DOL, EEOICPA Bulletin No. 06-10, "Illnesses That Presently Have No Known Causal Link to Toxic Substances" (June 2, 2006); EEOICPA Bulletin No. 06-14, "Illnesses That Presently Have No Known Causal Link to Exposure to Toxic Substances under EEOICPA" (August 1, 2006).

57. DOL, EEOICPA Bulletin No. 08-38, "Rescinding Bulletins 06-10 and 06-14" (June 25, 2008).

58. Refer to the "Site Exposure Matrices Website Help Guide," http://www.sem.dol.gov/expanded/help.cfm (accessed July 19, 2011).

59. For conditions without an established link in the matrix, the DOL will look at other evidence of causation, including information submitted by claimants, district medical consultants, industrial hygienists, and so on. However, the evidence must be compelling and probative in order to support compensation.

60. "EEOICP Site Exposure Matrices Website—Expanded SEM Data Now Available for 69 DOE Sites," http://www.sem.dol.gov/index.cfm (accessed August 1, 2011).

61. Donald Elisburg, speaking for the American Federation of Labor and Congress of Industrial Organizations (AFL-CIO) and the Building Construction Trades Department (BCTD), on November 21, 2003, *To Conduct Oversight of the Implementation of the Energy Employees Occupational Illness Compensation Program Hearings,* to the Committee on Energy and Natural Resources, U.S. Senate, 108th Cong., 1st sess. 108-334 (Washington, D.C.: U.S. Government Printing Office, 2004); also see Donald Elisburg, speaking for the AFL-CIO and BCTD, on October 30, 2003, for the *Energy Employees Workers' Compensation: Examining the Department of Labor's Role in Helping Workers with Energy-related Occupational Illnesses and Diseases Hearings,* to the Subcommittee on Workforce Protections, Committee on Education and the Workforce, U.S. House of Representatives, 108th Cong., 1st sess. 108-41 (Washington, D.C.: U.S. Government Printing Office, 2004).

62. For more on Haz-Map, refer to Jay A. Brown, "An Internet Database for the Classification and Dissemination of Information about Hazardous Chemical and Occupational Diseases," http://www.haz-map.com/AJIMtextWeb.doc (accessed July 15, 2011).

63. In the field of occupational health, surveillance usually refers to the systematic collection and analysis of information on disease, injury, and hazards for the purposes of preventing injury or occupational illness. Haz-Map, by contrast, is collected from secondary sources in the medical literature rather than from actual health events.

64. Laura Frank, "Plan to Pay Sick Nuclear Workers Unfairly Rejects Many, Doctor Says," *ProPublica* (July 31, 2009).

65. The report GAO-10-302 states: "Occupational health physicians we interviewed criticized the scientific soundness of the site exposure matrix, noting that the absence of published research linking certain chemicals to diseases does not constitute evidence that such links do not exist" (32).

66. Studies exist on risk of chronic low-dose exposures, covering such subjects as cancer survivors, nuclear weapons test participants, medical workers, crew of aircraft, radium workers, underground hard-rock miners, such as uranium and gold miners, Chernobyl accident emergency and recovery workers, and workers in nuclear weapons and power industries. The DOE has conducted research on this in relation to several sites, including Oak Ridge, Los Alamos, and Hanford. See Richard Wakeford, "Radiation in the Workplace—A Review of Studies of the Risks of Occupational Exposure to Ionising Radiation," *Journal of Radiological Protection* 29 (2009): A61–A79; Elisabeth Cardis et al., "The 15-Country Collaborative Study of Cancer Risk among Radiation Workers in the Nuclear Industry: Estimates of Radiation-related Cancer Risk," *Radiation Research* 167, no. 4 (2007): 396; National Academy of Sciences, Committee to Assess the Health Risks from Exposure to Low Levels of Ionizing Radiation, *BEIR VII Report* (Washington, D.C.: National Academies Press, 2006); United Nations Scientific Committee on the Effects of Atomic Radiation, Effects of Ionizing Radiation, vol. 1, *UNSCEAR 2006*, Report to the General Assembly, with Scientific Annexes (New York: United Nations, 2008).

67. Nash, *Inescapable Ecologies*, 147–51.

68. Thomas W. Henderson, "Toxic Tort Litigation: Medical and Scientific Principles in Causation," *American Journal of Epidemiology* 132, no. 1 (1990): S77.

69. Richard Wakefield and E. Janet Tawn, "The Meaning of Low Dose and Low Dose-rate," *Journal of Radiological Protection* 30 (2010): 1–3; E. W. Webster, "The Linear No-threshold Debate: A Summary," *Medical Physics* 25, no. 3 (1998): 300.

70. Richard Miller, Government Accountability Project, "Fact Sheet on NIOSH Special Exposure Cohort Rulemaking," http://www4.bfn.org/nuclear/gapfact1.htm (accessed September 28, 2010).

71. *Dose Reconstruction Program: Energy Employees Occupational Illness Compensation Program Act*, CD-ROM, (HHS/CDC/NIOSH no. 2007-113D, November, 2006), 14 min.

72. Updated methods can be found on the Electronic Code of Federal Regulations Web site: http://ecfr.gpoaccess.gov (accessed August 1, 2011). See also EEOICPA Bulletin No. 03-23; GAO, "Energy Employees Compensation" GAO-10-302.

73. Elisburg, *To Conduct Oversight of the Implementation of the Energy Employees Occupational Illness Compensation Program Hearings*. Claimants can present arguments that NIOSH made errors in application, such as inappropriate assumptions about a particular physical or chemical form of a radioactive material used in a facility where the employee worked. However, NIOSH's overall methods to carry out EEOICPA cannot be disputed.

74. Some amount of backlog was inevitable, because people began filing claims before the methods and program of dose reconstruction had been established.

75. Parascandola, "Chances, Individuals and Toxic Torts," 147.

76. "Toxic torts" refer to those court cases in which persons assert causes of action and seek compensation for adverse health effects resulting from exposures to toxic substances. Examples include litigation involving Agent Orange, asbestos, swine flu vaccine, silicone implants, and other toxic agents.

77. "Compensation Regimes Applicable to Radiation Workers in OECD Countries," *Nuclear Law Bulletin* 66, no. 2 (December 2000), http://www.oecd-nea.org/law/nlb/Nlb-66/007-013.pdf (accessed August 2, 2011). "OECD" refers to the Organisation for Economic Co-operation and Development, an intergovernmental organization of industrialized countries.

78. See the Web site of the UK nuclear industry's Compensation Scheme for Radiation-Linked Diseases: http://www.csrld.org.uk/default.php (accessed August 15, 2011).

79. This section draws heavily on the work of several epidemiologists, law and risk analysts, and scholars who have critically examined the probability of causation/assigned share, and the application of epidemiology by law and compensation schemes. Any errors in the discussion are the author's. Refer to Steven Shavell, "Uncertainty over Causation and the Determination of Civil Liability," Working Paper no. 1219 (National Bureau of Economic Research Working Paper Series, 1983); Louis Anthony Cox Jr., "Statistical Issues in the Estimation of Assigned Shares for Carcinogenesis Liability," *Risk Analysis* 7, no. 1 (1987): 71–80; Sander Greenland and James M. Robins, "Conceptual Problems in the Definition and Interpretation of Attributable Fractions," *American Journal of Epidemiology* 128, no. 6 (1988): 1185–97; James Robins and Sander Greenland, "The Probability of Causation under a Stochastic Model for Individual Risk," *Biometrics* 45 (1989): 1125–38; James M. Robins and Sander Greenland, "Estimability and Estimation of Excess and Etiologic Fractions," *Statistics in Medicine* 8 (1989): 845–59; Sander Greenland, "Relation of Probability of Causation to Relative Risk and Doubling Dose: A Methodologic Error That Has Become a Social Problem," *American Journal of Public Health* 89, no. 8 (1999): 1166–69; Jan Beyea and Sander Greenland, "The Importance of Specifying the Underlying Biologic Model in Estimating the Probability of Causation," *Health Physics* 76, no. 3 (1999): 269–74; Sander Greenland and James M. Robins, "Epidemiology, Justice, and the Probability of Causation," *Jurimetrics* 40 (2000): 321–40; James Robins, "Should Compensation Schemes Be Based on the Probability of Causation or Expected Years of Life Lost?" *Journal of Law and Policy* 12, no. 2 (2004): 537–48.

80. The work of Mark Parascandola is essential to this chapter's understanding the broad epistemic issues of law-epidemiology (mis)translations. See Mark Parascandola, "Evidence and Association: Epistemic Confusion and Toxic Tort Law," *Philosophy of Science* 63, no. 3 (1996): S168–76, esp. S173; "What Is Wrong with the Probability of Causation?" *Jurimetrics* 39 (1998): 29–44; "Epistemic Risk: Empirical Science and the Fear of Being Wrong," *Law, Probability and Risk* 9, nos. 3–4 (2010): 201–14; "Chances, Individuals and Toxic Torts."

81. There are significant differences in scientific and legal definitions of causation, particularly with respect to the reasoning processes and proof required in defining causation in science and law. For example, science generally examines causes for the distribution and determinants of diseases and is focused on populations; for law, by contrast, causation is used to resolve disputes based on evidence involving individuals or individual events.

82. Parascandola, "What Is Wrong with the Probability of Causation?" 39.

83. Parascandola, "Chances, Individuals and Toxic Torts," 148.

84. Parascandola, "Evidence and Association," S168.

85. Ibid., S172.

86. Ibid., S174.

87. Greenland and Robins, "Conceptual Problems in the Definition and Interpretation of Attributable Fractions"; Greenland, "Relation of Probability of Causation to Relative Risk and Doubling Dose"; Parascandola, "What Is Wrong with the Probability of Causation?"

88. Greenland, "Relation of Probability of Causation to Relative Risk and Doubling Dose," 1166–67.

89. Parascandola, "What Is Wrong with the Probability of Causation?" 37.

90. The POC formula implicitly assumes that excess radiogenic risk is independent of the background and does not interact with or depend on other risks. One consequence of this is that rare diseases, which exhibit large rate ratios between exposed and background populations, are compensated, while more common diseases are not paid out—regardless of how many people developed them due to exposure—because they do not result in a high excess risk ratio when the exposed population is compared to background population.

91. Beyea and Greenland, "The Importance of Specifying the Underlying Biologic Model in Estimating the Probability of Causation."

92. Ibid., 273.

93. Successor to the Atomic Bomb Casualty Commission created in 1947, the Radiation Effects Research Foundation (RERF) in Hiroshima and Nagasaki has been funded by the governments of Japan and the United States—specifically the DOE—to conduct epidemiologic studies that generate data on cancer incidence, cancer mortality, and noncancer effects in relation to radiation dose. Refer to the DOE Office of Health, Safety and Security Web site on the RERF, http://www.hss.energy.gov/healthsafety/IHS/hstudies/japan_radiate.html (accessed August 2, 2011).

94. EPA, "EPA Radiogenic Cancer Risk Models and Projections for the U.S. Population," EPA 402-R-11-001 (April 2011), http://www.epa.gov/radiation/docs/bluebook/bbfinalversion.pdf (accessed August 15, 2011).

95. The debates over this data range from disagreements over the biologic effects/harm of radiation doses experienced for long durations versus short episodes, to concerns over the likely misclassifications and overall injustices involved in a study that was conducted under extreme conditions and based on traumatic memories that had to be recalled from five or more previous years. See M. Susan Lindee, *Suffering Made Real: American Science and the Survivors at Hiroshima* (Chicago: University of Chicago Press, 1994).

96. Confidence intervals are used in the determination of the POC in order to give claimants the benefit of the doubt. The probability that the worker's cancer was related to the covered radiation exposure is calculated at the 99 percent confidence interval, meaning that compensation can be granted to those cases with as little as 1 percent chance that the POC is at least 50 percent or higher. Although the 99 percent credibility limit is in theory a more generous estimate, it cannot rectify false assumptions, bad substitutions, or inaccurate data.

97. Epidemiological studies are often the only evidence available to assess causation in many toxic tort cases. Uncertainty is not an obstacle for the law; "statistical evidence merely makes such uncertainty more explicit by quantifying it" (Parascandola, "Chances, Individuals and Toxic Torts," 153). As Parascandola poignantly argues, what is needed are better epistemological practices that recognize the distinctions between different forms of evidence and what they can and cannot contribute in particular situations to fulfill policy goals. See Mark Parascandola, "Cell Phones, Aluminum, Agent Orange: No? Yes? Maybe?" *The Dana Foundation* (January 1, 2001), http://www.dana.org/news/cerebrum/detail.aspx?id=3034 (accessed August 2, 2011).

98. Refer to Laura Frank's series "Deadly Denial"; also view the documentary film by Scott Bison, *Rocky Flats: Legacy* (Bison Pictures, 2011), 22 min., 21 sec., http://vimeo.com/21466919 (accessed August 15, 2011).

99. Mike Kessler, "Out in the Cold," 5280 (November 2007): 134–41, 162–82; Representative Udall of New Mexico, *Energy Employees Occupational Illness Compensation Program*, 9–10.

100. NIOSH bulletin, "NIOSH Awards Contract for Support of the Energy Employees Occupational Illness Compensation Program Act (EEOICPA) Responsibilities" (May 5, 2009), http://www.cdc.gov/niosh/ocas/pdfs/dr/pr050509.pdf (accessed August 2, 2011); Kessler, "Out in the Cold," 141; also refer to the MJW Corporation Web site (http://www.mjwcorp.com) and the Dade Moeller Web site (http://www.moellerinc.com).

101. Representative Udall of New Mexico, *Energy Employees Occupational Illness Compensation Program*, 9.

102. Antoinette Bonsignore, of the Linde Ceramics SEC Action Group, on behalf of ANWAG, letter to Hilda Solis, Secretary of the DOL, June 8, 2009, included on Frank Munger's *Atomic City Underground* blog on *Knox News*, http://blogs.knoxnews.com/munger/2009/06/sick_worker_advocates_appalled.html (accessed August 15, 2011).

103. Michel Foucault, *Discipline and Punish: The Birth of the Prison*, trans. Alan Sheridan (New York: Vintage Books, 1995), 184.

104. Wolf, *Studio 12*, KBDI-TV, July 30, 2008.

105. This chapter magnifies some of the logics, contradictions, and absurdities underlying EEOICPA. Rather than explore the response of workers in detail, worker activities are broadly considered here in relation to the legislation and its effects.

106. Mateo Taussig-Rubbo, "Sacrifice and Sovereignty," in *States of Violence:*

War, Capital Punishment, and Letting Die, ed. Austin Sarat and Jennifer L. Culbert (New York: Cambridge University Press, 2009), 83–126.

107. Alex Law, "Melting the Cold War Permafrost: Restructuring Military Industry in Scotland," *Antipode* 31, no. 4 (2000): 435.

108. Ibid.

109. John Lovering, "Military Expenditure and the Restructuring of Capitalism: The Military Industry in Britain," *Cambridge Journal of Economics* 14, no. 4 (1990): 464.

110. Catherine Lutz, ed., *The Bases of Empire: The Global Struggle against U.S. Military Posts* (New York: New York University Press, 2009); Catherine Lutz, *Homefront: A Military City and the American Twentieth Century* (Boston: Beacon Press, 2002); Rachel Woodward, *Military Geographies* (Malden, Mass.: Blackwell, 2004).

111. Len Ackland, "The Other Cleanup at Rocky Flats," *Denver Post* (August 7, 2005).

112. E. P. Thompson, "Exterminism," *New Left Review* 121 (1980): 3–32. Thompson states that the United States does not *have* a military–industrial complex; it *is* a military–industrial complex.

113. Masco, *The Nuclear Borderlands*, 289–327.

114. Elizabeth Povinelli distinguishes quiet deaths, slow deaths, rotting worlds, and the everyday drift toward death from the spectacular deathscapes of borders, agencies, and intensities; see Elizabeth Povinelli, "The Child in the Broom Closet: States of Killing and Letting Die," in *States of Violence: War, Capital Punishment, and Letting Die*, ed. Austin Sarat and Jennifer L. Culbert (New York: Cambridge University Press, 2009), 169–91. In a related effort, Berlant aims to make visible endemic forms of death that are experienced as the background of ordinary life, and that are resistant to accounts of causality, subjectivity, and the production of life ("Slow Death"). Finally, Rob Nixon's book *Slow Violence* offers the concept of "slow violence" and the paradox of the "long-term emergency." See Rob Nixon, "Slow Violence and the Environmentalism of the Poor: An Interview with Rob Nixon," by Ashley Dawson, *Social Text* journal blog (August 31, 2011), http://www.socialtextjournal.org/blog/2011/08/slow-violence-and-the-environmentalism-of-the-poor-an-interview-with-rob-nixon.php (accessed September 1, 2011).

115. Cindy L. Linden, "Fleshly Evils: Clinical and Cultural (Il)Logics of the Chronic Pain Subject in Contemporary U.S. Society," in *The Future of Flesh: A Cultural Survey of the Body*, ed. Zoe Detsi-Diamanti, Katerina Kitsi-Mitakou, and Effie Yiannopoulou (New York: Palgrave Macmillan, 2009), 79–97.

116. NIOSH, "Classes of Employees Currently Included in the SEC," http://www.cdc.gov/niosh/ocas/ocassec.html (accessed November 5, 2012).

117. Laura Frank, "Flats 'Hero' Is Gone, but His Cause Lives On," *Denver Post* (March 29, 2009), http://www.denverpost.com/search/ci_12020437 (accessed August 15, 2011).

118. The Charlie Wolf Nuclear Workers Compensation Act, HR 1828 / S 757, 111th Cong., 1st sess. (March 31, 2009). The bill was introduced in 2009 (by Mark

Udall of Colorado to the Senate; Jared Polis of Colorado to the House). Also see the ANWAG petition "Cold War Heroes Petition Congress to Reform Sick Worker Compensation Program" (September 13, 2010), http://www.state.nv.us/nucwaste/news2010/pdf/anwag100913petition.pdf (accessed August 9, 2011).

119. Elisburg, *To Conduct Oversight of the Implementation of the Energy Employees Occupational Illness Compensation Program Hearings.*

120. GAO, "Energy Employees Compensation" GAO-10-302, 13.

121. The Coalition for a Healthy Environment (Oak Ridge, Tennessee), United Nuclear Weapons Workers (St. Louis, Missouri), and Grassroots Organization of Sick Workers (Craig, Colorado) merged to create this new organization. Among the numerous nuclear worker advocacy and information Web sites: the Cold War Patriots Web site, http://www.coldwarpatriots.org/ (accessed August 15, 2011), Frank Munger's *Atomic City Underground* blog on *KnoxNews*, http://blogs.knoxnews.com/munger/ (accessed August 15, 2011), the Nuclear Waste Project Office, http://www.state.nv.us/nucwaste/about.htm (accessed August 15, 2011), "The Exposed: Poisoned Nuclear Workers," Nuclear Worker Rally in Oak Ridge, Tennessee (June 25, 2008), on the "Between the Ice Ages" Web site, http://www.wildclearing.com/ice-ages/the-exposed.html (accessed August 15, 2011).

122. Bernd Frohmann, "Documentary Ethics, Ontology, Politics," *Journal of Archival Science* 8, no. 3 (2008): 171.

123. Nikolas Rose and Carlos Novas, "Biological Citizenship," in *Global Assemblages: Technology, Politics, and Ethics as Anthropological Problems*, ed. Aihwa Ong and Stephen J. Collier (Malden, Mass., and Oxford: Blackwell, 2005); Hester Parr, "New-body Geographies: The Embodied Spaces of Health and Medical Information on the Internet," *Environment and Planning D: Society and Space* 20, no. 1 (2002): 73–95.

124. Jacques Derrida captures this paradox—the remedy both repairs and poisons—in his reflections on the figure of the "pharmakon." See Jacques Derrida, "From 'Plato's Pharmacy' in *Dissemination*," in *A Derrida Reader: Between the Blinds*, ed. Peggy Kamuf (New York: Columbia University Press, 1991), 112–39. Judith Butler discusses "precarious lives" that can be said to characterize these former nuclear workers, a population "exposed to arbitrary state violence who often have no other option than to appeal to the very state from which they need protection" (Judith Butler, *Frames of War: When Is Life Grievable?* [London and New York: Verso, 2009], 25–29).

125. This withdrawal of the state does not refuse individual legal rights. As Foucault has argued, what is needed is a new practice of "rights" that suspends the difference between human being and citizen and overcomes a legal concept that permanently reinscribes the separation between natural existence and political life.

126. Couze Venn, "Rubbish, the Remnant, Etcetera," *Theory, Culture and Society* 23, nos. 2–3 (2006): 46.

127. Murray, "Thanatopolitics," 210. Murray explores the self as an activity of expropriation; this entails an ethics and politics of vulnerability and permeability, in contrast to individual "rights," sovereignty, or integrity that rest on

understandings of the body as self-enclosed, separate, autonomous units or containers. Bodies serve as residual traces of economic and state practices, sources of evidence and proof, witnesses, collective memory triggers, environmental knowledge contact points, sources of humor, and so on. See Lisa Blackman, "Bodily Integrity," *Body and Society* 16, no. 3 (2010): 1–9.

128. Nash, *Inescapable Ecologies*, 204.

129. Where the environment and the human body begin and end is open to investigation; humans are agents and objects of environmental change, and the environment shapes human flesh and physiology. To conceptualize "exposure" necessitates reworking the modern idea of the body as a set of discrete parts bounded by skin; the ecological body, as Nash puts it (ibid., 8), is a permeable folding of the self and environment.

130. Stacy Alaimo, *Bodily Natures: Science, Environment, and the Material Self* (Bloomington and Indianapolis: Indiana University Press, 2010); Phil Brown, "When the Public Knows Better: Popular Epidemiology," *Environment* 35, no. 8 (1993): 16–41; Phil Brown, Steve Kroll-Smith, and Valerie J. Gunter, "Knowledge, Citizens, and Organizations: An Overview of Environments, Diseases, and Social Conflict," in *Illness and the Environment: A Reader in Contested Medicine*, ed. Phil Brown, Steve Kroll-Smith, and Valerie J. Gunter (New York: New York University Press, 2000), 9–25; Jason Coburn, *Street Science: Community Knowledge and Environmental Health Science* (Cambridge: MIT Press, 2005); Giovanna Di Chiro, "Local Actions, Global Visions: Remaking Environmental Expertise," *Frontiers* 18, no. 2 (1997): 203–31.

131. Also refer to the groundbreaking Nuclear Guardianship project of Joanna Macy, "Rocky Flats: A Call to Guardianship" lecture series and workshops (2011), http://www.rockyflatsnuclearguardianship.org/ (accessed August 23, 2011); Dr. LeRoy Moore's blog, http://leroymoore.wordpress.com/ (accessed August 23, 2011); and the experimental mutimedia exhibit of the Rocky Flats 1969 Fire, "Explore the Nuclear Age," sponsored by the Center for Environmental Journalism at the University of Colorado-Boulder.

132. Denise Grady, "Radiation Is Everywhere, but How to Rate Harm?" *New York Times* (April 4, 2011), http://www.nytimes.com/2011/04/05/health/05radiation.html?pagewanted=all (accessed July 17, 2011).

133. Eli Clare, "Stolen Bodies, Reclaimed Bodies: Disability and Queerness," *Public Culture* 13, no. 3 (2001): 359–65; Nixon, "Slow Violence and the Environmentalism of the Poor."

134. Rei Terada, "The Instant of My Death, and: Demeure: Fiction and Testimony," a review of *The Instant of My Death*, by Maurice Blanchot, and *Demeure: Fiction and Testimony*, by Jacques Derrida, trans. Elizabeth Rottenberg, *SubStance* 30, no. 3 (2001): 132.

135. Jacques Derrida, in his last interview, by *Le Monde* (August 19, 2004), cited in Didier Fassin, "Ethics of Survival: A Democratic Approach to the Politics of Life," *Humanity* 1, no. 1 (2010): 82; also see Judith Butler, "On Never Having Learned How to Live," *Differences: A Journal of Feminist Cultural Studies* 16, no. 3 (2005): 17–24.

4. Transnatural Revue

1. Rob Blackhurst, "Reasons to Be Fearful," *New Statesman* (July 31, 2008), http://www.newstatesman.com/books/2008/07/nuclear-weapons-atomic-war (accessed October 28, 2011); Nathan Hodge and Sharon Weinberger, *A Nuclear Family Vacation: Travels in the World of Atomic Weaponry* (New York: Bloomsbury, 2008).

2. The Fukushima nuclear disaster involved equipment failures, meltdowns, and release of radioactive materials at the Fukushima nuclear power plant following an earthquake and tsunami on March 11, 2011. It is the largest nuclear accident since the 1986 Chernobyl disaster.

3. Ralph H. Lutts, "Chemical Fallout: Rachel Carson's *Silent Spring*, Radioactive Fallout, and the Environmental Movement," *Environmental Review* 9, no. 3 (1985): 210–25; Cheryll Glotfelty, "Cold War, *Silent Spring*: The Trope of War in Modern Environmentalism," in *And No Birds Sing: Rhetorical Analyses of Rachel Carson's* Silent Spring, ed. Craig Waddell (Carbondale and Edwardsville: Southern Illinois University Press, 2000), 157–73; Rachel Carson, *Silent Spring* (New York: First Mariner Books, 2002).

4. Jonathan Crary, *Suspensions of Perception: Attention, Spectacle, and Modern Culture* (Cambridge, Mass., and London: MIT Press, 2000), 72–79; Shiloh Krupar and Stefan Al, "Notes on the Society of the ~~Spectacle~~ Brand," in *The SAGE Handbook of Architectural Theory*, ed. C. Greig Crysler, Stephen Cairns, and Hilde Heynen (Thousand Oaks, Calif.: SAGE Publications, 2012), 247–63.

5. Marc Lafleur, "The Bomb and the Bombshell: The Body as Virtual Battlefront," *Interculture* 5, no. 1 (2008): 29.

6. Hugh Gusterson, "Nuclear Tourism," *Journal for Cultural Research* 8, no. 1 (2004): 23–31; Matt Wray, "A Blast from the Past: Preserving and Interpreting the Atomic Age," *American Quarterly* 58, no. 2 (2006): 467–83; Hugh Gusterson, "Armageddon Tourism," *American Scientist* (September–October 2008), http://www.americanscientist.org/bookshelf/pub/armageddon-tourism (accessed October 28, 2011); Tom Vanderbilt, *Survival City: Adventures among the Ruins of Atomic America* (New York: Princeton Architectural Press, 2002); Scott C. Zeman and Michael A. Amundson, eds., *Atomic Culture: How We Learned to Stop Worrying and Love the Bomb* (Boulder: University Press of Colorado, 2004); Gabrielle Decamous, "Nuclear Activities and Modern Catastrophes: Art Faces the Radioactive Waves," *Leonardo* 44, no. 2 (2011): 124–32.

7. This is evidenced by the tourism to the potential nuclear waste storage site Yucca Mountain in Nevada and the permanent waste storage site Waste Isolation Pilot Plant (WIPP) in New Mexico. For a humorous example, see John D'Agata, *About a Mountain* (New York and London: W. W. Norton, 2010).

8. The concept "stuplime" is from Sianne Ngai, *Ugly Feelings* (Cambridge, Mass., and London: Harvard University Press, 2005), 248–97. For "nuclear sublime," see Alan Nadel, *Containment Culture: American Narratives, Postmodernism, and the Atomic Age* (Durham, N.C., and London: Duke University Press, 1995).

9. Krupar and Al, "Notes on the Society of the ~~Spectacle~~ Brand"; Marita Sturken, *Tourists of History: Memory, Kitsch, and Consumerism from Oklahoma City to Ground Zero* (Durham, N.C., and London: Duke University Press, 2007).

10. Ulrich Beck, *Ecological Politics in an Age of Risk,* trans. Amos Weisz (Cambridge and Malden, Mass.: Polity Press, 1995).

11. Rob Nixon, *Slow Violence and the Environmentalism of the Poor* (Cambridge, Mass., and London: Harvard University Press, 2011), 65.

12. Ann Laura Stoler, "Imperial Debris: Reflections on Ruins and Ruination," *Cultural Anthropology* 23, no. 2 (2008): 200; Nixon, *Slow Violence and the Environmentalism of the Poor,* 65.

13. Raymond Williams, "Dominant, Residual, and Emergent," in *Marxism and Literature* (Oxford and New York: Oxford University Press, 1977), 121–27; Walter Benjamin, *The Arcades Project,* trans. Howard Eiland and Kevin McLaughlin (Cambridge, Mass., and London: Belknap Press of Harvard University Press, 1999); Jean Baudrillard, "The Remainder," in *Simulacra and Simulation,* trans. Sheila Faria Glaser, http://www9.georgetown.edu/faculty/irvinem/theory/Baudrillard-Simulacra_and_Simulation.pdf (accessed July 17, 2011), 95.

14. Thomas James Barnes, "Residues of the Cold War: Emergent Waste Consciousness in Postwar American Culture and Fiction," Ph.D. dissertation, University of Western Ontario, London, Ontario, Canada, 2011.

15. See, for example, the work of Mierle Laderman Ukeles, Robert Smithson, Nancy Holt, Mel Chin, and Mark Dion. Also see Mira Engler, *Designing America's Waste Landscapes* (Baltimore and London: Johns Hopkins University Press, 2004); Lucy R. Lippard, *On the Beaten Track: Tourism, Art, and Place* (New York: New Press, 1999); Miwon Kwon, *One Place after Another: Site-Specific Art and Locational Identity* (Cambridge, Mass., and London: MIT Press, 2002).

16. Leslie Marmon Silko, *Ceremony* (New York: New American Library, 1977); Martin Cruz Smith, *Night Wing* (New York: Ballantine Books, 1990); A. A. Carr, *Eye Killers* (Norman and London: University of Oklahoma Press, 1995); Ellen Meloy, *The Last Cheater's Waltz: Beauty and Violence in the Desert Southwest* (Tucson: University of Arizona Press, 1999); Jake Page, *Cavern* (Albuquerque: University of New Mexico Press, 2002). Also see John Beck, *Dirty Wars: Landscape, Power, and Waste in Western American Literature* (Lincoln and London: University of Nebraska Press, 2009), 177–201; David Mogen, Scott P. Sanders, and Joanne B. Karpinski, eds., *Frontier Gothic: Terror and Wonder at the Frontier in American Literature* (Rutherford, N.J.: Fairleigh Dickinson University Press, 1993).

17. Engler, *Designing America's Waste Landscapes,* 229.

18. Phaedra C. Pezzullo, "Toxic Tours: Communicating the 'Presence' of Chemical Contamination," in *Communication and Public Participation in Environmental Decision-Making,* ed. Stephen P. Depoe, John W. Delicath, and Marie-France Aepli Elsenbeer (Albany: State University of New York Press, 2004), 235–54. Toxic tours are an example of what this chapter refers to as "counter-spectacle," which draws on ideas associated with "interventions," "performance activisms," "ruptural performances," and "ethical spectacles." See Stephen Duncombe, *Dream: Re-imagining Progressive Politics in an Age of Fantasy* (New York: New Press, 2007); Tony Perucci, "What the Fuck Is That? The Poetics of Ruptural Performance," *Liminalities: A Journal of Performance Studies* 5, no. 3 (2009): 1–18.

19. John Wills, "'Welcome to the Atomic Park': American Nuclear Landscapes

and the 'Unnaturally Natural,'" *Environment and History* 7 (2001): 449–72; Peter Galison's work on nuclear wastelands and wilderness, as reviewed by Corydon Ireland, "Wasteland and Wilderness," *Harvard Science* (October 8, 2009), http://news.harvard.edu/gazette/story/2009/10/wasteland-and-wilderness/ (accessed October 29, 2011); Sheila Pinkel, "Thermonuclear Gardens: Information Artworks about the U.S. Military–Industrial Complex," *Leonardo* 34, no. 4 (2001): 319–26.

20. Ecocide is a neologism for environmental devastation on a scale and intensity comparable to genocide.

21. Artists that have documented and/or guided tours through U.S. nuclear testing sites and contamination, military installations, and bombing ranges include the Center for Land Use Interpretation; Trevor Paglen; Sarah Kanouse; Eve Andree Laramee; Smudge (Jamie Kruse and Elizabeth Ellsworth); the Demilit collective (Javier Arbona, Bryan Finoki, and Nick Sowers); Sheila Pinkel; the National Toxic Land/Labor Conservation Service (Sarah Kanouse and Shiloh Krupar).

22. Phaedra C. Pezzullo, *Toxic Tourism: Rhetorics of Pollution, Travel, and Environmental Justice* (Tuscaloosa: University of Alabama Press, 2007).

23. Engler, *Designing America's Waste Landscapes*, xvi.

24. Peter Goin, *Nuclear Landscapes (Creating the North American Landscape)* (Baltimore and London: Johns Hopkins University Press, 1991); Carole Gallagher, *American Ground Zero: The Secret Nuclear War* (Cambridge: MIT Press, 1993); Richard Misrach, *Bravo 20: The Bombing of the American West* (Baltimore and London: Johns Hopkins University Press, 1990); and Patrick Nagatani, *Nuclear Enchantment* (Albuquerque: University of New Mexico Press, 1991). Also see Mike Davis, "Dead West: Ecocide in Marlboro Country," *New Left Review* 200 (July/August 1993): 49–73; Alan Berger, *Reclaiming the American West* (New York: Princeton Architectural Press, 2002); Alan Berger, *Drosscape: Wasting Land in Urban America* (New York: Princeton Architectural Press, 2006).

25. Beck, *Dirty Wars*.

26. Donna J. Haraway, *When Species Meet* (Minneapolis and London: University of Minnesota Press, 2008), 20.

27. Kathryn Yussof, "Biopolitical Economies and the Political Aesthetics of Climate Change," *Theory, Culture and Society* 27, nos. 2–3 (2010): 73 and 77.

28. This interpretation combines Jacques Rancière's concept of political aesthetics, a model of aesthetics that is ontologically social, with Michel Foucault's work on ethics as an arts of existence. See Jacques Rancière, *The Politics of Aesthetics*, trans. Gabriel Rockhill (New York: Continuum, 2004); Timothy O'Leary, *Foucault: The Art of Ethics* (London and New York: Continuum, 2002).

29. Yusoff, "Biopolitical Economies and the Political Aesthetics of Climate Change," 79; Haraway, *When Species Meet;* Steve Hinchliffe, "Reconstituting Nature Conservation: Towards a Careful Political Ecology," *Geoforum* 39 (2008): 88–97.

30. Yusoff, "Biopolitical Economies and the Political Aesthetics of Climate Change," 86.

31. Ibid., 79; Rancière, *The Politics of Aesthetics*, 9; George Pavlich, "Nietzsche,

Critique and the Promise of Not Being Thus . . . ," *International Journal for the Semiotics of Law* 13, no. 4 (2000): 357–75.

32. This chapter's "transnatural aesthetics" approach was inspired by several "nuclear artists": Ed Grothus of the Los Alamos "Black Hole" salvage yard; the atomic art sculptures and masks of Santa Fe–based Tony Price; Japan-based artist Kenji Yanobe's "postapocalyptic" playgrounds, atom suits, and atom cars; Natalie Jeremijenko's feral robotic dogs that sniff/detect pollution and radiation; and, while not overtly humorous, scientific illustrator Cornelia Hesse-Honegger's depictions of mutated insects around various nuclear facilities.

33. Nicholas Holm advises, "Humour must therefore be grasped as a site of constant tension between, on the one hand, its utopic promise to prise open gaps within the sensible and, on the other, its ability to reconfirm the existing consensus of sense and nonsense under the guise of free play" (Nicholas Holm, "The Distribution of the Nonsensical and the Political Aesthetics of Humour," *Transformations*, no. 19 [2011], http://www.transformationsjournal.org/journal/issue_19/article_04.shtml [accessed October 28, 2011]).

34. Richard Schechner, *Performance Theory* (New York and London: Routledge, 1988), 141.

35. Misrach, *Bravo 20;* Beck, *Dirty Wars.*

36. Refer to Nuclia Waste's story, http://www.Nucliawaste.com (accessed September 20, 2011).

37. Vinay Gidwani, *Capital, Interrupted: Agrarian Development and the Politics of Work in India* (Minneapolis and London: University of Minnesota Press, 2008), 18–19.

38. Joseph Masco, "Mutant Ecologies: Radioactive Life in Post–Cold War New Mexico," *Cultural Anthropology* 19, no. 4 (2004): 526.

39. The concept of the "transnatural" has recently been employed in art circles. See, for example, the exhibition Becoming TransNatural (March 4–April 1, 2011, Amsterdam, the Netherlands) and the 2012 Transnatural Festival (September 7–October 7, 2012, Amsterdam, the Netherlands), http://www.transnatural.nl/ (accessed September 20, 2011); artist Lucy McRae, "Becoming TransNatural Campaign," http://www.lucymcrae.net/2011/02/becoming-transnatural-exhibition.html (accessed September 20, 2011). In the field of ecocriticism, Simal Begoña discusses the "transnatural" challenges that environmental criticism currently faces; see Simal Begoña, "The Junkyard in the Jungle: Transnational, Transnatural Nature in Karen Tei Yamashita's *Through the Arc of the Rain Forest*," *Journal of Transnational American Studies* 2, no. 1 (2010): 1–25.

40. Gay Hawkins, *The Ethics of Waste: How We Relate to Rubbish* (Lanham, Md., and Oxford: Rowman and Littlefield, 2006), 11.

41. Mary Douglas, *Purity and Danger: An Analysis of Concepts of Pollution and Taboo* (London and New York: Routledge, 1966).

42. Jody Baker, "Modeling Industrial Thresholds: Waste at the Confluence of Social and Ecological Turbulence," *Cultronix*, no. 1 (1994), http://eserver.org/cultronix/baker (accessed September 20, 2011).

43. Judith Butler, *Bodies That Matter: On the Discursive Limits of "Sex"* (New York and London: Routledge, 1993), 241.

44. John Law, "Enacting Naturecultures: A Note from STS" (Lancaster, U.K.: Centre for Science Studies, Lancaster University, 2004), http://www.lancs.ac.uk/fass/sociology/papers/law-enacting-naturecultures.pdf (accessed September 20, 2011).

45. Jennifer Gabrys, "Sink: The Dirt of Systems," *Environment and Planning D: Society and Space* 27, no. 4 (2009): 676.

46. Donna J. Haraway, *Modest_Witness@Second_Millennium.FemaleMan©_Meets_OncoMouse™* (New York and London: Routledge, 1997), 62. Pu 239 is the transuranium element known as plutonium.

47. Donna Haraway, "The Promise of Monsters: A Regenerative Politics for Inappropriate/d Others," in *Cultural Studies*, ed. Lawrence Grossberg, Cary Nelson, and Paula Treichler (New York and London: Routledge, 1992), 295–37.

48. Eve Kosofsky Sedgwick, *Touching Feeling: Affect, Pedagogy, Performativity* (Durham, N.C., and London: Duke University Press, 2003), 149–50.

49. Masco, "Mutant Ecologies"; Barbara Freeman, "*Frankenstein* with Kant: A Theory of Monstrosity, or the Monstrosity of Theory," *SubStance* 16, no. 1, issue 52 (1987): 21–31.

50. Bruno Latour, *We Have Never Been Modern*, trans. Catherine Porter (Cambridge: Harvard University Press, 1993).

51. Haraway, "The Promise of Monsters"; Rosalind Garland-Thomson, "Integrating Disability, Transforming Feminist Theory," *NWSA Journal* 14, no. 3 (2002): 1–32; Shiloh Krupar, "The Biopsic Adventures of Mammary Glam: Breast Cancer Detection and the Promise of Cancer Glamor," *Social Semiotics* 22, no. 1 (2012): 47–82.

52. Gwendolyn Blue and Melanie Rock, "Trans-biopolitics: Complexity in Interspecies Relations," *Health* 15, no. 4 (2011): 353–68, esp. 354 and 357.

53. Stacy Alaimo, *Bodily Natures: Science, Environment, and the Material Self* (Bloomington and Indianapolis: Indiana University Press, 2010), 11. Also refer to Mel Chen's work on animacy theory in *Animacies: Biopolitics, Racial Mattering and Queer Affect* (Durham, N.C., and London: Duke University Press, 2012), esp. 189–221.

54. Alaimo, *Bodily Natures*, 3.

55. Lorraine Code, *Ecological Thinking: The Politics of Epistemic Location* (Oxford: Oxford University Press, 2006), 19.

56. Michel Foucault, *Technologies of the Self*, ed. Luther H. Martin, Huck Gutman, and Patrick H. Hutton (Amherst: University of Massachusetts Press, 1988); Michel Foucault, *Ethics: Subjectivity and Truth*, ed. Paul Rabinow, trans. Robert Hurley (New York: New Press, 1994).

57. Éric Darier, "Foucault against Environmental Ethics," in *Discourses of the Environment*, ed. Éric Darier (Oxford and Malden, Mass.: Blackwell, 1999), 217–40.

58. Ibid., esp. 234.

59. Catriona Sandilands, "Eco Homo: Queering the Ecological Body Politic," *Social Philosophy Today* 19 (2004): 17–39.

60. Blue and Rock, "Trans-biopolitics," 353.

61. Sarah Whatmore, "Materialist Returns: Practising Cultural Geography in and for a More-Than-Human World," *cultural geographies* 13 (2006): 600–609; Noel Castree, "A Post-environmental Ethics?" *Ethics, Place and Environment* 6, no. 1 (2003): 3–12.

62. Jane Bennett, "The Force of Things: Steps to an Ecology of Matter," *Political Theory* 32 (2004): 365.

63. Noreen Giffney, "Queer Apocal(o)ptic/ism: The Death Drive and the Human," in *Queering the Non/Human*, ed. Noreen Giffney and Myra J. Hird (Hampshire, U.K., and Burlington, Vt.: Ashgate, 2008), 55–78; Tobin Siebers, "Disability in Theory: From Social Constructionism to the New Realism of the Body," *American Literary History* 13, no. 4 (2001): 737–54.

64. Kathryn Yussof contends that "aesthetics must be considered as part of the practice of politics *and* a space that configures the realm of what is possible in that politics. This is to suggest aesthetics as a form of ethics or an 'aesthetics of existence'" (Yussof, "Biopolitical Economies and the Political Aesthetics of Climate Change," 73).

65. Catriona Mortimer-Sandilands and Bruce Erickson, "Introduction: A Genealogy of Queer Ecologies," in *Queer Ecologies: Sex, Nature, Politics, Desire*, ed. Catriona Mortimer-Sandilands and Bruce Erickson (Bloomington and Indianapolis: Indiana University Press, 2010), 35.

66. Giovanna Di Chiro, "Polluted Politics? Confronting Toxic Discourse, Sex Panic, and Eco-normativity," in Mortimer-Sandilands and Erickson, *Queer Ecologies*, 199–230; Katie Hogan, "Undoing Nature: Coalition Building as Queer Environmentalism," in Mortimer-Sandilands and Erickson, *Queer Ecologies*, 231–53; Timothy W. Luke, *Ecocritique: Contesting the Politics of Nature, Economy, and Culture* (Minneapolis and London: University of Minnesota Press, 1997).

67. Nixon, *Slow Violence and the Environmentalism of the Poor*, 236–37.

68. Mortimer-Sandilands and Erickson, *Queer Ecologies*, 35–36.

69. Yusoff, "Biopolitical Economies and the Political Aesthetics of Climate Change," 93.

70. "Transuranic" refers to chemical elements with atomic numbers greater than 92, the atomic number of uranium.

71. Nuclia Waste, http://www.Nucliawaste.com (accessed September 20, 2011).

72. Daniel C. Brouwer, "Risibility Politics: Camp Humor in HIV/AIDS Zines," in *Public Modalities: Rhetoric, Culture, Media, and the Shape of Public Life*, ed. Daniel C. Brouwer and Robert Asen (Tuscaloosa: University of Alabama Press, 2010), 227 and 229.

73. http://www.Nucliawaste.com (accessed September 20, 2011).

74. Mortimer-Sandilands and Erickson, *Queer Ecologies*, 39.

75. Moe Meyer, ed., *The Politics and Poetics of Camp* (London and New York: Routledge, 1994); Susan Sontag, "Notes on Camp," in *A Susan Sontag Reader* (New York: Vintage Books, 1983), 105–19; Linda Hutcheon, *A Theory of Parody: The Teachings of Twentieth-Century Art Forms* (New York: Methuen, 1985); Judith Butler, *Gender Trouble: Feminism and the Subversion of Identity* (New York and

London: Routledge, 1999); Barbara Kirshenblatt-Gimblett, "Disputing Taste," *Destination Culture: Tourism, Museums, and Heritage* (Berkeley and Los Angeles: University of California Press, 1998), 259–81.

76. Mortimer-Sandilands and Erickson, *Queer Ecologies,* 37.

77. John Scanlan, *On Garbage* (London: Reaktion, 2005).

78. Brouwer, "Risibility Politics," 219–39.

79. M. Jimmie Killingsworth and Jacqueline S. Palmer, "The Discourse of Environmentalist Hysteria," in *Proceedings of the Conference on Communication and Our Environment,* ed. James G. Cantrill and M. Jimmie Killingsworth (Marquette: Northern Michigan University Printing Services, 1993), 321–35.

80. John Downey and Natalie Fenton, "New Media, Counter Publicity and the Public Sphere," *New Media and Society* 5, no. 2 (2003): 185–202; Daniel C. Brouwer and Robert Asen, eds., *Public Modalities: Rhetoric, Culture, Media, and the Shape of Public Life* (Tuscaloosa: University of Alabama Press, 2010).

81. Sandilands, "Eco Homo," 25.

82. For example, Noel Gough and Annette Gough, "Tales from Camp Wilde: Queer(y)ing Environmental Education Research," *Canadian Journal of Environmental Education* 8, no. 1 (2003): 44–66.

83. Catriona Sandilands, "Lavender's Green? Some Thoughts on Queer(y)ing Environmental Politics," *UnderCurrents: Critical Environmental Studies* 6 (May 1994): 22.

84. Noreen Giffney and Myra J. Hird, "Introduction: Queering the Non/Human," in *Queering the Non/Human,* ed. Noreen Giffney and Myra J. Hird (Hampshire, U.K., and Burlington, Vt.: Ashgate, 2008), 1–16; Andrew Ross, "Uses of Camp," *Yale Journal of Criticism* 2, no. 2 (1988): 1.

85. Lin Nelson, "Promise Her Everything: The Nuclear Power Industry's Agenda for Women," *Feminist Studies* 10, no. 2 (1984): 293.

86. Catriona Mortimer-Sandilands, "Unnatural Passions?: Notes Toward a Queer Ecology," *Invisible Culture: An Electronic Journal for Visual Culture* 9 (2005), http://www.rochester.edu/in_visible_culture/Issue_9/sandilands.html (accessed September 20, 2011).

87. Catherine Lutz, "Making War at Home in the United States: Militarization and the Current Crisis," *American Anthropologist* 104, no. 3 (2002): 723–35, esp. 724; Andrew Ross, "The Future Is a Risky Business," in *FutureNatural: Nature, Science, Culture,* ed. George Robertson et al. (London and New York: Routledge, 1996), 7–20.

88. Bryan C. Taylor, "Home Zero: Images of Home and Field in Nuclear-Cultural Studies," *Western Journal of Communication* 61, no. 2 (1997): 209–34; Gillian Brown, "Nuclear Domesticity: Sequence and Survival," *Yale Journal of Criticism* 2, no. 1 (1988): 179–91.

89. See, for example, the interviews included in Hillary Rosner, "At the Foot of the Rockies, Cleaning a Radioactive Wasteland," *New York Times* (June 7, 2005), http://www.nytimes.com/2005/06/07/science/earth/07flat.html?pagewanted=print (accessed July 25, 2011). Nancy Tuor, then president and chief executive of Kaiser-Hill (the company hired by the U.S. Department of

Energy to remediate Rocky Flats), compared the cleaning of the plutonium-processing Building 771—reputed to have been one of the most contaminated buildings on earth—to "cleaning the bathroom."

90. Carol Cohn, "Slick 'Ems, Glick 'Ems, Christmas Trees, and Cookie Cutters: Nuclear Language and How We Learned to Pat the Bomb," *Bulletin of the Atomic Scientists* 43, no. 5 (1987): 17–24; Amy Kaplan, "Manifest Domesticity," *American Literature* 70, no. 3 (1998): 581–606.

91. Sedgwick, *Touching Feeling*, 149–50.

92. Gabrys, "Sink," 676; Masco, "Mutant Ecologies," 517–50.

93. NIMBY is an acronym for "not in my backyard," a rallying cry often heard in response to the "placing" of toxic waste or other environmental hazards near residential homes.

94. Ed Cohen, *A Body Worth Defending: Immunity, Biopolitics, and the Apotheosis of the Modern Body* (Durham, N.C., and London: Duke University Press, 2009).

95. Sandilands, "Eco Homo," 32.

96. Masco, "Mutant Ecologies."

97. Elizabeth Grosz, *Space, Time and Perversion: Essays on the Politics of Bodies* (New York: Routledge, 1995), 174.

98. Mortimer-Sandilands, "Unnatural Passions?"

99. Stoler, "Imperial Debris," 200 and 211.

100. Barnes, "Residues of the Cold War," 166 and 186.

101. Ibid., 168.

102. Ibid., 8, 49–50.

103. For a relevant examination of postindustrial diseases, sexuality, and disability that criticizes the supposed "transcendence" of the technologized body, see Sharon Betcher, "Putting My Foot (Prosthesis, Crutches, Phantom) Down: Considering Technology as Transcendence in the Writings of Donna Haraway," *Women's Studies Quarterly* 29, nos. 3/4 (2001): 35–53.

104. Barnes, "Residues of the Cold War," 211.

105. Lawrence Buell, "Toxic Discourse," *Critical Inquiry* 24, no. 3 (1998): 642 and 648.

106. Ibid., 654 and 659.

107. Peter Coviello, "Apocalypse from Now On," in *Queer Frontiers: Millennial Geographies, Genders, and Generations* (Madison and London: University of Wisconsin Press, 2000), 39–63.

108. Hawkins, *The Ethics of Waste*, 7–13.

109. James Acord was the inspiration for the character of Reever in James Flint's *The Book of Ash* (London and New York: Penguin, 2004).

110. Kat Austen, "An Artist Who Had a Radioactive Imagination," *New Scientist* (May 2, 2011), http://www.newscientist.com/blogs/culturelab/2011/05/an-artist-who-had-a-radioactive-imagination.html (accessed October 28, 2011).

111. William Haver, "Queer Research; or, How to Practice Invention to the Brink of Intelligibility," in *The Eight Technologies of Otherness*, ed. Sue Golding (London and New York: Routledge, 1997), 277–92.

112. James Acord passed away January 9, 2011. See Jon Michaud, "Postscript:

James Acord, Alchemist for the Nuclear Age," *New Yorker* (January 20, 2011), http://www.newyorker.com/online/blogs/backissues/2011/01/postscript-james-acord.html (accessed October 25, 2011); *James L. Acord* memorial blog, http://jameslacord.com/author/egalore/ (accessed October 25, 2011); Philip Munger, "James L. Acord Passes," *Progressive Alaska* blog (January 9, 2011), http://progres sivealaska.blogspot.com/2011/01/james-l-acord-passes.html (accessed October 2, 2011).

113. Austen, "An Artist Who Had a Radioactive Imagination."

114. Decamous, "Nuclear Activities and Modern Catastrophes," 129; Brian Freer, "Ambushed at 50," *Plazm Magazine* (1995), http://www.plazm.com/maga zine/features/all-articles/ambushed-at-50?p=1 (accessed October 12, 2011).

115. The Three Mile Island disaster was a partial core meltdown of a pressurized water reactor of the Three Mile Island Nuclear Generating Station near Harrisburg, Pennsylvania, in 1979.

116. For more biographical information, refer to Philip Schuyler, "Moving to Richland—I," *New Yorker* (October 14, 1991): 59–64, 68, 72–73, 76–80, 82, 84–96; Philip Schuyler, "Moving to Richland—II," *New Yorker* (October 21, 1991): 62–64, 73–81, 84–88, 90, 97–107.

117. James Acord, presentation at the Influencers 2010 "Art, Guerilla Communication, Radical Entertainment" (Center of Contemporary Culture of Barcelona, Spain, February 4–6, 2010), video of talk found at http://theinfluencers.org/en/james-acord (accessed October 28, 2011); "James Acord: Atomic Artist," interview by *Nuclear News* (November 2002), http://www.ans.org/pubs/magazines/nn/docs/2002-11-3.pdf (accessed October 28, 2011).

118. For an overview of the politics of transmutation within nuclear science and engineering internationally, see Arjun Makhijani, Hisham Zerriffi, and Annie Makhijani, "Magical Thinking: Another Go at Transmutation," *Bulletin of the Atomic Scientists* 57, no. 2 (2001): 34–41. Although transmutation of high-level nuclear waste was previously seen as a dead end because it was considered too expensive and "too congenial to proliferation," it has returned as a topic of research—even lauded as the solution to nuclear waste. The idea is to convert long-lived radionuclides into radionuclides with relatively short half-lives. The current interest in transmutation can be linked to the fraught public relations over geologic repositories for nuclear waste and to the efforts of the commercial nuclear power industry to rally political and economic support.

119. A major product of the fission of uranium 235, technetium 99 is used as a gamma ray–free source of beta particles. Uranium 235 is an isotope of uranium making up a small percentage of naturally occurring uranium; it is the only fissile isotope that is a primordial nuclide or found in significant quantity in nature. Unlike the predominant isotope uranium 238, it can sustain a fission chain reaction. Uranium 235 has a half-life of 700 million years.

120. Schuyler, "Moving to Richland—I," 72.

121. "James Acord: Atomic Artist," *Nuclear News*, 52–53.

122. Acord presentation at the Influencers 2010.

123. Michele Stenehjem Gerber, *On the Home Front: The Cold War Legacy of the Hanford Nuclear Site* (Lincoln and London: University of Nebraska Press, 1992), 16–22; John M. Findlay and Bruce Hevly, *Atomic Frontier Days: Hanford and the American West* (Seattle: University of Washington Press, 2011).

124. The B Reactor at Hanford was the first large-scale plutonium production reactor in the world. It was designed and built by DuPont based on an experimental design by Enrico Fermi.

125. This plutonium then went on to other facilities across the United States, where it was machined/integrated into nuclear weapons. The making of the atomic bomb took place across an integrated geographic "factory line" of facilities that each performed some crucial aspect of bomb production.

126. Hanford represents approximately two-thirds of the nation's high-level radioactive waste by volume. See "From Bombs to $800 Handbags: Trouble Stalks America's Biggest Clean-Up," *Economist* (March 17, 2011), http://www.economist.com/node/18396103?story_id=18396103 (accessed October 10, 2011); the Physicians for Social Responsibility Web site on Hanford, http://www.psr.org/chapters/washington/hanford/hanford-and-environmental.html (accessed October 23, 2011).

127. This figure expands when land on the other side of the Columbia River is taken into account.

128. One section that is partly open to the public is the Hanford Reach National Monument, created in 2000 from former security buffer land at the Hanford site. It is a national monument with two main habitats, desert and river, and is famous for elk and large numbers of salmon. See Susan Zwinger and Skip Smith, *The Hanford Reach: A Land of Contrasts* (Tucson: University of Arizona Press, 2004).

129. Schuyler, "Moving to Richland—I."

130. James Flint, "Looking for Acord," written for the *Observer* (unpublished, July 1998), http://ds.dial.pipex.com/town/park/di21/art_tech_ec_files/acord.htm (accessed October 28, 2011).

131. Acord, presentation at the Influencers 2010.

132. Ngai, *Ugly Feelings*, 276.

133. Refer to Sianne Ngai's comments on "slapstick" in ibid., 292–97; Nicholas Holm on the organization of the "nonsensible" in his article "The Distribution of the Nonsensical and the Political Aesthetics of Humour"; and Florian Keller on "overorthodoxy" in *Andy Kaufman: Wrestling with the American Dream* (Minneapolis and London: University of Minnesota Press, 2005).

134. James Acord, "Q & A: The Art of Transmutation," interview by Daniel Cressey, *Nature* 458, no. 2 (2009): 577.

135. Gilles Deleuze and Félix Guattari, *A Thousand Plateaus: Capitalism and Schizophrenia*, trans. Brian Massumi (Minneapolis and London: University of Minnesota Press, 1987), 411; Manuel De Landa "Deleuze, Diagrams, and the Open-ended Becoming," in *Becomings: Explorations in Time, Memory, and Futures*, ed. Elizabeth Grosz (Ithaca, N.Y., and London: Cornell University Press, 1999), 29–41, esp. 37–38; Andrew Barry, "Materialist Politics: Metallurgy," in *Political*

Matter: Technoscience, Democracy, and Public Life, ed. Bruce Braun and Sarah J. Whatmore (Minneapolis and London: University of Minnesota Press, 2010), 89–117.

136. Brian Holmes explains "extradisciplinary" art as follows: "The extradisciplinary ambition is to carry out rigorous investigations on terrains as far away from art as finance, biotech, geography, urbanism, psychiatry, the electromagnetic spectrum, etc., to bring forth on those terrains the 'free play of the faculties' and the intersubjective experimentation that are characteristic of modern art, but also to try to identify, inside those same domains, the spectacular or instrumental uses so often made of the subversive liberty of aesthetic play" (Brian Holmes, "Extradisciplinary Investigations: Towards a New Critique of Institutions," *Transversal,* no. 1 [2006], http://eipcp.net/transversal/0106/holmes/en [accessed February 14, 2012]).

137. Haraway, "The Promise of Monsters," 297; Richard A. Rogers, "Overcoming the Objectification of Nature in Constitutive Theories: Toward a Transhuman, Materialist Theory of Communication," *Western Journal of Communication* 62, no. 3 (1998): 244–72.

138. Rogers, "Overcoming the Objectification of Nature in Constitutive Theories," 245.

139. "Applied sciences" refers both to the contemporary field and the early sciences that explored the mundane arenas where materials displayed their complexity, such as in kitchens, docks, and buildings.

140. Acord, presentation at the Influencers 2010.

141. Acord suggests a project akin to that of Walter Benjamin, who sought to redeem certain aspects of modernity crystallized in the outmoded and residual Arcades. In the case of Acord, it is the residual nuclear reactors that might unleash other potentials than the history of progress, destruction, and war. The term "rotten perfection" is from Richard Rogers's discussion of Rocky Flats as the hubris of humans (Rogers, "Overcoming the Objectification of Nature in Constitutive Theories," 268).

142. Cyril Stanley Smith, "Matter versus Materials: A Historical View," *Science* 162, no. 3854 (November 8, 1968): 639.

143. Manuel De Landa, "Uniformity and Variability: An Essay in the Philosophy of Matter," *Posthuman Destinies* blog (August 1, 2010), http://www.sciy.org/2010/08/01/uniformity-and-variability-an-essay-in-the-philosophy-of-matter-by-manuel-de-landa/ (accessed October 21, 2011).

144. Manuel De Landa, "Material Complexity," *Digital Tectonics,* 18, http://newmediaabington.pbworks.com/f/delanda_material_complexity.pdf (accessed October 21, 2011).

145. Michel Foucault, "18 February 1976," in *Society Must Be Defended: Lectures at the Collège de France, 1975–1976,* ed. Mauro Bertani and Alessandro Fontana, trans. David Macey (New York: Picador, 1997), 163.

146. David Biello, "A Need for New Warheads?" *Scientific American* 297 (November 2007): 80–85, esp. 83–84.

147. A nuclear reactor, by general definition, can be anything that promotes a nuclear process.

148. Uranium has been used in the decorative arts, providing a yellow glow to glass and a bright-orange quality to glazes.

149. Flint, "Looking for Acord."

150. Acord, presentation at the Influencers 2010.

151. Nadel, *Containment Culture,* 24–25.

152. Fred Moody, "The Monstrance and Me," *Pacific Northwest Magazine* (2001), http://seattletimes.nwsource.com/pacificnw/2001/0114/cover.html (accessed October 28, 2011).

153. Chris Arnot, "Sculpting with Nukes," *Guardian* (October 25, 1999), http://www.guardian.co.uk/culture/1999/oct/26/artsfeatures/print (accessed October 10, 2011).

154. Regina Hackett, *Another Bouncing Ball* blog, http://www.artsjournal.com/anotherbb/2009/06/_james_l_acord_is.html (accessed October 10, 2011).

155. Beck, *Dirty Wars,* 10–11.

156. "Southwestern Acquires Unusual Sculpture," *Southwestern Newsroom* (July 25, 2011), http://www.southwestern.edu/live/news/5453-southwestern-acquires-ununsual-sculpture (accessed October 25, 2011).

Conclusion

1. "Proving ground" is usually the name for a military installation or federally withdrawn land area where such testing takes place.

2. John Beck, *Dirty Wars: Landscape, Power, and Waste in Western American Literature* (Lincoln and London: University of Nebraska Press, 2009), 5.

3. Ibid., 7. Matthew Wolf-Meyer argues that U.S. colonial and industrial projects linger in the form of cancer hot spots and exposures. Exposure has come to create and define many communities, requiring that we understand race and class in this region to be experienced through historically/geographically specific risks and disease susceptibilities. See Matthew Wolf-Meyer, "The Mutable Nation: Cancer, Race, and American 'Biological Citizenship'" (working paper presented at the Institute for Advanced Study/Quadrant Program, University of Minnesota, Minneapolis, November 2010).

4. Michel Foucault, "18 February 1976," in *Society Must Be Defended: Lectures at the Collège de France, 1975–1976,* ed. Mauro Bertani and Alessandro Fontana, trans. David Macey (New York: Picador, 2003), 159–60.

5. Beck, *Dirty Wars,* 38–39.

6. Michel Foucault, *The History of Sexuality, Volume 1: An Introduction,* trans. Robert Hurley (New York, Vintage Books, 1990), 137.

7. Beck, *Dirty Wars,* 37.

8. Leerom Medovoi, "Global Society Must Be Defended: Biopolitics without Boundaries," *Social Text* 25, no. 2 (2007): 57; Paul Rutherford, "Ecological Modernization and Environmental Risk," in *Discourses of the Environment,* ed. Éric

Darier (Oxford and Malden, Mass.: Blackwell, 1999), 95–118; Leerom Medovoi, "A Contribution to the Critique of Political Ecology: Sustainability as Disavowal," *New Formations: A Journal of Culture/Theory/Politics*, no. 69 (2010): 129–43.

9. Michel Foucault, "21 January 1976," in *Society Must Be Defended*, 50.

10. Rosalyn Diprose, "Toward an Ethico-Politics of the Posthuman: Foucault and Merleau-Ponty," *Parrhesia* 8 (2009): 7–8. Also refer to Diana Coole and Samantha Frost, "Introducing the New Materialisms," in *New Materialisms: Ontology, Agency, and Politics*, ed. Diana Coole and Samantha Frost (Durham, N.C., and London: Duke University Press, 2010), 32–33.

11. Foucault, "21 January 1976," 54–55.

12. John Lowering, "The Production and Consumption of the 'Means of Violence': Implications of the Reconfiguration of the State, Economic Internationalisation, and the End of the Cold War," *Geoforum* 25, no. 4 (1994): 471–86; Catherine Lutz, "Making War at Home in the United States: Militarization and the Current Crisis," *American Anthropologist* 104, no. 3 (2002): 723–35.

13. See Wolf-Meyer, "The Mutable Nation."

14. François Ewald, "Insurance and Risk," in *The Foucault Effect: Studies in Governmentality, with Two Lectures by and an Interview with Michel Foucault*, ed. Graham Burchell, Colin Gordon, and Peter Miller (Chicago: University of Chicago Press, 1991), 197–210; Deborah Lupton, *The Imperative of Health: Public Health and the Regulated Body* (Thousand Oaks, Calif.: SAGE Publications, 1995), 89; Deborah Lupton, *Risk* (London and New York: Routledge, 1999).

15. Ewald, "Insurance and Risk," 199.

16. Brian Massumi, "National Enterprise Emergency: Steps toward an Ecology of Powers," *Theory, Culture and Society* 26, no. 6 (2009): 153–85, esp. 158–59.

17. Shiloh Krupar and Stefan Al, "Notes on the Society of the ~~Spectacle~~ Brand," in *The SAGE Handbook of Architectural Theory*, ed. C. Greig Crysler, Stephen Cairns, and Hilde Heynen (Thousand Oaks, Calif.: SAGE Publications, 2012), 247–63.

18. Mark Halsey, "Against 'Green' Criminology," *British Journal of Criminology* 44 (2004): 833–53; Mark Halsey, *Deleuze and Environmental Damage: Violence of the Text* (Burlington, Vt.: Ashgate, 2006).

19. Lauren Berlant, "Slow Death (Sovereignty, Obesity, Lateral Agency)," *Critical Inquiry* 33, no. 4 (2007): 754–80.

20. Elizabeth A. Povinelli, "The Child in the Broom Closet: States of Killing and Letting Die," in *States of Violence: War, Capital Punishment, and Letting Die*, ed. Austin Sarat and Jennifer L. Culbert (New York: Cambridge University Press, 2009), 169–91.

21. Beck, *Dirty Wars*, 47.

22. Michel Foucault, "Confronting Governments: Human Rights," in *The Chomsky–Foucault Debate on Human Nature* (New York: New Press, 2006), 212.

23. Louisa Cadman, "How (Not) to Be Governed: Foucault, Critique, and the Political," *Environment and Planning D: Society and Space* 28, no. 3 (2010): 539.

24. Ibid., 553.

25. Michel Foucault, "The Subject and Power," *Critical Inquiry* 8, no. 4 (1982): 785.

26. Beck, *Dirty Wars*, 210.

27. Eve Kosofsky Sedgwick, *Touching Feeling: Affect, Pedagogy, Performativity* (Durham, N.C., and London: Duke University Press, 2003), 140.

28. S. Lochlann Jain, "The Mortality Effect: Counting the Dead in the Cancer Trial," *Public Culture* 22, no. 1 (2010): 89–117.

29. Lawrence Buell, "Toxic Discourse," *Critical Inquiry* 24, no. 3 (1998): 662.

30. Barbara Adam, "Radiated Identities: In Pursuit of the Temporal Complexity of Conceptual Cultural Practices," in *Timescapes of Modernity: The Environment and Invisible Hazards* (London and New York: Routledge, 1998), 146.

31. Diprose, "Toward an Ethico-Politics of the Posthuman," 15.

32. George Pavlich, "Critical Genres and Radical Criminology in Britain," *British Journal of Criminology* 41 (2001): 158.

33. Judith Revel explains: "The dissymmetry between the powers over life and life's power of invention . . . can be seen as an 'ontology', a term that appears more frequently in Foucault's later writings. Yet this ontology is both an 'ontology of actuality' and an 'ontology of ourself', which means that it is a matter of thinking at the same time determinations and freedom, the objectivization of singularities and the paradoxical power that the latter have, in spite of everything, to constitute themselves as subjectivities" (Judith Revel, "Identity, Nature, Life: Three Biopolitical Deconstructions," *Theory, Culture and Society* 26, no. 6 [2009]: 52–53). Also see Timothy C. Campbell, *Improper Life: Technology and Biopolitics from Heidegger to Agamben* (Minneapolis and London: University of Minnesota Press, 2011), 119–56.

34. Judith Butler, *Frames of War: When Is Life Grievable?* (London and New York: Verso, 2009), 51–52.

35. Halsey, "Against 'Green' Criminology," 850.

36. Steve Hinchliffe, "Reconstituting Nature Conservation: Towards a Careful Political Ecology," *Geoforum* 39, no. 1 (2008): 96. Also see Michel Callon's notion of "hot situations" in his chapter "An Essay on Framing and Overflowing: Economic Externalities Revisited by Sociology," in *The Law of the Markets*, ed. Michel Callon (Oxford and Malden, Mass.: Blackwell, 1998), 244–69.

37. Michael Hardt, *Gilles Deleuze: An Apprenticeship in Philosophy* (Minneapolis and London: University of Minnesota Press, 1993), 120–21.

38. Revel, "Identity, Nature, Life," 45–54. The goal is not to decenter ethics for aesthetics, but to imagine new ways to bring politics and aesthetics together critically, without assigning rules or conceptual legislation.

39. Paul Rabinow, "Introduction: The History of Systems of Thought," in Michel Foucault, *Ethics: Subjectivity and Truth*, ed. Paul Rabinow, trans. Robert Hurley (New York: New Press, 1994), xix.

40. The activity of "styling," in relation to the conduct of one's life and relation to truth, might propel new life-forms and public-forming interventions, which are open, unpredictable, robust, objective, real, fragile, and marked by controversy.

See George Pavlich, "Nietzsche, Critique and the Promise of Not Being Thus . . . ," *International Journal for the Semiotics of Law* 13, no. 4 (2000): 357–75; Mariana Valverde, "Experience and Truth Telling in Post-humanist World: A Foucauldian Contribution to Feminist Ethical Reflections," in *Feminism and the Final Foucault,* ed. Dianna Taylor and Karen Vintges (Urbana and Chicago: University of Illinois Press, 2004), 67–90; Bernd Frohmann, "Documentary Ethics, Ontology, Politics," *Journal of Archival Science* 8, no. 3 (2008): 165–80; Timothy O'Leary, *Foucault: The Art of Ethics* (London and New York: Continuum, 2002).

41. "Hot spotting" is inspired by the creative forms and methodologies of figural writing, institutional critique, and experimental ethnographies by Walter Benjamin, Donna J. Haraway, Allan Pred, Andrea Fraser, Kathleen Stewart, Hugh Gusterson, John Law, Jon McKenzie, Leslie Salzinger, Michael Taussig, Nigel Thrift, Karen E. Till, Anna Lowenthal Tsing, Ramona Fernandez, and Gregory Ulmer.

42. George Kamberelis, "Ingestion, Elimination, Sex, and Song: Trickster as Premodern Avatar of Postmodern Research Practice," *Qualitative Inquiry* 9, no. 5 (2003): 673–704, esp. 674.

43. Ladelle McWhorter, "Practicing Practicing," in Taylor and Vintges, *Feminism and the Final Foucault,* 143–62.

44. Frohmann, "Documentary Ethics, Ontology, Politics," 175; also see Allen Feldman, "Faux Documentary and the Memory of Realism," *American Anthropologist* 100, no. 2 (1998): 494–502.

45. Michel Callon, "Writing and (Re)writing Devices as Tools for Managing Complexity," in *Complexities: Social Studies of Knowledge Practices,* ed. John Law and Annemarie Mol (Durham, N.C., and London: Duke University Press, 2002), 191–217.

46. Michel Foucault, "The History of Sexuality," in *Power/Knowledge: Selected Interviews and Other Writings, 1972–1977,* ed. Colin Gordon, trans. Colin Gordon, Leo Marshall, John Mepham, and Kate Soper (New York: Pantheon Books, 1980), 193.

47. Sarah Whatmore, "Materialist Returns: Practicing Cultural Geography in and for a More-Than-Human World," *cultural geographies* 13 (2006): 602.

48. Adriana Petryna, *Life Exposed: Biological Citizens after Chernobyl* (Princeton, N.J., and Oxford: Princeton University Press, 2002).

49. Halsey, "Against 'Green' Criminology," 848; Stacy Alaimo, *Bodily Natures: Science, Environment, and the Material Self* (Bloomington and Indianapolis: Indiana University Press, 2010), 16–17.

50. Stuart J. Murray, "Thanatopolitics: On the Use of Death for Mobilizing Political Life," *Polygraph* 18 (2006): 195.

51. See Berlant, "Slow Death."

52. Cary Wolfe, *Critical Environments: Postmodern Theory and the Pragmatics of the Outside* (Minneapolis and London: University of Minnesota Press, 1998), xxii.

53. Michel Foucault, "The Masked Philosopher," cited in Martha Cooper and Carole Blair, "Foucault's Ethics," *Qualitative Inquiry* 8, no. 4 (2008): 526. A different translation of this quotation can be found in Michel Foucault, *Foucault Live:*

Collected Interviews, 1961–1984, ed. Sylvère Lotringer (New York: Semiotext[e], 1996), 305. Also refer to Ed Cohen, "The Courage of Curiosity, or the Heart of Truth (A Mash-Up)," *Criticism* 52, no. 2 (2010): 201.

54. Pavlich, "Nietzsche, Critique and the Promise of Not Being Thus . . . ," 374.

55. Shiloh Krupar, "The Fictional World of Absurdist Drama according to its Influence on Me," *Environment and Planning D: Society and Space* 29, no. 3 (2011): 558.

56. Friedrich Nietzsche, *The Gay Science*, ed. and trans. Walter Kaufmann (New York: Vintage, 1974), 246.

57. See Atul Gawande, "The Hot Spotters," *New Yorker* (January 24, 2011), http://www.newyorker.com/reporting/2011/01/24/110124fa_fact_gawande (accessed October 2, 2011).

58. Nadine Ehlers, "Risking 'Safety': Breast Cancer Prognosis and the Strategic Enterprise of Life" (working paper, Women's and Gender Studies Program, Georgetown University, 2010). Ehlers states: "This imperative to live—to optimize life—is one of the central organizing principles of modern biopolitical risk society" (1).

59. "Plutonium Memorial Design Competition," *Bulletin of the Atomic Scientists* 58, no. 3 (2002): 39; *Mike and Maaike* blog, Michael Simonian's *24110*, http://www.mikeandmaaike.com/#p_24110-monument (accessed October 23, 2011); *Just Urbanism* blog, "Bulletin of the Atomic Scientists' Plutonium Memorial Competition," http://www.justurbanism.com/2007/10/bulletin_of_the_atomic_scientists_plutonium_m/ (accessed October 23, 2011). Also see *dpr-barcelona* blog, "From Salt Mines to Floating Stars as Nuclear Waste Disposals" (July 12, 2010), http://dprbcn.wordpress.com/2010/07/12/from-salt-mines-to-floating-stars/ (accessed October 23, 2011).

60. Nuclear Energy Institute, "EPA's Radiation Protection Standard for Yucca Mountain Is Consistent with Global Norms" (October 1, 2008), http://www.nei.org/newsandevents/newsreleases/epas_radiation_protection_standard_for_yucca_mountain_is_consistent_with_global_norms (accessed October 23, 2011); Nuclear Energy Institution, "Radiation and Human Health," http://nuclear.inl.gov/docs/factsheets/radiation_human_health_1003.pdf (accessed October 23, 2011).

61. Mira Engler, *Designing America's Waste Landscapes* (Baltimore and London: Johns Hopkins University Press, 2004), 232.

62. Derek Gregory, *The Colonial Present* (Malden, Mass., and Oxford: Blackwell, 2004); also see Rob Nixon, *Slow Violence and the Environmentalism of the Poor* (Cambridge, Mass., and London: Harvard University Press, 2011).

63. Sculptor James Acord, featured in chapter 4, worked on memorials involving nuclear waste materials. Also refer to the monument project for the Waste Isolation Pilot Plant outside Carlsbad, New Mexico; see Peter C. van Wyck, *Signs of Danger: Waste, Trauma, and Nuclear Threat* (Minneapolis and London: University of Minnesota Press, 2005).

64. For examples, refer to Amy Standen, "U.S. Military Boosts Clean Energy, with Startup Help," KQED, October 24, 2011.

INDEX

Shiloh R. Krupar is assistant professor of culture and politics in the Edmund A. Walsh School of Foreign Service at Georgetown University.